# Biotechnology in Agriculture and Forestry 36

## *Somaclonal Variation in Crop Improvement II*

Edited by Y.P.S. Bajaj

With 118 Figures and 86 Tables

Springer

Professor Dr. Y.P.S. Bajaj
A-137
New Friends Colony
New Delhi 110065, India

ISBN 3-540-60549-5 Springer-Verlag Berlin Heidelberg New York

Library of Congress Cataloging-in-Publication Data. Main entry under title: Trees. (Biotechnology in agriculture and forestry: (1) Bibliography: p. Includes index. I. Tree crops Propagation–In vitro. 2. Trees–Propagation–In vitro. I. Bajaj, Y.P.S., 1936– . II. Series. SB170:T725 1985 634.9 85-17309. ISBN 3-540-60549-5


Printed in Germany



Cover design: Design & Production, Heidelberg

Typesetting: Thomson Press (India) Ltd., New Delhi

SPIN: 10471198 31/3137/SPS – 5 4 3 2 1 0 – Printed on acid-free paper

*Dedicated to*
*Professor Jean Semal*
*of the Faculté des Sciences Agronomiques,*
*Gembloux (Belgium), whose laboratory I had*
*the privilege of visiting in 1985*

# Preface

Somaclonal variation in plant cell cultures has been the focus of interest recently for the induction of much needed genetic variability in crops. It also enables one to add to or intensify only one feature of an established variety possessing a combination of most of the useful agronomic traits. Over the past 5 years, much information has accumulated on the in vitro induction of genetic variability in a number of plants of economic importance. Taking these developments into consideration, the present book, like the previous volume, *Somaclonal Variation in Crop Improvement I*, published in 1990, is special in its approach. It comprises 24 chapters dealing with somaclonal variants showing resistance to salt/drought, herbicides, viruses, *Alternaria*, *Fusarium*, *Glomerella*, *Verticillium*, *Phytophthora*, fall armyworm, etc. in a number of plant species. The book has been divided into two sections:

*Section I.* Somaclonal variation in agricultural crops (wheat, rice, maize, sorghum, potato, tomato, *Lotus*, *Stylosanthes*, banana, strawberry, citrus, colt cherry).

*Section II.* Somaclonal variation in medicinal and aromatic plants (*Atropa*, *Carthamus*, *Hypericum*, *Lavatera*, *Nicotiana*, *Primula*, *Rauwolfia*, *Scilla*, and *Zinnia*).

This book will be of great assistance to research workers, teachers, and advanced students of plant biotechnology, tissue culture, pathology, horticulture, pharmacy, and especially plant breeding.

New Delhi, March 1996

Professor Y.P.S. Bajaj
Series Editor

# Contents

## Section I Somaclonal Variation in Agricultural Crops – Cereals, Potato, Fruits, Legumes

I.9 Somaclonal Variation for Salt Tolerance in Tomato and Potato
M. Tal (With 2 Figures)



I.10 Somaclonal Variation in *Lotus corniculatus* L. (Birdsfoot Trefoil)
M. Niizeki (With 6 Figures)



I.11 Somaclonal Variation in *Stylosanthes* Species
I.D. Godwin (With 5 Figures)



I.12 Somaclonal Variation in Banana and Plantain (*Musa* Species)
O. Reuveni, Y. Israeli, and E. Lahav (With 2 Figures)

I.13 Somaclonal Variation for Resistance to *Fusarium*- and *Glomerella*-Caused Diseases in Strawberry (*Fragaria* × *ananassa*)
H. Toyoda (With 5 Figures)



I.14 In Vitro Selection for Salt Tolerance in *Citrus* Rootstocks
J. Bouharmont and N. Beloualy (With 5 Figures)



I.15 In Vitro Selection for Salt/Drought Tolerance in Colt Cherry (*Prunus avium* × *pseudocerasus*)
S.J. Ochatt (With 6 Figures)



**Section II Somaclonal Variation in Medicinal and Aromatic Plants**

II.1 In Vitro Induction of Herbicide Resistance in *Atropa belladonna* L.
M. Yamazaki and K. Saito (With 3 Figures)

# List of Contributors

AHMED, K.Z., Department of Genetics, Faculty of Agriculture, University of Minia, Minia 61517, Egypt

BANG, J.W., Department of Biology, College of Natural Sciences, Chungnam National University, Daejon 305-764, Korea

BELOUALY, N., Laboratoire de Cytogénétique, Université Catholique de Louvain, Place Croix-du-Sud 4, 1348 Louvain-la-Neuve, Belgium (Present address: SODEA, BP. 6280, Rabat, Morocco)

BINH, D.Q., Former Scientist of Department of Genetics and Plant Breeding, University of Agricultural Sciences, 2103 Gödöllő, Hungary (Present address: Institute of Biology, CNRS, Nghiado-Tuliem, Hanoi, Vietnam)

BOUHARMONT, J., Laboratoire de Cytogénétique, Université Catholique de Louvain, Place Croix-du-Sud 4, 1348 Louvain-la-Neuve, Belgium

BRUŇÁKOVÁ, K., Department of Experimental Botany and Genetics, Faculty of Science, P.J. Šafárik University, Mánesova 23, 04154 Košice, Slovakia

BUTLER, L.G., Department of Biochemistry, Purdue University, West Lafayette, IN 47907-1153, USA

CAI, T., Department of Biochemistry, Purdue University, West Lafayette, IN 47907-1153, USA (Present address: Department of Biotechnology Research, Pioneer Hi-Bred International Inc., 7300 N.W. 62nd Avenue, P.O. Box 1004, Johnston, IA 50131-1004, USA)

CASSELLS, A.C., Department of Plant Science, University College, Cork, Ireland

ČELLÁROVÁ, E., Department of Experimental Botany and Genetics, Faculty of Science, P.J. Šafárik University, Mánesova 23, 04154 Košice, Slovakia

Choi, H.W., Department of Biology, College of Natural Sciences,
Chungnam National University, Daejon 305-764, Korea

Deng, Y.H., Institute of Crop Breeding and Cultivation,
Chinese Academy of Agricultural Sciences, Beijing, 100081, China

Ducreux, G., Morphogénèse Végétale Expérimentale,
Bât. 360, Université Paris Sud, 91405 Orsay Cedex, France

Duncan, R.R., Department of Crop and Soil Science,
University of Georgia, Georgia Station, Griffin, GA 30223-1797, USA

Godwin, I.D., Department of Agriculture, The University of Queensland,
Brisbane QLD, 4072, Australia

Haicour, R., Morphogénèse Végétale Expérimentale,
Bât 360, Université Paris Sud, 91405 Orsay Cedex, France

Heszky, L.E., Department of Genetics and Plant Breeding,
University of Agricultural Sciences, 2103 Gödöllő, Hungary

Iriondo, J.M., Departamento de Biología Vegetal, ETSI Agrónomos,
Universidad Politécnica, 28040 Madrid, Spain

Isenhour, D.J., Department of Research Specialists,
Pioneer Hi-Bred International, Box 85, Johnston, IA 50131-0085, USA

Israeli, Y., Jordan Valley Banana Experiment Station,
Zemach 15132, Israel

Jadari, R., Institut Agronomique et Vétérinaire Hassan II,
BP 6202, Morocco

Kanda, M., Laboratory of Floriculture,
Chiba Horticultural Experiment Station, 1762 Yamamoto, Tateyama,
Chiba 294, Japan

Kneusel, R.E., Biologisches Institut II, Lehrstuhl für Biochemie
der Pflanzen, Albert-Ludwigs-Universität, Schänzlestr. 1,
79104 Freiburg, Germany

Kunakh, V.A., Institute of Molecular Biology and Genetics,
National Academy of Sciences of Ukraine, Kiev 252143, Ukraine

Kunothai-Muhsin, A., Southeast Asian Regional Centre
for Tropical Biology (Seameo-Biotrop), J1 Raya Km 6, P.O. Box 116,
Bogor, Indonesia

Lahav, E., Institute of Horticulture, Agricultural Research Organization, The Volcani Center, P.O. Box 6, Bet Dagan 50250, Israel

Ling, D.H., Department of Genetics, South China Institute of Botany, Academia Sinica, Guangzhou 510650, China

Matern, U., Biologisches Institut II, Lehrstuhl für Biochemie der Pflanzen, Albert-Ludwigs-Universität, Schänzlestr. 1, 79104 Freiburg, Germany

Mesterházy, Á., Cereal Research Institute, P.O. Box 391, 6701 Szeged, Hungary

Niizeki, M., Laboratory of Bioscience and Biotechnology, Faculty of Agriculture, Hirosaki University, Hirosaki, Aomori-ken 036, Japan

Ochatt, S.J., INRA, Station d'Amélioration des Espèces Fruitières et Ornementales, B.P. 57, 49071 Beaucouzé Cedex, France

Oleschenko, L.T., Department of Phytopathogenic Viruses, Zabolotny Institute of Microbiology and Virology, National Academy of Sciences of Ukraine, Zabolotny Street 154, Kiev 252143, Ukraine

Pérez, C., Departamento de Biología Vegetal, ETSI Agrónomos, Universidad Politécnica, 28040 Madrid, Spain

Reuveni, O., Institute of Horticulture, Agricultural Research Organization, The Volcani Center, P.O. Box 6, Bet Dagan 50250, Israel

Rossignol, L., Morphogénèse Végétale Expérimentale, Bât. 360, Université Paris Sud, 91405 Orsay Cedex, France

Sági, F., Cereal Research Institute, P.O. Box 391, 6701 Szeged, Hungary

Saito, K., Faculty of Pharmaceutical Sciences, Laboratory of Molecular Biology and Biotechnology, Research Center of Medicinal Resources, Chiba University, Yayoi-cho 1-33, Inage-ku, Chiba 263, Japan

Scherbatenko, I.S., Department of Phytopathogenic Viruses, Zabolotny Institute of Microbiology and Virology, National Academy of Sciences of Ukraine, Zabolotny Street 154, Kiev 252143, Ukraine

SEN, P.T., Department of Plant Science, University College, Cork, Ireland

SIHACHAKR, D., Morphogénèse Végétale Expérimentale, Bât. 360,
Université Paris Sud, 91405 Orsay Cedex, France

SIMON-KISS, I., Rice Breeding Section, Irrigation Research Institute,
5541 Szarvas, Hungary

STIEVE, S.M., Department of Horticulture, University of Wisconsin,
1575 Linden Drive, Madison, WI 53706, USA

STIMART, D.P., Department of Horticulture, University of Wisconsin,
1575 Linden Drive, Madison, WI 53706, USA

TAL, M., Department of Life Sciences, Ben Gurion University
of the Negev, P.O. Box 653, Beer Sheva 84105, Israel

TOYODA, H., Laboratory of Plant Pathology, Faculty of Agriculture,
Kinki University, 3327-204 Nakamachi, Nara 631, Japan

WISEMAN, B.R., USDA-ARS, Insect Biology and Population Management
Research Laboratory, P.O. Box 748, Tifton, GA 31793-0748, USA

YAMAZAKI, M., Faculty of Pharmaceutical Sciences,
Laboratory of Molecular Biology and Biotechnology,
Research Center of Medicinal Resources, Chiba University,
Yayoi-cho 1-33, Inage-ku, Chiba 263, Japan

ZHOU, H.S., Institute of Crop Breeding and Cultivation,
Chinese Academy of Agricultural Sciences, Beijing, 100081, China

# Section I
# Somaclonal Variation in Agricultural Crops – Cereals, Potato, Fruits, Legumes

# I.1 In Vitro Production of *Fusarium*-Resistant Wheat Plants

K.Z. AHMED[1], Á. MESTERHÁZY[2], and F. SÁGI[2]

## 1 Introduction

Bread wheat can suffer from a number of diseases caused by various fungi, bacteria, and viruses. Some of the pathogenic fungi can invade the seeds and produce toxic metabolites harmful for both humans and livestock. Thus, it is understandable that the seed should be protected from fungal infection by all possible methods. Among them, the breeding for resistance and biotechnology can be regarded as the most acceptable.

Somaclonal variation of wheat plants regenerated from callus cultures of various explants has been extensively studied (Ahloowalia 1982; Maddock et al. 1983; Larkin et al. 1984; Ahloowalia and Sherington 1985; Galiba et al. 1985; Bajaj 1986; Maddock and Semple 1986; Chen et al. 1987; Ryan et al. 1987; see also Bajaj 1990; Hanson et al. 1994). In wheat somaclones, stalk length, size and structure (fertility) of the ear, leaf angle, heading time, yield, and yield components are the traits varying most frequently; however, variations in grain protein content and composition (Larkin et al. 1984; Cooper et al. 1986), baking quality characters (Ryan et al. 1987), $\beta$-amylase isozymes (Ryan and Scowcroft 1987), and frost tolerance (Lazar et al. 1988; Galiba and Sutka 1989) were also recorded. It has been shown that reduced stalk length in regenerated somaclones of the Mv4 winter wheat cultivar is due to a mutation in the Rht locus (Sági et al. 1993). There is a growing body of evidence that for these variations, in vitro-induced chromosomal changes (Maddock et al. 1983; Karp and Maddock 1984; Ahloowalia and Sherington 1985; Davies et al. 1986) and DNA rearrangements (Rode et al. 1987; Morère-Le Paven et al. 1992; Chowdhury et al. 1994) are at least partly responsible. According to our knowledge, among the somaclones of wheat, *Fusarium* resistance has not yet been reported.

### 1.1 Fusaria and Fusariose

In the voluminous group of fungi attacking wheat and damaging wheat seeds, species belonging to the genus *Fusarium* became especially widespread, some-

[1] Dept. of Genetics, Faculty of Agriculture, University of Minia, Minia 61517, Egypt
[2] Cereal Research Institute, P.O. Box 391, H-6701 Szeged, Hungary

Biotechnology in Agriculture and Forestry, Vol. 36
Somaclonal Variation in Crop Improvement II (ed. by Y.P.S. Bajaj)

times causing epidemics and serious health problems. On wheat, *Fusarium graminearum* Schwabe (perfect form *Gibberella zeae* Petch.), *F. culmorum* Sacc., *F. nivale* (*Microdochium nivale*), *F. avenaceum* (*G. avencea*), and *F. sporotrichioides* occur most frequently. From diseased plants we could identify *F. graminearum* and *F. culmorum* as dominant species (Mesterházy 1984). They cause the so-called fusariose (seedling blight, foot rot, scab, or head blight). The most dangerous for health of these being *Fusarium* head blight (Fig. 1), which directly threatens the developing seeds, especially under stress conditions (excess humidity and high temperature, unbalanced mineral supply, lodging, etc.).

The mycotoxins produced by the pathogenic *Fusarium* species are numerous and of differing chemical nature. Among them the highly toxic trichothecenes such as deoxyninalenol or vomitoxin (DON), T-2 toxin, causing hemorrhagia, neural disorders, and damage to the immune system and the less toxic zearalenone and its derivatives, affecting animal and human reproduction due to their estrogen activity, are the most important ones. Beside these, more than 70 toxic compounds produced by *Fusaria* have been reported (Chelkowski 1989). Their identification and determination can be carried out by efficient and sensitive analytical methods such as high pressure liquid chromatography (HPLC) and mass spectrometry (MS) (Tanaka et al. 1985; Mirocha et al. 1992; Mesterházy and Bartók 1993).

In naturally contaminated wheat samples, *Fusarium* toxin level can vary within wide limits. In Hungary, 80% of the wheat samples were contaminated by

A

B

**Fig. 1A, B.** *Fusarium* scab of wheat ears. **A** Scabby grains in infected ear in comparison with a control ear. **B** Scabby and healthy grains from infected and control ears

DON in 1987. Amounts ranging from traces to 8.50 ppm DON were detected in wheat samples, as described by Snijders (1990b); some data are listed in Table 1. Cleaning can remove only a part of the scabby grains with the highest DON concentrations. Experiments show that milling can decrease the toxin content in flour, but that the scabby grains mostly remain in bran on milling, so that bran and bran products from scabby wheat can be highly poisonous (Scott et al. 1984; Abbas et al. 1985; Lee et al. 1987; Lepschy v. Gleissenthal et al. 1989). Breadmaking hardly decomposes DON and other heat-stable *Fusarium* toxins (Scott et al. 1984; Abbas et al. 1985), but noodles and cooked spaghetti retained only 42–53% of the DON originally present in the flour and durum wheat semolina, respectively (Nowick et al. 1988).

*Fusarium* trichothecene toxins are inhibitory to growth and protein synthesis of the wheat plant also. DON in low concentrations inhibits coleoptile and root growth, and 14 ppm DON can completely stop seedling growth (Snijders 1988; Wakulinski 1989; Shimada and Otani 1990). Similarly, the T-2 toxin minimizes elongation of the wheat coleoptile (Cole et al. 1981). In this respect, it is interesting that wheat shoot apices free from any *Fusarium* infection were able to synthesize zearalenone (Wang and Meng 1990).

## 1.2 Pathogenesis and Epidemiology

*Fusarium* species invade wheat all over the world. Generally, a disease outbreak needs susceptible genotypes, presence of the pathogen, and favorable ecological conditions. Under dry conditions, crown rot and foot rot are predominant, as in Australia, the Mediterranean basin, and the Russian and American plains. Under wet and warm conditions at flowering time and subsequently, the monsoon summer wheat regions like India, South Japan, China, and tropical and subtropical parts of the world are endangered. *F. graminearum* is dominant in the warmer, *F. culmorum* in the cooler regions. *F. nivale*, which is not a *Fusarium* species now, attacks wheat in areas where long snow cover is usual.

It is important that *F. graminearum* and *F. culmorum* do not have vertical races. The often experienced isolate-genotype interaction originates from methodical and mostly ecological differences.

*Fusariam* species can infect wheat plants at any developmental stage (Mesterházy 1989). Most sensitive, however, are the very young tissues like coleoptile,

**Table 1.** Amounts of some *Fusarium* toxins found in naturally infected seed samples of wheat

| Toxin | Concentration (mg/kg) | Reference |
|---|---|---|
| DON (deoxynivalenol) | 0.09– 0.74 | Tanaka et al. (1985) |
| | 0.20–30.40 | Chelkowski (1989) |
| | <0.30– 9.76 | Gilbert (1989) |
| | 0.07–43.80 | Lepschy v. Gleissenthal et al. (1989) |
| ZEN (zearalenone) | 0.10– 8.00 | Gilbert (1989) |
| Moniliformin | <0.01 | Gilbert (1989) |

initial root system, embryos, and grains in early developmental stages. Higher susceptibility also occurs in senescent tissues like roots or ripened heads. The highest yield losses occur when a warm, rainy period prevails during and just after flowering. The infected embryos die very early, or mostly small, scabby grains develop. Even seemingly healthy grains often bear latent infection. Late epidemics after formation of grains can result in a high ratio of scabby grains of nearly normal size, the yield loss is less, but the product may be unsuitable for human and animal consumption even after mechanical cleaning. Fertilization, soil conditions, and agrotechnical factors can have contradictory effects, sometimes favoring, sometimes disfavoring disease development. Of course, the severity of epidemics depends largely on the resistance level of the genotypes, so that it is not possible to forecast an epidemic, based alone on the weather conditions.

The source of inoculum is mostly plant debris overwintering on the soil surface. Infected wheat or corn represents a real risk of starting a severe epidemic. When infected, volunteer cereals in set-aside fields can also threaten wheats grown in the vicinity.

Yield losses due to natural epidemics may be as high as 60–70% (Munteanu et al. 1972); the losses, however, are generally much less. Under artificial infection, the losses in susceptible or highly susceptible genotypes may reach even 100% (Mesterházy 1988). As scabby grains may, depending on the cultivar, attain even 50–60 ppm DON (Mesterházy, unpubl.), it is clear that the ratio of scabby grains determines the DON content in the grain mass (Mesterházy and Bartók 1993).

## 1.3 Available Genetic Variability and Resistance Breeding in Wheat

The worldwide search for sources of *Fusarium* resistance has been successful. The highest resistance level has been achieved in countries in the spring wheat regions, where head blight is annually present at the epidemic level, allowing effective selection based approximately on natural infection pressure. This, supplemented with intensive breeding, resulted in genotypes with excellent resistance to scab, such as Sumey-3 from China, Nobeoka Bozu from Japan, and other spring wheats (Gu 1983; Mesterházy 1983, 1989, Snijders 1990a; Lemmens et al. 1993). They exhibit no immunity, but do not suffer significantly even if infected artificially with the most pathogenic isolates. This resistance level ensures full protection even under conditions that favor severe epidemic outbreaks.

In winter wheat areas, a regular natural selection pressure does not exist. Therefore, the maximal resistance level is much lower than that found in the best spring wheats, and many highly susceptible genotypes are grown commercially. Consequently, irregular epidemics can develop into severe epidemics causing high yield losses. However, less severe epidemics can cause indirect losses via the inferior quality due to toxin contamination.

The genetic background of *Fusarium* resistance is different. The cv. Sumey-3 has two or three major genes, in Nobeoka Bozu a polygenic background has been found (Yu 1982, Gocho 1985, Zhou et al. 1987). Studies in other genotypes refer mostly to the polygenic model, and transgressive segregation seems to be normal

(Snijders 1990c). We should mention that resistance to seedling blight does not always correlate with scab or foot rot resistance, and the same is also true for foot rot and scab resistance. Therefore, to create plants with a complex resistance to all three forms needs a special approach (Mesterházy 1988, 1989).

It is, however, a significant feature that, generally, resistance to *F. graminearum* and *F. culmorum* is common, as confirmed by numerous tests (Mesterházy 1987, 1989; Snijders et al. 1993). But, there are several genotypes for which this correlation is not applicable (Szunics and Szunics 1992). The reason for this (genetic, methodical, or other specific reactions) is not known. There are also data on the relationships between resistance to *F. avenaceum*, *F. poae*, *F. sporotrichioides*, *F. graminearum*, and *F. culmorum* (Hollins 1993). We have similar data from Eeuwijk et al. (1994) for wheat, for triticale from Arseniuk et al. (1992), and for maize (Mesterházy and Kovács 1986). Therefore, the very often used term *Fusarium* resistance in wheat seems to be correct.

Intensive breeding programs are now underway in Mexico (CIMMYT), China, Japan, Hungary, Germany, Austria, France, and other countries, to incorporate resistance to scab. We have the sources of resistance, methods of screening, and selection under artificial infection pressure, but no commercial success has as yet been achieved in the winter wheat region. The work is not easy; confirmation of selection results and the selection itself needs skilled personnel and time. Thus, in vitro work may have a chance to achieve this goal earlier.

## 2 Review of In Vitro Studies

Since a breakthrough in breeding for *Fusarium* resistance is difficult to attain besides the traditional methods based on the host-pathogen interaction, the possible role of in vitro techniques using toxic fungal metabolites as supplements in nutrient media of different cell cultures must be clarified (Wenzel 1985; van der Bulk 1991). This approach was first applied by Gengenbach and Green (1975) for selecting T cytoplasm maize calli resistant to *Helminthosporium* T toxin. Later, Gengenbach and Rines (1986) regenerated maize plants resistant to *H. maydis* race T, demonstrating the possibilities of biotechnological methods. This technique has been adopted in selection of in vitro cultures of wheat (Ahmed et al. 1991; Li and Huang 1992; Fadel and Wenzel 1993) and various crop plants for *Fusarium* resistance, partly also at the haploid level (anther culture). However, *Fusarium*-resistant plants could only rarely be regenerated from resistant callus or cells (Gengenbach and Rines 1986; Wenzel and Foroughi-Wehr 1990; Ahmed et al. 1991). This limited success raises the question: how reliable are the *Fusarium* toxins as inducers of stable resistance?

The answer is relatively simple and positive for those diseases where a highly specific pathotoxin acts as a determining agent of pathogenicity, as in the case of some *Pseudomonas*, *Xanthomonas*, or *Helminthosporium* (*Bipolaris*) species (Mathysse 1983; Király 1986). Concerning the *Fusarium* species, some data are available on the correlation between toxin sensitivity and resistance to fusaric acid and lycomarasmine produced by *F. oxysporum* or to naphtazine, a metabolite of *F. solani*, but their exact role in disease development could not yet be clarified (Király 1986). Far more unknown and complicated are the relationships with *F. graminearum* and *F. culmorum*, which release many toxins. Resistant wheat varieties were less susceptible to DON, for instance, than susceptible ones (Snijders 1988), and the sensitivity of wheat mesophyll chloroplasts to DON correlates with varietal resistance to *F. graminearum* (X. M. Huang et al. 1991a, b). Wang and Miller (1988) found that in vitro growth inhibition by DON and field resistance of wheat correlate negatively. Nevertheless, according to Liu et al. (1990), this toxin behaves in low concentrations ($10^{-6}$ to $10^{-3}$ M) as a growth substance in wheat callus cultures, stimulating root and shoot regeneration. Since the treated regenerants did not develop resistance, DON was regarded as a compound without selective activity. By contrast, Shimada and Otani (1990) claim that DON might be useful for selecting wheat cells resistant to *F. graminearum*, while nivalenol and fungal culture filtrate do not. Furthermore, as DON is an inhibitor of protein synthesis, it may prepare the way for fungus invasion, when the defense mechanisms of the plant need protein synthesis. Here the data of Miller et al. (1985) are important, as they claim that resistant wheat varieties are able to metabolize DON, but susceptible varieties either much less or not at all.

In wheat, an unselected $R_2$ somaclone of the variety Tobari 66 has been found to be more resistant to *Fusarium* (Ahmed et al. 1991), and among the $F_1$ hydrids from the cross of *Fusarium*-susceptible Marquis and resistant Wangshu Bai wheats, nearly 50% of the haploid lines tolerated $10^{-6}$ M DON and 3-acetyl-DON. After artificial inoculation, these lines also showed *Fusarium* resistance in the field (Simmonds et al. 1993). Consequently, somaclonal variation not induced by in vitro selection pressure can be similarly useful in generating *Fusarium* resistance in wheat. In this respect, the anther culture-derived haploid system seems to be particularly successful, probably due to some minor (recessive) genes present even in the susceptible genotypes (Simmonds et al. 1993).

## 3 Induced or Noninduced Somaclonal Variation for *Fusarium* Resistance?

Induction of toxin-resistant calli on selective media and subsequent regeneration of putative resistant plants has already proved successful (Wenzel 1985; Daub 1986; Gengenbach and Rines 1986; Van den Bulk 1991). However, the lack of a sound theoretical basis (e.g., the unclear role of the toxins in pathogenesis, questionable transfer of resistance from calli or microspore-embryoids to the regenerated plants), makes its practical application in creating scab-resistant lines uncertain. Production of unselected somaclones is easier, requiring less time and energy, but the success of obtaining resistant individuals in this way is per definitionem (Larkin and Scowcroft 1981) random and unpredictable.

Other approaches, like utilization of gametoclonal variation and selection in cell or protoplast cultures, seem to be not feasible, due to inefficiency (high frequency of albino haploids and sporadic regeneration, respectively). Consequently, it is reasonable to compare the potential of selected and unselected somaclonal variation in generating *Fusarium* resistance in wheat.

### 3.1 In Vitro Selection Techniques and Resistance Test

For in vitro selection of somatic wheat calli resistant to toxic *Fusarium* metabolites, double-layer and culture filtrate techniques were used. The double-layer technique was a modification of the method described by Lepoivre et al. (1986). For the culture filtrate technique, the toxin-containing suspensions were prepared according to the procedure published by Mesterházy (1977). Details of the methods are given in Section 5, Protocol.

With the double-layer system, calli of five spring (Lerma Rojo, Sakha 8, Sakha 69, Siete Cerros, Tobari 66) and four winter wheats (GK Bence, GK Kincsö, GK Mini Manó, GK Ságvári) were selected. In experiments with culture filtrate, calli of five spring and one winter wheat cultivars (Giza 155, Giza 157, Giza 160, Sakha 8, Sakha 69 and GK Ságvári) were used.

The regenerated plants ($R_0$) produced by both techniques were transplanted into pots filled with sterile soil and transferred in a greenhouse to grow to maturity. $R_1$ to $R_3$ selfed generations were evaluated for seedling blight resistance using a seedling test (Mesterházy 1978). Besides the plants of the original cultivars, individuals of two *Fusarium*-resistant wheat genotypes (74–2 and 84–4) served as controls. $R_0$ plants were saved for seed production and not subjected to resistance test. $R_2$ to $R_3$ plants were regarded as more resistant or susceptible than the original variety, when they differed significantly in at least two of the four seedling traits compared (seed germination, mortality, shoot length, and dry matter content).

### 3.2 Callus Growth and Plant Regeneration

Independently of callus origin and type, the growth of these inoculi was strongly inhibited on the upper MS layer and somewhat less on the MS medium containing *Fusarium* culture filtrate (Table 2). After 4 weeks, the majority of the

**Table 2.** Fresh weight changes of GK Ságvári winter wheat calli after 4 weeks of culture on control and toxin-containing MS media (30 calli each), in grams. (Ahmed et al. 1991)

| Mean callus weight (g) | MS | MS upper layer without and with *F. graminearum* | | MS + 30% Czapek-Dox broth | MS + 30% *F. graminearum* filtrate |
|---|---|---|---|---|---|
| Initial | 1.18 | 1.10 | 1.31 | 1.16 | 1.07 |
| After 4 weeks | 1.73 | 1.48 | 1.11 | 1.60 | 1.14 |
| Difference, g | 0.55 | 0.38 | −0.20[a] | 0.44 | 0.07[a] |
| Difference, % | 46.61 | 34.55 | −15.27 | 37.93 | 6.54 |

[a] Significant at $p = 0.1$ probability level, compared to the corresponding control values.

calli turned brown and died in both culture systems. Inhibition of growth and decay of the calli, however, was influenced also by the fungus isolate and the wheat cultivar. The regenerating ability of the selected calli also decreased or was completely lost. Selected calli of spring wheats differentiated into plantlets, while those of the winter wheat varieties developed roots only, if at all, on the regenerating medium. Results of a representative experiment with double-layer technique are summarized in Table 3. The culture filtrate experiment produced similar results. In both in vitro selection systems, Sakha 8 excelled with the highest percentage of resistant calli, followed by Lerma Rojo 64 (in the double-layer technique only) and Sakha 69. At the adult plant level, Lerma Rojo 64 proved to be *Fusarium*-susceptible; however, Sakha 8 and Sakha 69 exhibited good resistance. Several plants regenerated from resistant calli had low viability, abnormal habit, and were partially or completely sterile.

Variation of the genotypic response in vitro selection for disease resistance is well documented in the literature. When selecting calli and cell cultures for *Fusarium* resistance, similar observations were made (Liu et al. 1990; Yan et al. 1990). The varying effect of different *Fusarium* isolates through change in toxin efficiency has also been mentioned (Yan et al. 1990). The detrimental effect of in vitro selection by *Fusarium* toxins on the regenerating ability of callus cultures, as pointed out by Yan et al. (1990), has also been confirmed by our findings. As found by Yan et al. (1990), in wheat, the survival rate of calli from genotypes more resistant to scab was better on the toxic than that of the more susceptible genotypes.

### 3.3 *Fusarium* Resistance of the Progenies—Results and Reflections

In the seedling test, the first two selfed generations ($R_1$ and $R_2$) of selected callus origin gave disappointing results. Among 19 $R_1$ and 20 $R_2$ lines tested, only two $R_1$ lines (10.5%) proved to be more resistant than the original variety, and none of the $R_2$ lines showed improved disease resistance. $R_1$ and $R_2$ progeny of the

**Table 3.** Selection of *Fusarium* toxin-resistant wheat calli with the double-layer technique and their plant regeneration as compared to those of the controls. (Ahmed et al. 1991)[a]

| Variety | Total calli | Resistant calli | | Regenerated plants | |
|---|---|---|---|---|---|
| | No. | No. | % | No. | % |
| Lerma Rojo 64 | 50 | 9 | 18.0 | 11 | 22.0 |
| " control | 10 | | | 2 | 20.0 |
| Sakha 8 | 45 | 14 | 31.0 | 8 | 17.8 |
| " control | 12 | | | 6 | 50.0 |
| Sakha 69 | 60 | 6 | 10.0 | 9 | 15.0 |
| " control | 18 | | | 9 | 50.0 |
| Tobari 66 | 38 | 4 | 10.5 | 2 | 5.3 |
| " control | 15 | | | 5 | 33.3 |

[a] Resistant calli of GK Bence, GK Kincsö, GK Mini Manó, and GK Ságvári failed to regenerate plants.

primary regenerants from unselected wheat calli were not more successful either, producing a single *Fusarium*-resistant line only (8%). However, in the $R_3$ generation, the relationships changed positively. From the 28 selected $R_3$ lines, 9 (32.1%) became more resistant than the original cultivar, and among the unselected $R_3$ lines, 7 (46.7%) showed better seedling resistance. The majority of the $R_1$ and $R_2$ lines derived from both selected and unselected calli were susceptible or exhibited a disease reaction similar to that of the donor variety. The susceptible lines disappeared from the $R_3$ populations, but the frequency of the lines with similar disease reaction increased (Table 4).

Resistance parameters of some positive somaclones are presented in Table 5. Obviously, seed germination was least affected by *Fusarium* inoculation, and this parameter could be only slightly improved by the in vitro techniques. In the other three traits, especially in shoot growth and dry matter content, much greater progress could be achieved. As far as their *Fusarium* resistance is concerned, some of the lines of in vitro origin were almost as good (Giza 155 CF/24A) as or equal (Sakha 69 CF-13–7) to the reference cultivars. The example of the unselected F2/18-B Tobari 66 line shows that the *Fusarium* resistance induced in vitro is heritable, and consequently mediated by somaclonal variation.

Figure 2 demonstrates how vigorously plants of another unselected, resistant line (Sakha 8 CF/8-3, $R_3$) grew in perlite inoculated with *F. graminearum* and *F.*

**Table 4.** Inventory of the *Fusarium* disease reaction in $R_1$ to $R_3$ progeny of primary regenerants originated from selected and unselected calli. (Ahmed et al., unpubl.)

| Cultivar | Line Selected | Line Unselected | No. of lines | No. of lines and their resistance level Susceptible | Unchanged | Resistant |
|---|---|---|---|---|---|---|
| Sakha 8 | $R_1$, $R_2$ | | 14 | 7 | 6 | 1 |
| | | $R_2$ | 4 | 4 | 0 | 0 |
| Sakha 69 | $R_1$, $R_2$ | | 13 | 8 | 5 | 0 |
| | | $R_2$ | 4 | 3 | 1 | 0 |
| Giza 155 | $R_1$ | | 4 | 0 | 3 | 1 |
| Giza 157 | $R_1$ | | 3 | 0 | 3 | 0 |
| Tobari 66 | | $R_2$ | 2 | 0 | 1 | 1 |
| Lerma R. 64 | $R_2$ | | 5 | 0 | 5 | 0 |
| Total | | | 49 | 22 | 24 | 3 |
| % | | | 100 | 44.9 | 48.9 | 6.1 |
| Sakha | $R_3$ | | 11 | 0 | 6 | 5 |
| | | $R_3$ | 11 | 0 | 7 | 4 |
| Sakha 69 | $R_3$ | | 6 | 0 | 4 | 2 |
| | | $R_3$ | 1 | 0 | 0 | 1 |
| Giza 155 | $R_3$ | | 1 | 0 | 0 | 1 |
| Giza 157 | $R_3$ | | 3 | 0 | 3 | 0 |
| Giza 160 | $R_3$ | | 1 | 0 | 0 | 1 |
| Tobari 66 | $R_3$ | | 1 | 0 | 1 | 0 |
| | | $R_3$ | 3 | 0 | 1 | 2 |
| Lerma R. 64 | $R_3$ | | 5 | 0 | 5 | 0 |
| Total | | | 43 | 0 | 27 | 16 |
| % | | | 100 | 0 | 62.8 | 37.2 |

**Table 5.** Resistance parameters of some *Fusarium*-resistant wheat somaclones in percent of the controls. Control values are given in % of those of the uninfected plants. (Ahmed et al. 1991)

| Cultivar | Line, progeny Selected | Unselected | Seed germination | Seedling mortality | Shoot length | Dry matter content |
|---|---|---|---|---|---|---|
| Giza 155 | | Control | 80.0 | 52.5 | 44.6 | 50.7 |
| | CF/24A, $R_1$ | | 92.5 | 22.5[a] | 112.1[a] | 92.8[a] |
| | CF/24A, $R_3$ | | 92.5 | 24.5[a] | 83.4[a] | 87.8[a] |
| Sakha 8 | | Control | 63.2 | 52.6 | 38.3 | 44.5 |
| | CF/3-2, $R_3$ | | 90.0[a] | 20.0[a] | 69.7[a] | 88.6[a] |
| | CF/8-3, $R_3$ | | 88.9[a] | 15.6[a] | 92.3[a] | 77.1[a] |
| Sakha 69 | | Control | 82.5 | 32.5 | 55.9 | 55.9 |
| | CF/13-7, $R_3$ | | 100.0 | 4.2[a] | 94.7[a] | 90.1[a] |
| Tobari 66 | | Control | 55.6 | 58.5 | 34.0 | 38.4 |
| | | F2/18-B, $R_2$ | 73.7 | 32.9[a] | 94.1[a] | 103.3[a] |
| | | F2/18-B, $R_3$ | 87.5[a] | 15.0[a] | 88.1[a] | 86.0[a] |
| 74-4 | | Reference | 99.2 | 16.5 | 93.2 | 102.7 |
| 84-4 | | Reference | 100.4 | 14.4 | 102.6 | 96.6 |

[a] Values are significantly different from the control values at p = 5% probability level.

**Fig. 2.** *Fusarium* resistance of Sakha 8 and its CF/8-3 $R_3$ line derived from selected callus. *From left to right* Sakha 8 and CF/8-3 $R_3$ not inoculated controls, Sakha 8 + *F. graminearum*, CF/8-3 + *F. graminearum*, Sakha 8 + *F. culmorum*, and CF/8-3 + *F. culmorum*

*culmorum*, while the majority of the seeds from the original variety failed to germinate; the seedlings could either not emerge, or died.

As can be seen from Table 6, Sakha 8 and tobari 66 can be regarded as the most successful varieties, producing the highest number of lines with improved *Fusarium* resistance among the cultivars involved in the experiments. From comparison of the resistance grade of the original cultivars with the number of resistant lines produced in vitro, it is clear that the extent of the resistance does not correlate with the number of resistant regenerants. From two cultivars (Giza

**Table 6.** Grouping of the wheat varieties according to their performance in production of lines with different *Fusarium* resistance by using in vitro techniques. (Ahmed et al., unpubl.)

| Cultivar | Resistance grade | Total no. of lines | Lines and their disease reaction | | | | | |
|---|---|---|---|---|---|---|---|---|
| | | | Susceptible | | Unchanged | | Resistant | |
| | | | No. | % | No. | % | No. | % |
| Tobari 66 | NR | 6 | 0 | 0.0 | 3 | 50.0 | 3 | 50.0 |
| Giza 155 | NR | 5 | 3 | 60.0 | 1 | 20.0 | 1 | 20.0 |
| Giza 160 | NR | 1 | 0 | 0.0 | 0 | 0.0 | 1 | 100.0 |
| Lerma R. 64 | NR | 10 | 0 | 0.0 | 10 | 100.0 | 0 | 0.0 |
| Sakha 8 | I | 40 | 11 | 27.5 | 19 | 47.5 | 10 | 25.0 |
| Giza 157 | I | 6 | 0 | 0.0 | 6 | 100.0 | 0 | 0.0 |
| Sakha 69 | R | 24 | 11 | 45.8 | 10 | 41.7 | 3 | 12.5 |

NR = not resistant (susceptible); I = intermediary; R = resistant.

157, Lerma Rojo 64), resistant lines could not be obtained. It is probable that these cultivars are recalcitrant to the variation induced by in vitro culturing.

Comparing the *Fusarium* resistance of lines originated from selected and unselected calli, the latter seem to be more efficient in this respect (Table 7). Nevertheless, this statement should be confirmed by further experiments.

Between the two in vitro selection methods used, the culture filtrate technique can be preferred as it is less time-consuming and its induction potential is not lower than that of the double-layer method in spite of its less drastic selection pressure.

As determined by reversed-phase high performance liquid chromatography (Mesterházy and Bartók 1993), DON was found occasionally, and if present, only in trace amounts in the selective media.

For testing field resistance, R3 seeds will be multiplied in the greenhouse and sown in the autumn 1995.

The dependence of in vitro selection success on genotype is a well-known phenomenon, described in several host-fungus relations (Chawla and Wenzel 1987; Pauly et al. 1987). However, it cannot be interpreted simply by the resistance or susceptibility of the donor genotypes, since their calli used for the in

**Table 7.** Comparative performance of the varieties in producing lines with varied *Fusarium* resistance from selected and unselected calli. (Ahmed et al. 1991)

| Cultivar | Callus group | Total no. of lines | Lines and their disease reaction | | | | | |
|---|---|---|---|---|---|---|---|---|
| | | | Susceptible | | Unchanged | | Resistant | |
| | | | No. | % | No. | % | No. | % |
| Sakha 8 | Selected | 22 | 6 | 27.3 | 11 | 50.0 | 5 | 22.7 |
| | Unselected | 15 | 4 | 26.7 | 7 | 46.7 | 4 | 26.7 |
| Sakha 69 | Selected | 17 | 8 | 47.1 | 7 | 41.2 | 2 | 11.7 |
| | Unselected | 5 | 3 | 60.0 | 1 | 20.0 | 1 | 20.0 |
| Tobari 66 | Selected | 1 | 0 | 0.0 | 1 | 0.0 | 0 | 0.0 |
| | Unselected | 3 | 0 | 0.0 | 1 | 33.3 | 2 | 66.7 |

vitro selection consist of a cell population morphologically, cytologically, and functionally quite different from that of the explant, leading to a divergent in vitro response. Furthermore, the source of the regenerated plants may be a cell or a cell group which escaped the selective toxic metabolites, or was surrounded by protecting, resistant cells (Hammerschlag 1988), or alternatively, has been partially damaged/mutagenized by the selective agent(s) (Gengenbach and Rines 1986). These cells can then easily regenerate less viable or abnormal plants, if they can regenerate at all. The induced resistance present in the callus can also be lost during regeneration (Ahmed et al. 1991) or not transferred into the differentiating tissues (Daub 1986). In the regenerants, as observed in wheat, the induced characters can partly segregate or can disappear and reappear transitionally in the subsequent generations, due probably to dominant-recessive gene shifts or transpositional events (Larkin et al. 1984; Ahloowalia and Sherington 1985).

Thus, when selected for *Fusarium* resistance in tissue culture, the final outcome can be an instable $R_1$ wheat population with a heterogenic *Fusarium* reaction and a low number of resistant lines, as found also in our study. The $R_2$ may represent a transitional generation, from which the epigenetically resistant, toxin- but not pathogen-resistant, and less vigorous lines will probably be selected out, while new resistant individuals emerge due to reparation of cytological defect and to gene shifts, since recombinations cannot take place during selfing. In $R_3$ the stabilization process seems to be continued, resulting in the further appearance of resistant and the elimination of the remaining weaker and susceptible lines (Table 4).

Although Yu et al. (1990) and D.C. Huang et al. (1991) found no difference between resistance variation in somaclones and sexual $F_2$ generation in two wheat crosses, according to our observations, genetic alterations advantageous to *Fusarium* resistance can be evoked in wheat by somaclonal variation, and since in this case the complex interactions between cells and pathotoxins on the one hand, and among the pathotoxins themselves on the other do not occur to reduce regeneration rate of the resistant plants, a more resistant plant output can be obtained (Table 7). Near or complete absence of DON from the selective media cannot be made responsible for the limited selection success, since the *Fusarium* isolates can produce a number of toxins other than DON (see Sect. 1.1), and the selection potential of DON is questionable (Liu et al. 1990).

Another aspect must be considered when explaining the above results. Since the work of Mesterházy (1987), we know that *Fusarium* resistance in seedlings and ears should not be the same. Also resistance of the calli may show specificity, and it is not known whether it correlates with seedling or adult stage resistance or both. It is possible that clarification of these problems can contribute to the more efficient transfer of resistance from calli to seedlings and ears.

## 4 Summary and Conclusions

Pathogenic *Fusarium* fungi, especially *F. graminearum* and *F. culmorum*, often cause serious yield and quality losses in wheat. Through contamination with toxic metabolites, they can render the end product, the grains, unsuitable for human consumption and animal feeding. Thus, incorporation of genetic resistance into both bread and forage wheat is an important task. In spite of international breeding efforts, no desirable sortiment of *Fusarium*-resistant winter wheat could as yet be created. Therefore, the development of resistant lines via in vitro techniques seemed to be a real alternative.

Somatic callus cultures of 15 wheat varieties were established at the Cereal Research Institute, Szeged, Hungary and partially selected in vitro by means of the double-layer and culture filtrate techniques, i.e., by using nutrient media supplemented with pathotoxins from isolates of *F. graminearum* and *F. culmorum*. Altogether, 179 primary ($R_0$) lines were obtained and their further selected progenies ($R_1$–$R_3$) were screened for *Fusarium* seedling resistance by seed inoculation in a greenhouse test. From a total of 92 lines, 19 (20.6%) showed *Fusarium* resistance improved over the donor varieties, 22 (23.9%) were less resistant than the original cultivars, and the reaction of 51 lines (55.4%) remained unchanged. The resistance level of some regenerants was equal to that of the *Fusarium*-resistant reference cultivars and remained constant in the $R_3$ generation. The frequency of the in vitro-derived resistant lines depended on the variety, but there was no correlation between resistance grade of the donors and the number of the related resistant regenerants. Selected calli produced somewhat less resistant lines than nonselected. In the selective media, DON could be detected in trace amounts only, but this does not mean that these media were ineffective, as other, undetermined toxins, could be present.

In conclusion, both in vitro selection techniques ensured a selection pressure high enough to obtain wheat calli resistant to *Fusarium* toxins with acceptable frequency, under conditions definitively cheaper than by application of commercially available trichothecenes or other toxic compounds. It is advisable to run putative resistant calli through further selection cycles and not exclude their susceptible descendants from the experiment, since in the following generations some of them can develop into resistant lines. Plants regenerated from unselected calli are also worth screening for *Fusarium* resistance, and in a tissue culture-linked *Fusarium* resistance project, under continuous control of toxicity of selective media, combined use of in vitro selection and somaclonal variation can be recommended.

## 5 Protocol

Initiate and grow callus cultures preferably from immature, 10–12-day-old embryos of immature, 1–2-cm-long inflorescences on solid MS medium (Murashige and Skoog 1962) containing 1–2 mg/l 2,4-D at 26 °C in the dark. Practically all regenerable callus cultures are suitable. For double-layer culture, inoculate a ca. 1-cm-thick potato dextrose agar medium filled in cylindrical glasses (e.g., baby food jars) with mycelium of *F. graminearum* or *F. culmorum*. Incubate the culture at 26 °C for 4 days in a 18/6 h light-dark cycle and then at 5 °C for 3 days in continuous light for fungus growth and toxin production. Autoclave the flasks at 120 °C for 15 min to kill the fungal cells (the toxic compounds are heat-stable). After cooling, overlayer the fungus medium with an equal amount of MS medium supplemented with 1 mg/l 2,4-D and leave to stand at room temperature in the dark for 7 days, allowing the toxins to diffuse up into the MS layer. Place 4–5-week-old calli on the upper medium and keep them on it for 4–5 weeks at 26 °C in 16 h photoperiods. Prepare the control cultures similarly, but without fungus inoculation. Transfer the healthy looking, surviving calli to MS medium containing 0 or 0.1 mg/l 2,4-D for regeneration.

For the culture filtrate technique, add 30% cell-free culture liquid (homogenized and filtered Czapek-Dox medium after 1 week aerated fungus culture and 2–6 wkeeks maintenance in a refrigerator) to MS medium supplemented with 0.1 mg/l 2,4-D and 7% agar before autoclaving. Place 4–5-week-old calli on the toxic MS medium (20 calli per 100 × 15 mm Petri dish containing 40 ml medium) and incubate the cultures for 4–5 weeks at 26 °C in 16-h photoperiods. Then transfer the healthy calli to fresh toxic medium for two further selection cycles and place the resistant calli on MS regenerating medium as above. Replace culture filtrate with pure Czapek-Dox medium for the controls and proceed similarly.

*Acknowledgments.* The authors wish to acknowledge the support of the OTKA (National Scientific Research Foundation) project 585.

## References

Abbas HK, Mirocha CJ, Pawlosky RJ, Pusch D (1985) Effect of cleaning, milling and baking on deoxynivalenol in wheat. Appl Environ Microbiol 50: 82–486

Ahloowalia BS (1982) Plant regeneration from callus cultures of wheat. Crop Sci 22: 405–410

Ahloowalia BS, Sherington J (1985) Transmission of somaclonal variation in wheat. Euphytica 34: 525–537

Ahmed KZ, Mesterházy Á, Sági F (1991) In vitro techniques for selecting wheat (*Triticum aestivum* L.) for *Fusarium* resistance. Double-layer culture technique. Euphytica 57: 251–257

Arseniuk E, Góral T, Czembor HJ (1992) Pathogenicity and differential response in *Fusarium spp.* -X *Tritico-secale* Wittmack (*Triticum aestivum* L.) *Secale cereale* pathosystems. Hodowla Rosl Aklim Nasienn 37: 17–24

Bajaj YPS (1986) In vitro regeneration of diverse plants and the cryopreservation of germplasm in wheat (*Triticum aestivum* L.). Cereal Res Commun 14: 305–311

Bajaj YPS (ed) (1990) Biotechnology in agriculture and forestry, vol 13. Wheat. Springer, Berlin Heidelberg New York

Chawla HS, Wenzel G (1987) In vitro selection in barley and wheat for resistance against *Helminthosporium sativum.* Plant Breeding 49: 159–163

Chelkowski J (1989) Formation of mycotoxins produced by *Fusaria* in heads of wheat, triticale and rye In: Chelkowski J (ed) *Fusarium* mycotoxins, taxonomy and pathogenicity, vol 2 Elsevier Amsterdam, pp 63–84

Chen THH, Lazar MD, Scoles GJ, Gusta LV, Kartha KK (1987) Somaclonal variation in a population of winter wheat. J Plant Physiol 130: 27–36

Chowdhury MKU, Vasil V, Vasil IK (1994) Molecular analysis of plants regenerated from embryogenic cultures of wheat (*Triticum aestivum* L.). Theor Appl Genet 87: 821–828

Cole RJ, Dorner JW, Cox RH, Cunfer BM, Cutler HG, Stuart BP (1981) The isolation and identification of several trichothecene mycotoxins from *Fusarium heterosporum.* J Nat Prod 44: 324–330

Cooper DB, Sears RG, Lookhart GL, Jones BL (1986) Heritable somaclonal variation in gliadin proteins of wheat plants derived from immature embryo callus culture. Theor Appl Genet 71: 784–790

Daub EM (1986) Tissue culture and the selection of resistance to pathogens. Annu Rev Phytopathol 24: 159–186

Davies PA, Pallotta MA, Ryan SA, Scowcroft WR, Larkin LJ (1986) Somaclonal variation in wheat: genetic and cytogenetic characterization of alcohol dehydrogenase 1 mutants. Theor Appl Genet 72: 644–653

van Eeuwijk FA, Mesterházy Á, Kling Ch I, Ruckenbauer P, Saur L, Bürstmayr H, Lemmens M, Maurin M, Snijders CHA (1994) Assessing non-specificity of resistance in wheat to head blight caused by inoculation with European strains of *Fusarium culmorum, F. gramicearum* and *F. nivale,* using a multiplicative model for interaction. Theor Appl Genet (accepted for publication)

Fadel F, Wenzel G (1993) In vitro selection for tolerance in $F_1$ microspore populations of wheat. Plant Breeding 110: 85–95

Galiba G, Sutka J (1989) Frost resistance of somaclones derived from *Triticum aestivum* L. winter wheat calli. Plant Breeding 102: 101–104

Galiba G, Kertész K, Sutka J, Sági L (1985) Differences in somaclonal variation in three winter wheat (*Triticum aestivum* L.) varieties. Cereal Res Commun 13: 343–350

Gengenbach BG, Green CE (1975) Selection of T cytoplasm maize callus cultures resistant to *Helminthosporium maydis* race T pathotoxin. Crop Sci 15: 645–649

Gengenbach BG, Rines HW (1986) Use of pathotoxins in selection of disease-resistant mutants in tissue culture. Iowa State J Res 60: 449–476

Gilbert J (1989) Current views on the occurrence and significance of *Fusarium* toxins. J Appl Bacteriol Symp Suppl 895–985

Gocho H (1985) Wheat breeding for scab resistance. Wheat Inf Serv 60: 41 (Abstr)

Gu JQ (1983) Study of the genetics of resistance to wheat scab. Sci Agric 18: 500–523

Hammerschlag FA (1988) Selection of peach cells for insensitivity to culture filtrates of *Xanthomonas campestris* f. sp. *pruni* and regeneration of resistant plants. Theor Appl Genet 76: 865–869

Hanson K, Hucl P, Baker RJ (1994) Comparative field performance of tissue culture-derived lines and breeder lines of HY320 spring wheat. Plant Breeding 112: 183–191

Hollins W (1993) *Fusarium* resistance breeding in PBI: identification of resistance, resistance sources, maturity problems and adult glasshouse tests. Meeting of the European Wheat *Fusarium* Network Vienna Feb (Abstr)

Huang DC, Liu ZZ, Sun XJ, Zhang FQ, Tang YL (1991) Somaclonal variation in resistance to scab (*Gibberella zeae* Petch.) in wheat. Acta Agric Shanghai Suppl 7: 60–64

Huang XM, Liu ZZ, Wang ZY, Lu SH, Yao QH, Yang YM (1991a) The effect of deoxynivalenol on the viability of mesophyll protoplasts of wheat cultivars with different scab resistance. Acta Agric; Sin 7: 8

Huang XM, Liu ZZ, Yao QH, Wang ZY, Lu SH, Yao QH, Yang YM (1991b) The different sensitivities to DON of mesophyll protoplasts of wheat cultivars differing in sensitivity to scab. Acta Agric Sin 7: 8–15

Karp A, Maddock SE (1984) Chromosome variation in wheat plants regenerated from cultured immature embryos. Theor Appl Genet 67: 249–255

Király Z (1986) Toxins. In: Goodman RN, Király Z, Wood KR (eds) The biochemistry and physiology of plant disease. University of Missouri Press, Columbia, pp 589–641

Larkin JP, Scowcroft WR (1981) Somaclonal variation—a novel source of variability from cell cultures for plant improvement. Theor Appl Genet 60: 197–214

Larkin PJ, Ryan SA, Brettell SIR, Scowcroft WR (1984) Heritable somaclonal variation in wheat. Theor Appl Genet 67: 443–455

Lazar MD, Chen THH, Gusta LV, Kartha KK (1988) Somaclonal variation for freezing tolerance in a population derived from Norstar winter wheat. Theor Appl Genet 75: 480–484

Lee VS, Jang HS, Tanaka T, Oh YJ, Cho CM, Ueno Y (1987) Effect of milling on decontamination of *Fusarium* mycotoxins nivalenol, deoxynivalenol and zearalenone in Korean wheat. J Agric Food Chem 35: 126–129

Lemmens M, Bürstmayr H, Ruckenbauer P (1993) Variation in *Fusarium* head blight susceptibility of international and Austrian wheat breeding material. Bodenkultur 44: 65–78

Lepoivre P, Viseur J, Duhem K, Carels N (1986) Double-layer culture technique as a tool for selection of calluses resistant to toxic material from plant pathogenic fungi. In: Semal J (ed) Somaclonal variations and crop improvement. Nijhoff, Dordrecht, pp 45–52

Lepschy v Gleissenthal J, Dietrich R, Märtbauer E, Schuster M, Süss A, Terplan G (1989) A survey on the occurrence of *Fusarium* mycotoxins in Bavarian cereals from the 1987 harvest. Z Lebensm Unters Forsch 188: 521–526

Li J, Huang WF (1992) Preliminary study on the induction of resistant mutants to the pathotoxin of *Fusarium graminearum* Schw. in wheat by tissue culture. J Hebei Agric Univ 15: 47–51

Liu ZZ, Chen QQ, Yao QH, Huang XM, Lu SH, Wei CM, Sun XJ, Zhang FQ, Tang YL (1990) Growth hormone-like action of deoxynivalenol toxin on induction and differentiation of wheat calluses. Acta Agric Shanghai Suppl 7: 1–7

Maddock SE, Semple JT (1986) Field assessment of somaclonal variation in wheat. J Exp Bot 37: 1065–1078

Maddock SE, Lancaster VA, Risiott R, Franklin J (1983) Plant regeneration from cultured embryos and inflorescences of 25 cultivars of wheat (*Triticum aestivum*). J Exp Bot 34: 915–926
Mathysse AG (1983) The use of tissue cultures in the study of crown gall and other bacterial diseases. In: Helgeson JP, Deverall BJ (eds) Use of tissue culture and protoplasts in plant pathology. Academic Press New York, pp 39–68
Mesterházy Á (1977) Reaction of winter wheat varieties to four *Fusarium* species. Phytopathol Z 90: 104–112
Mesterházy Á (1978) Comparative analysis of artificial inoculation methods with *Fusarium* spp. on winter wheat varieties. Phytopathol Z 93: 12–25
Mesterházy Á (1983) Breeding wheat for resistance to *Fusarium graminearum* and *F. culmorum*. Z Pflanzenzuecht 91: 295–311
Mesterházy Á (1984) *Fusarium* species of wheat in South Hungary, 1970–1983. Cereal Res Commun 12: 167–170
Mesterházy Á (1987) Selection of head blight resistant wheats through improved seedling resistance. Plant Breeding 98: 25–36
Mesterházy Á (1988) Expression of resistance to *Fusarium graminearum* and *F. culmorum* under various experimental conditions. J Phytopathol 133: 304–310
Mesterházy Á (1989) Progress in breeding of wheat and corn not susceptible to infection by fusaria. In: Chelkowski J (ed) *Fusarium* – Mycotoxins, taxonomy and pathogenicity. Elsevier, Amsterdam pp 357–386
Mesterházy Á, Bartók T (1993) Resistance, pathogenicity and *Fusarium* spp. influencing toxin (DON) contamination of wheat varieties. 3rd Eur Seminar Fusarium—mycotoxins, taxonomy, pathogenicity and host resistance, Radzików, Poland. Hodowla Rosl Aklim Nasienn 37: 9–15
Mesterházy Á, Kovács K (1986) Breeding corn against fusarial stalk rot, ear rot and seedling blight. Acta Phytopathol Entomol Hung 21: 231–249
Miller SA, Williams GR, Medina-Filho H, Evans D (1985) A somaclonal variant of tomato resistant to race 2 of *Fusarium oxysporum f. sp. lycopersici*. Phytopathology 75: 1354
Mirocha CJ, Chen J, Xie W, Xu Y (1992) Biology and chemistry of fumomisin and aal toxins. 3rd Eur Seminar *Fusarium*—mycotoxins, taxonomy, pathogenicity and host resistance, Radzików (Abstr)
Morère-Le Paven MC, De Buyser J, Henry Y, Corre F, Hartmann C, Rode A (1992) Multiple patterns of mtDNA reorganization in plants regenerated from different in vitro cultured explants of a single wheat variety. Theor Appl Genet 85: 9–14
Munteanu I, Muresan t, Tataru V (1972) *Fusarium* wilt in wheat and integrated disease control in Romania. Acta Phytopathol Acad Sci Hung 21: 17–29
Murashige T, Skoog F (1962) A revised medium for rapid growth and bioassays with tobacco tissue cultures. Physiol Plant 15: 473–497
Nowick TW, Gaba DG, Dexter JE, Matsuo RR, Clear RM (1988) Retention of the *Fusarium* mycotoxin deoxynivalenol in wheat during processing and cooking of spaghetti and noodles. J Cereal Sci 8: 189–202
Pauly MH, Shane WW, Gengenbach BG (1987) Selection for bacterial blight phytotoxin resistance in wheat tissue culture. Crop Sci 27: 340–344
Rode A, Hartmann C, Falconet D, Lejeune B, Quétier F, Benslimane A, Henry Y, De Buyser J (1987) Extensive mitochondrial DNA variation in somatic tissue cultures initiated from immature embryos. Curr Genet 12: 369–376
Ryan SA, Scowcroft WR (1987) A somaclonal variant of wheat in additional $\beta$-amylase isozymes. Theor Appl Genet 73: 459–464
Ryan SA, Larkin PJ, Ellison FW (1987) Somaclonal variation in some agronomic and quality characters in wheat. Theor Appl Genet 74: 77–82
Sági H, Börner A, Ács E, Bartók T, Sági F (1993) Genetic identification, agronomic performance and technological quality of tissue culture-induced dwarfs of the Mv 4 winter wheat. Cereal Res Commun 21: 309–315
Scott PM, Kanhere SR, Dexter JE, Brennan PW, Trenholm HL (1984) Distribution of the trichothecene mycotoxin deoxynivalenol (vomitoxin) during the milling of naturally contaminated hard red spring wheat and its fate in baked products. Food Add Contam 1: 313–323
Shimada T, Otani M (1990) Effects of *Fusarium* mycotoxins on the growth of shoots and roots at germination in some Japanese wheat cultivars. Cereal Res Commun 18: 229–232

Simmonds J, Fregean-Reid J, Pandeya R, Sampson D, Fedak G (1993) Potential of anther culture in breedig for *Fusarium* head blight resistance in wheat. Cereal Res Commun 21: 141–147
Snijders CHA (1988) The phytotoxic action of deoxynivalenol and zearalenone on wheat seedlings. Proc Jpn Assoc Mycotox Suppl 1: 103–104
Snijders CHA (1990a) Genetic variation for resistance to *Fusarium* head blight in bread wheat. Euphytica 50: 171–179
Snijders CHA (1990b) *Fusarium* head blight and mycotoxin contamination of wheat, a review. Neth J Plant Pathol 96: 187–198
Snijders CHA (1990c) The inheritance to head blight caused by *Fusarium culmorum* in winter wheat. Euphytica 50: 11–18
Snijders CHA, van Eeuwijk FA, Mesterházy Á, Kling Ch I, Ruckenbauer P, Saur L, Maurin N (1993) Resistance of wheat to head blight caused by inoculation with European strains of *Fusarium culmorum, F. graminearum* and *F. nivale*. 6th Int Congr Plant pathology, July 28–Aug 6, Montreal (Abstr)
Szunics Lu, Szunics L (1992) Methods for infecting wheat ear with *Fusarium* and the susceptibility of the varieties. Növénytermelés 41: 201–210 (in Hungarian)
Tanaka T, Hasegawa A, Matsuki Y, Ishii K, Ueno Y (1985) Improved methodology for the simultaneous detection of the trichothecene mycotoxins deoxynivalenol and nivalenol in cereals. Food Add Contam 2: 125–137
Van den Bulk RW (1991) Application of cell and tissue culture and in vitro selection for disease resistance breeding—a review. Euphytica 65: 269–285
Wakulinski W (1989) Phytotoxicity of *Fusarium* metabolites in relation to pathogenicity. In: Chelkowski J (ed) *Fusarium* mycotoxins, taxonomy and pathogenicity, vol 2. Elsevier, Amsterdam, pp 257–268
Wang H, Meng FJ (1990) Formation of endogenous zearalenone and its inhibition by malathion in winter wheat during vernalization. Acta Phytopathol Sin 16: 197–200
Wang YZ, Miller JD (1988) Effects of *Fusarium graminearum* metabolites on wheat tissue in relation to *Fusarium* head blight resistance. J Phytopathol 122: 118–125
Wenzel G (1985) Strategies in unconventional breeding for disease resistance. Annu Rev Phytopathol 23: 149–172
Wenzel G, Foroughi-Wehr B (1990) Progeny tests of barley, wheat and potato regenerated from cell cultures after in vitro selection for disease resistance. Theor Appl Genet 80: 359–365
Yan Z, Cuilan Z, Yuwen W, Shuxin R (1990) Scab resistant wheat breeding by cell engineering. Proc Int Symp Wheat breeding—prospects and future approaches, Albena, pp 136–138
Yu YJ (1982) Monosomic analysis for wheat scab resistance and yield components in the wheat cultivar Soo-moo-3. Cereal Res Commun 10: 185–189
Yu YJ, Liao PG, Pu Z, Yu FQ (1990) Genetic analysis of resistance to scab of regenerated plants ($R_2$) from wheat immature embryo culture. Acta Genet Sin 17: 461–468
Zhou CF, Xua SS, Qian CM, Yao GC, Shen JX (1987) Studies on breeding wheat for scab resistance. Acad Agric Sin 20: 19–25

# I.2 In Vitro Production of Male Sterile Rice Plants

D.H. LING[1]

## 1 Introduction

Huge economic and social benefits from the application of hybrid rice production have been obtained in China since 1976. The total area of hybrid rice plantation has now reached 16 million ha/year and constitutes 50% of the total rice area in China. Compared with conventional varieties, the yield of hybrid rice has increased by an average of around 1500 kg/ha (Fu and Gong 1994). Outside China (United States, Philippines, Thailand, Japan, Indonesia, Korea, etc.), hybrid rice production has also been successful on a small scale (Yuan and Mao 1991). However, the female parents of most combinations used in hybrid rice production in China were almost all derived from the same cytoplasm, Wild Rice Abortive cytoplast (WA-type cytoplasm). The long time span (almost 20 years), the large growing area (16 million ha/year), and the use of a single source of the cytoplasm of male sterile (ms) may cause the hybrid rice to be vulnerable to diseases or insect epidemics. Moreover, a single source of cytoplasm for the male sterile lines limits the choice of hybrid combination. Researchers in hybrid rice breeding have therefore been seeting a new source of cytoplasm (non-WA type cytoplasm) for a long time.

To develop a male sterile line to be used for hybrid seed production, traditionally the following methods were used: screening by test cross; cross breeding including backcrossing; radiation breeding, etc. In recent years, new methods of biotechnology have been developed in various crops to induce male sterility, such as anther culture (Kaul 1988), protoplast fusion (Akamatsu et al. 1988; Sakai and Imamura 1988), and genetic transformation (Mariani et al. 1990; Aarts et al. 1993). During the past few years, rice somaclones derived from somatic cell cultures have yielded male sterile mutants (Ling et al. 1987b; Ling 1991). A new male sterile line was developed from these ms mutants. This chapter describes the male sterile mutants in somaclones and the characters of a CMS line 54257/162-5A in detail. Both parents of female 54257 and male 162–5 of the ms line were somaclones and derived from $IR_{54}$ and $IR_{52}$, respectively.

[1] South China Institute of Botany, Academia Sinica, Guangzhou 510650, China

Biotechnology in Agriculture and Forestry, Vol. 36
Somaclonal Variation in Crop Improvement II (ed. by Y.P.S. Bajaj)

## 2 Sterile Mutants from Somaclones in Rice

Sterile mutant/variation are quite common in somaclones from in vitro culture of rice. In our case, the mutants which caused sterility in the somaclones were chromosome female sterile and male sterile mutants.

### 2.1 Chromosome Mutants

These mutants were usually found in the first generation of regenerated plants. Because of the chromosome problem in the plant, they are usually male, but also female sterile, so that when pollinated with normal pollen from other plants, almost no seeds were set. This included chromosome number and structure mutants.

#### *2.1.1 Chromosome Number Mutants*

From 319 plants regenerated from somatic cell culture of $IR_{36}$ and $IR_{54}$ between 1984 and 1985, 10 plants were tetraploid (Table 1); in these plants the number of chromosomes in root tip cells was 48 (Fig. 1) and the chromosome configuration of PMC in meiosis was 12 IV or 11 IV + 2 II (Fig. 2). Because the chromosome pairing is abnormal, only few seeds were set, so they were sterile (Ling et al. 1987a).

One triploid was found from anther culture of indica rice in 1980. During meiosis, no chromosomal pairing was observed, and 36 univalants were observed in metophase I. This was an asyndetic triploid mutant, and no seed was set (Ling et al. 1981).

#### *2.1.2 Chromosome Structure Mutants*

In 1985, one sterile plant was found among 122 regenerated plants ($R_1$) from mature seed culture in $IR_{54}$. Compared with the parent $IR_{54}$, the sterile $R_1$ plant was shorter in stature, with fewer tillers. The anthers of the mutant were smaller,

**Table 1.** Ploidy and fertility in plants regenerated from somatic cell culture in indica rice

| Year and variety | | Diploid and fertility | | | Tetraploid | Total |
|---|---|---|---|---|---|---|
| | | Fully fertile | Semifertile | Sterile | | |
| 1984 | $IR_{54}$ | 116 (95.1) | 12 (9.8) | 1 (0.8) | 3 (2.4) | 132 |
| | $IR_{36}$ | 89 (87.2) | 10 (9.8) | 1 (0.98) | 2 (1.9) | 102 |
| 1985 | $IR_{54}$ | 74 (80.4) | 9 (9.7) | 4 (4.3) | 5 (5.4) | 92 |
| | $IR_{36}$ | 9 (81.8) | 1 (9.1) | 1 (9.1) | 0 | 11 |

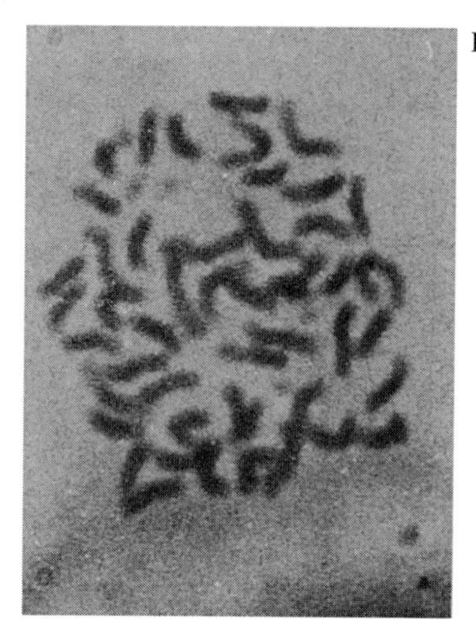

**Fig. 1.** Tetraploid in mitosis, 2n = 48. (Ling et al. 1987a)

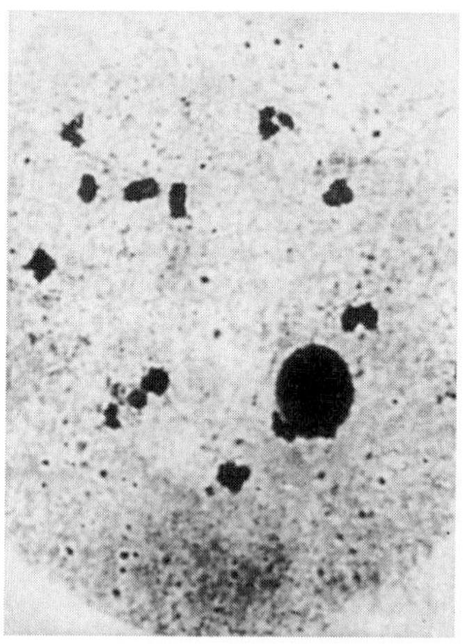

**Fig. 2.** Tetraploid in meiosis, showing 12 IV. (Ling et al. 1987a)

did not split, and were thus unable to shed its almost empty pollen during flowering (Fig. 4). From more than 200 panicles which were vegetatively produced by separating tillers, only three seeds were obtained in self-pollination condition. The chromosome number in the root tip cell was normal, 2n = 24 (Fig. 3).

Analysis of chromosome configuration in meiosis revealed a drastic disturbance. Only 5.7% of the PMCs showed the normal configuration of 12II (Table 2, Fig. 5). Multivalents of different complexity were frequent, such as tetravalent (1IV + 10II, 2IV + 8II), hexavalents (1VI + 9II, 1VI + 1IV + 7II), octovalents (1VIII + 8II, 1VIII + 1IV + 6II), and decavalents (1X + 7II; Fig. 5). Chromosome arrangements with one decavalent prevailed (50.9%). Multivalents of even higher orders were not found. The decavalents were open in half the cases (52%), almost half were ring-shaped (48%), some of the latter group being looped or 8-shaped. Chromosome configurations at diakinesis thus indicated that

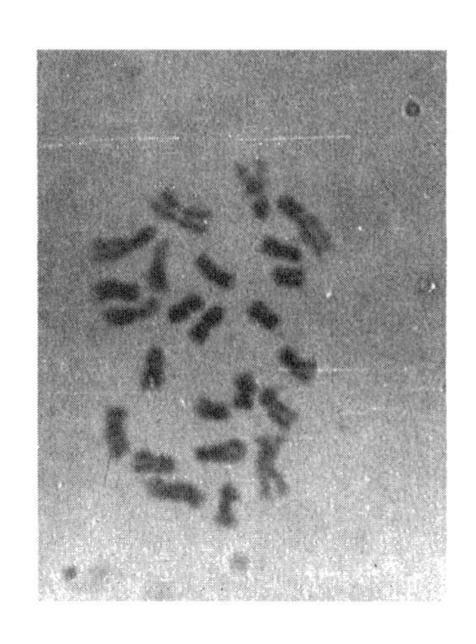

**Fig. 3.** The chromosome in root tip cell in MRT, 2n = 24. (Ling et al. 1987a)

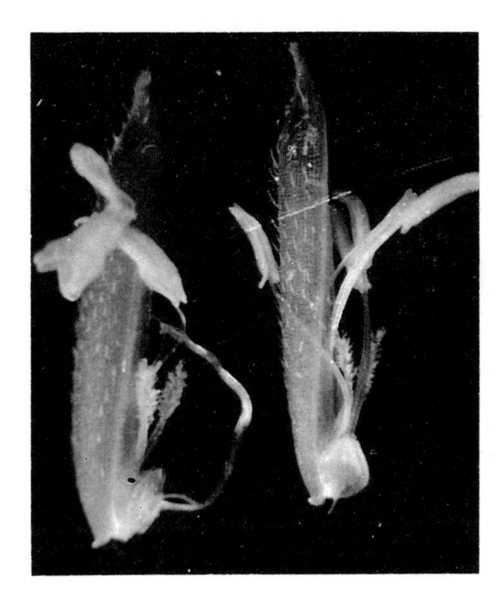

**Fig. 4.** The MRT mutant (*right*) and its donor variety, $IR_{54}$. The anthers in the mutant did not split. (Ling 1987)

a multiple reciprocal translocation (MRT) had occurred with five non-homologous chromosomes involved.

The chromosome distribution at metaphase II was more closely followed up in the MRT mutant (Table 2). In only 11.6% of the PMCs were the chromosomes evenly distributed (12/12) to the two daughter cells. The most common distribution pattern proved to be 13/11, amounting to 50.8%. Tetrads developed normally. A pollination test, by which the normal pollen from the parent variety, $IR_{54}$, was used for checking the function of the egg cells of the mutant, also showed high disturbance on the female side because only a few seeds were set (Ling et al. 1987a).

**Table 2.** Chromosome configurations at diakinesis of MRT sterile rice mutant induced from somaclones

| Chromosome configuration | Total no. of plants | 12II | 10II +1IV | 9II +1VI | 8II+ 1VIII | 8II +2IV | 7II +1IV +1VI | 7II +1X[a] | 6II +31V | 6II+ 1IV+ 1VIII | 6II +2VI |
|---|---|---|---|---|---|---|---|---|---|---|---|
| Observed frequency (%) | 438 | 25<br>5.7 | 13<br>3.0 | 29<br>6.6 | 36<br>8.2 | 27<br>6.2 | 61<br>13.5 | 223<br>50.9 | 10<br>2.3 | 9<br>2.1 | 5<br>1.1 |

[a] Of the 223 decavalents observed, 107 (48%) were ring-shaped and 116 (52%) open.

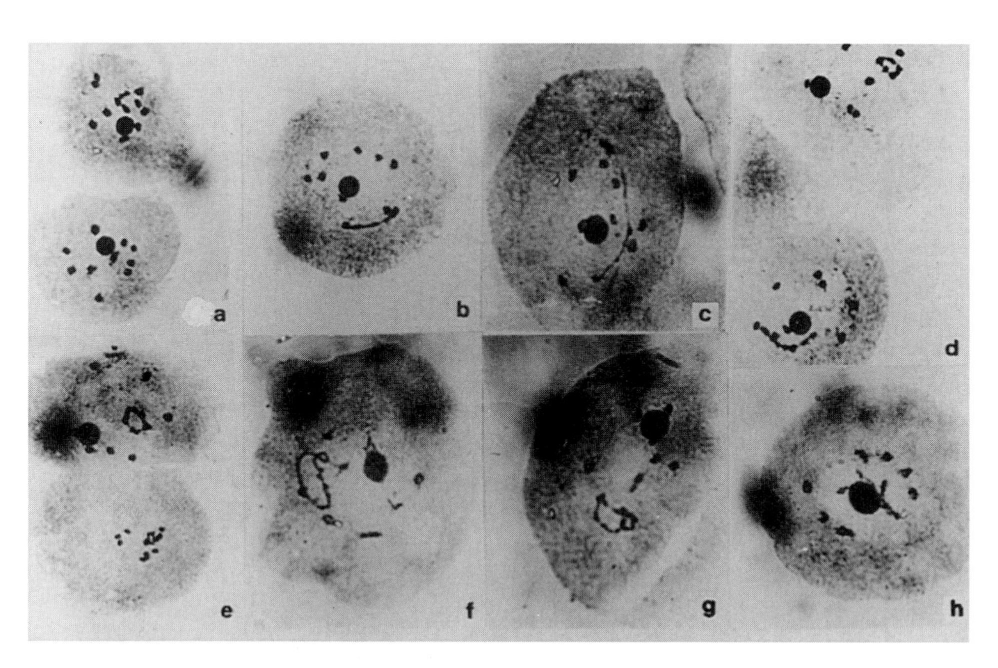


**Fig. 5a-h.** Chromosome configurations in meiosis of an MRT mutant. **a** 12II (*base*) and 10II + 1IV (*top*). **b** and **h** 8II + II. **c** 9II + 1VI. **d-g** 7II + 1X

### *2.1.3 Asynapsis and Desynapsis Mutants from In Vitro Culture*

Sterile somaclones from in vitro culture of rice were caused by asynaptic and desynaptic mutants which were found in 1980 and 1987, respectively. They were gene mutation, and disturbed chromosome pairing in meiosis.

From 23 first-generation pollen plants from anther culture of an indica variety, Qing Er-ai, two asynaptic triploid plants were found in 1980. In diplotene, diakinesis, and metaphase, the chromosomes did not pair and showed 36 univalents. The distribution of lagging chromosomes continued in anaphase I. The following abnormal phenomena in meiosis were observed: (1) monstrositas (abnormality) of spindle: divergent spindle in V-shape, curved spindle in C-shape or poly-pole spindle and poly-spindles. (2) Syncytium consisting of from two to more than ten PMCs. (3) Tetrad division, where usually no division occurs at the tetrad stage. (4) Abnormal microspore (from monads to octoads at the tetrad stage: Ling et al. 1981).

The desynnaptic mutant was derived from young panicle culture of $IR_{50}$. In this mutant, the chromosome pairing was normal in early prophase of meiosis (pachytene, diplotene), but after late prophase, the pairing of some chromosomes disappeared and a varying number of univalent stragglers and laggards could be observed in diakinesis, metaphase I, and anaphase I. Normal chromosome pairing (12II) comprised only 17.7% of all cells observed (Ling et al. 1988). In both mutants the imbalance of chromosome distribution in meiosis is of desynapsis and desynapsis resulted in sterility.

## 2.2 Female Sterile Mutant from In Vitro Culture

One female sterile mutant (Fig. 6) was found in a somaclone of the R3 generation which was derived from a mature seed culture of $IR_{50}$. The pistil of the mutant lacked stigma and style and consisted of ovary only (Fig. 7). Most of the ovaries in the mutant were without embryo sac, which was full of parenchyma cells (Fig. 8), so the mutant was completely sterile. Some (30%) of the spikelets possessed two or more ovaries (Fig. 8). No seed was set in the mutant plant. The anthers of the mutant were fully or partially degenerated (Fig. 7). None of the spikelets possessed all six anthers. In spikelets of the mutant, the maximum number of anthers was four (3.9% of the total number observed), and 25% of the spikelets were without any anther at all. In the last case, the stamen consisted of six filaments only. The degenerated anthers were empty, but in the normal anther of the mutants, the pollen grains were stained by I-IK solution and were functional. In the spikelets, the lemma was overdeveloped while the palea was underdeveloped and small, so that the spikelets could not close after flowering and the anthers emerged out of the spikelet before flowering (Fig. 6). Hybrid seeds were obtained when the pollen of the mutant plant (as male parent) were pollinated to $IR_{50}$ (as female parent). The hybrid $F_1$ of ($IR_{50}$/mutant) was expressed as fertile, as the normal plant and the seed-setting frequency was between 47.8 and 88.5%. Out of 458 plants in the $F_2$ generation, 22 female sterile plants were segregated and the result of $X^2$ of the segregation ratio showed that the female sterile mutant was controlled by two recessive genes (Ling et al. 1991a).

## 2.3 Male Sterile Mutant from Somaclones

The sterile mutants/variations in somaclones included chromosome, female, male, and other problems of sterility. To identify whether a sterile plant found in the somalones is male sterile or not, first, a test cross (TC) is made with pollen from a normal variety as male parent to the sterile somaclones (as female parent). Only the one in which hybrid seeds set after pollination was a male sterile mutant.

6

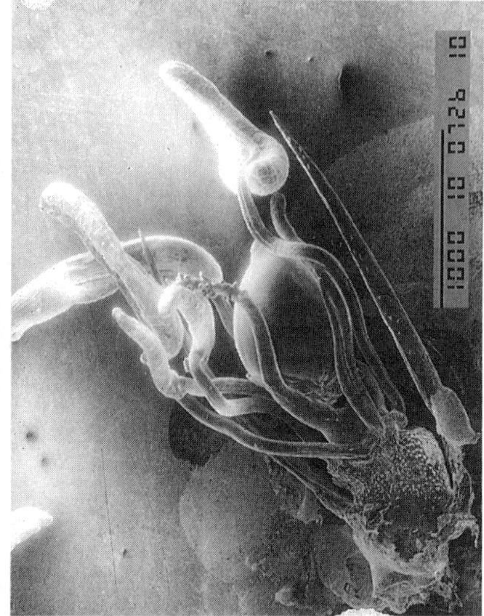
7

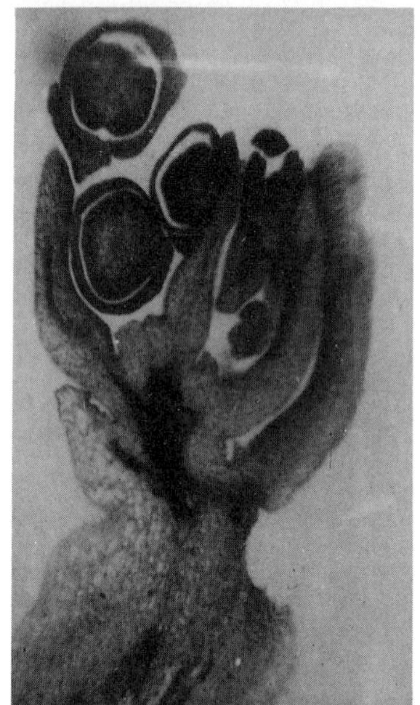
8

**Fig. 6.** The panicle of female sterile mutant (*left*) and its donor variety, $IR_{50}$. (Ling et al. 1991b)

**Fig. 7.** Male and female organs in the female sterile mutant. (Ling et al. 1991b)

**Fig. 8.** Section of a spikelet from the female sterile mutant. There are four pistils in a spikelet and all the embryo sacs are full of parenchyma cells. (Ling et al. 1991b)

Three different kinds of male sterile mutants in somaclones of rice were found: (1) pollen-free (Fig. 9); (2) pollen-abortive (Fig. 10); (3) antherless (Fig. 11). The major characters of these three mutants were described earlier (Ling 1991). Only the one in which sterility was both maintained and restored can be used in hybrid seed production. Until now, the number of male sterile mutants found and

**Fig. 9.** Anthers in the pollen-free ms mutant with no pollen contained in the anthers. (Ling et al. 1987b)

**Fig. 10.** Abortive pollen in 515A, in which all pollen were not stained by I-KI solution, cf. Figs. 14 and 15. (Ling et al. 1991b)

**Fig. 11.** The spikelet of the antherless mutant in which all six anthers degenerated and became stigma-like organs

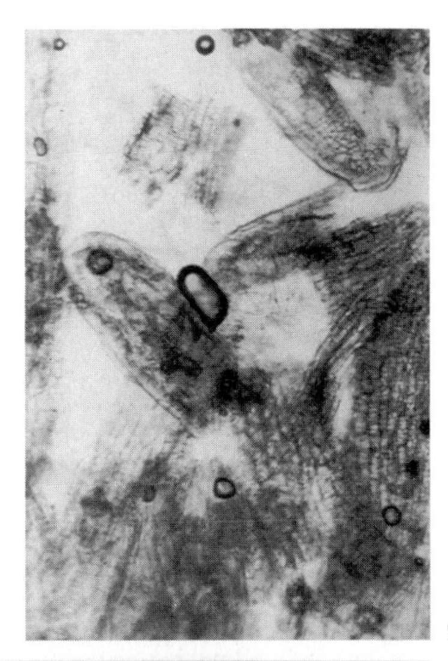

9

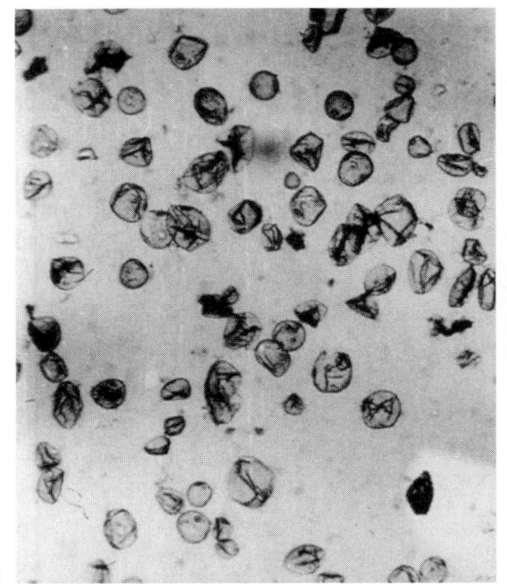

10

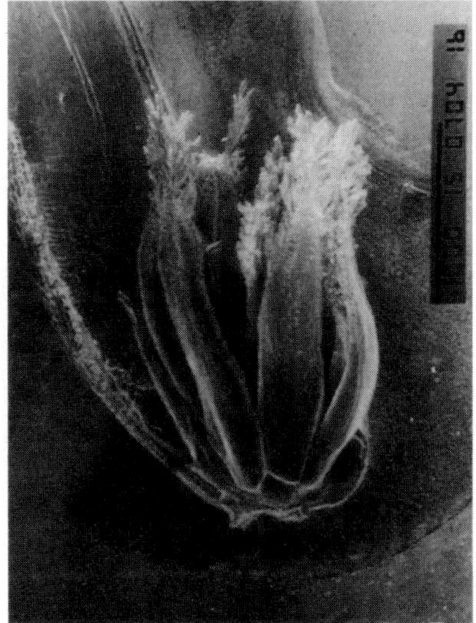

11

test-crossed was around 35 (Table 3). They were derived from both $R_1$ and $R_2$. Among them, 8 were pollen-free, 25 pollen-abortive, and $2_x$ antherless. Of the 35 ms mutants listed in Table 3, all could be restored. In other words, we found no ms mutant in which male sterility could not be restored.

### *2.3.1 Pollen-Free Male Sterile Mutants from Somaclones*

The anthers in this mutant were empty and contained no pollen (Fig. 9). All the TC $F_1$ hybrids in pollen-free mutants were fertile (Table 3). To find out how to

**Table 3.** Type of ms mutant in somaclones of rice and the fertility in their test-crossing hybrid $F_1$[a]

| Type of ms mutant | Total no. of mutants | Fertility expression in TC hybrid $F_1$ | | |
|---|---|---|---|---|
| | | F. only | S. only | F. and S |
| Pollen-free | 8 | 8 | 0 | 0 |
| Pollen-abortive | 25 | 19 | 0 | 6 |
| Antherless | 2 | 2 | 0 | 0 |

[a] The data list only mutants used for test crossing and with results in test crossing hybrid $F_1$ from 1985–1994. F. only = All the TC hybrids are fertile. S. only = All the TC hybrids are sterile. F. and S. = Both fertile and sterile plants in TC $F_1$.

maintain and restore this kind of mutant, a test cross was made, so that some mutants were continuously test-crossed for 2–3 years (using vegetative ratoowing), for example, the mutant 91–20, which was derived from the $R_2$ generation of mature seed culture in $IR_{54}$. We selected more than ten varieties as TC male parents to cross it. The male parent included the IR line (an established restorer line), an established maintainer in the WA-type of the CMS line (such as Zhen Shang 97B, Er Jiu aiB) and maintainers and restorers of other CMS lines. The result showed that all of the TC $F_1$ hybrids were fertile and no sterile plant was found, demonstrating that the sterility of the mutant 91–20 could not be maintained. Until now no sterile offspring in the TC $F_1$ from pollen-free ms mutants has been found. In other words, we found no variety that could maintain the pollen-free ms mutant. From heredity of male sterility, we know that if male sterility was cooperatively controlled by both nuclear and cytoplasm genes, it can be both maintained and restored; if controlled by the cytoplasm gene only, the ms cannot be restored, and if controlled by the nuclear gene only, the ms cannot be maintained. So the ms mutant 91–20 may be nuclear gene-controlled. It is well known that male sterility cannot be maintained if it is controlled by a nuclear gene only (Yuan and Mao 1991).

### *2.3.2 Antherless Somaclones*

This is a male sterile mutant in which the anthers had degenerated (Ling et al. 1991b; Fig. 11). A few TC hybrid plants were obtained in the antherless mutant, although plenty of spikelets in the mutant were pollinated with pollen from several varieties. Fertilization and seed setting in the antherless mutant were very low. The low seed setting on hand pollination showed that the female organ of the mutant was also abnormal. The TC hybrid $F_1$ plants from a single male parent (Zhen Shang 97B) were fertile. Obviously, the sterility of this kind of mutant could not be maintained.

### *2.3.3 Pollen-Abortive ms Mutants in Somaclones*

In pollen abortive male sterile mutants, sterility could be maintained in some varieties, and be restored in others. From 25 mutants obtained from 1985–1993, only 6 were found whose sterility could be maintained by some varieties, and restored by others. There were no sterile plants in the TC hybrid $F_1$ from the other 19 mutants (Table 3). Because of shortage of labor, the number of varieties used for the test cross in most mutants (including pollen-free male sterile mutants) was limited. It was thus very difficult to draw the conclusion that these 19 mutants (Table 3) could not be maintained at all.

Table 4 shows the six mutants in which male sterility could be both maintained and restored. The male parents used for the test crossing and their restoration and maintenance are also listed in the table.

According to whether the male sterility of the mutants was to be continuously maintained or not, two groups of mutants could be divided, in Table 4.

1. Mutants in which male sterility was not continuously maintained: the male sterility of two mutants. 9S1 and 7P3-1 (Nos. 2 and 3 in Table 4), were maintained for three and two generations, respectively. Mutant 9S1 was derived from the $R_1$ generation by young panicle culture of $TN_1$ in 1987. Male sterility could be maintained by $IR_{24}$ and restored by both Zhen Shang 97B and Er Jui-ai B. Interestingly, $IR_{24}$ is the restorer, and Zhen Shang 97 B and Er Jui-ai B are the maintainers in CMS lines of WA-type cytoplasm. Thus the relationship of M and R in this mutant was exactly opposite to the established WA-type CMS line. Unfortunately, the male sterility of the mutant was maintained for only two generations. In the third backcrossing generation ($IR_{24}$ as the male parent), hybrid $F_1$ plants became fertile.

The male sterility of the mutant 7P3-1 was maintained for only two generations. In fact, in the first BC generation, the $F_1$ plants were segregated into fertile and sterile (Table 4). When the sterile plant was selected to backcross continuously to $IR_{24}$ and Qing Er-ai, no male sterile plant was found in the third BC generation.

Q925 and 3936 are ms mutants which were found recently and their sterility could be maintained and restored. It is interesting that the sterility of the two mutants could be maintained by IR varieties ($IR_{28}$ and $IR_{26389}$), Er Jiu-aiB, and Zhen Shang 97B (Table 4). It is well known that IR varieties and Er Jiu aiB/Zhen Shang 97B are restorers and maintainers, respectively, in the CMS line with WA-type cytoplasm. It is difficult to draw any conclusion as yet because it is only the second backcross generation.

2. Mutants in which male sterility could continuously be maintained. Male sterility in the mutant 54257 was continuously maintained for more than ten generations and in 2P1, four generations (Nos. 1 and 4 in Table 4). The relevant maintainer, the sterile expression of both population and individual in each BC $F_1$ of the two mutants (54257 and 2P1) are listed in Tables 5 and 6, respectively. The BC hybrid $F_1$ populations of both 54257 and 2P1 were large enough (more than 800 in 54257/162–5 and 41 in 2P1/Er Jui-aiB) to identify the stability of an ms mutant. The results in Tables 5 and 6 show that the BC $F_1$ from both mutants were completely male sterile, so they are stable for establishing a CMS line. Meanwhile, their sterility could be restored by some varieties.

**Table 4.** The expression of fertility of TC $F_1$ in some pollen-abortive mutants

| Code no. of mutant | Origin of mutant | | | Male parent (MP) and expression of fertility in TC $F_1$ | | | | | | | | | | | | |
|---|---|---|---|---|---|---|---|---|---|---|---|---|---|---|---|---|
| | Variety | Generation | Year | | | | | | | | | | | | | |
| 1 54257 | $IR_{54}$ | $R_2$ | 1985 | MP | $IR_{36}$ | $IR_{54}$ | $IR_{24}$ | 162–5 | GC | EJ | ZS | SE | WX25 | T64 | T222 | |
| | | | | $F_1$ | R | R | R | M | M | M | M | R | R | R | R | |
| 2 9S1 | TN1 | $R_1$ | 1987 | MP | $IR_{24}$ | ZS | EJ | | | | | | | | | |
| | | | | $F_1$ | M | R | R | | | | | | | | | |
| 3 7P3-1 | QR-Ai | $R_2$ | 1988 | MP | $IR_{24}$ | | $IR_{52}$ | EJ | QR-Ai | | | | | | | |
| | | | | $F_1$ | 13R:3M | | R | R | 41R:1M | | | | | | | |
| 4 2P1 | $IR_{25}$ | $R_2$ | 1989 | MP | $IR_{26}$ | $IR_{24}$ | 114H | Ce222 | Ce64 | WX25 | MH63 | EJ | 7017 | ZS | 162–5 | TZ |
| | | | | $F_1$ | R | HR | R | R | R | R | R | M | M | M | M | M |
| 5 Q925 | QR-Ai | $R_1$ | 1993 | MP | $IR_{28}$ | $IR_{26389}$ | EJ | ZS | NT | GC | Xian98 | 210 | | | | |
| | | | | $F_1$ | M | M | M | M | M | M | R | R | | | | |
| 6 3936 | 3550 | $R_1$ | 1993 | MP | EJ | CT232 | IR28 | GC | | | | | | | | |
| | | | | $F_1$ | M | M | R | R | | | | | | | | |

R = restore (fertile); HR = half restore (half-fertile); M = maintain (sterile); MP = male parent; CT = Chen Te; EJ = Er Jiu-ai; GC = Gui Chao 2#, MH = Min Huai63; 114H = 114 Huai QR-Ai = Qing Er-Ai; SE = Shuang Er-zhan; Tz = Tian Zhen B; WX25 = Wai Xian 25; ZS = Zhen Sang 97B.

**Table 5.** Sterile expression of $F_1$ and BC offspring when 2P1 was used as female parent

| Generation | Population fertility | | | Seed fertility | | |
|---|---|---|---|---|---|---|
| | Total no. of plants | No. of ms plants | ms (%) | Total no. of spikelets | No. of seeds set | Frequency of seed setting (%) |
| | xEr Jui-aiB | | | | | |
| $F_1$ | 18 | 18 | 100 | 84.5 | 5.0 | 5.9 |
| $B_1F_1$ | 53 | 53 | 100 | 62.9 | 1.5 | 2.4 |
| $B_2F_1$ | 89 | 73 | 82 | 145.1 | 7.1 | 4.9 |
| $B_3F_1$ | 45 | 44 | 97 | 157.4 | 7.5 | 4.4 |
| $B_4F_1$ | 41 | 41 | 100 | | | |
| | $xIR_{26}$ | | | | | |
| $F_1$ | 18 | 18 | 100 | | | |
| $B_1F_1$ | 61 | 6 | 8 | 108.9 | 82.4 | 75.7 |

**Table 6.** Pollen and seed fertility of the hybrid (54257/162–5) $F_1$ and its backcrossing offspring

| Generation | Population fertility | | | Pollen staining (%) | | | Seed setting (%) | |
|---|---|---|---|---|---|---|---|---|
| | Total plants | No. of ms plants | ms (%) | Non | Half | Full | Free | Self |
| x162–5 | | | | | | | | |
| $F_1$ | 16 | 16 | 100 | 50 | 33 | 17 | 0.2 | 0 |
| $B_1F_1$ | 23 | 23 | 100 | 50 | 50 | 0 | 3.25 | 0 |
| $B_2F_1$ | 157 | 137 | 90.1 | 29.4 | 47 | 23.5 | 2.6 | 0 |
| $B_3F_1$ | 162 | 156 | 96.3 | 25 | 36 | 39 | 8.2 | 0 |
| $B_4F_1$ | 28 | 28 | 100 | 17.8 | 67.9 | 14.3 | 7.5 | 0 |
| $B_5F_1$ | 852 | 852 | 100 | 28.4 | 58.9 | 12.7 | 2.9 | 0 |
| $B_6F_1$ | 332 | 332 | 100 | 29.5 | 40.4 | 30.1 | 5.8 | 0 |
| $B_7F_1$ | 16 | 16 | 100 | 31.3 | 50 | 18.7 | 0 | 0 |
| $B_8F_1$ | 20 | 20 | 100 | 35.0 | 45 | 20 | 1.9 | 0 |
| xWai Xian | 25 | | | | | | | |
| $F_1$ | 156 | 0 | 0 | 0 | 0 | 100 | 90.5 | – |

Non = The pollen not stained by 1% I-IK solution.
Half = The frequency of pollen staining was 11–50%.
Full = The frequency of pollen staining was more than 51%.

The mutant 2P1 was derived from the $R_2$ generation from young panicle culture of $IR_{26}$ in 1990. From 235 $R_2$ plants, 18 sterile plants were segregated. The pollen grains were not stained by 1% I-IK solution and no or only a few seeds were set under conditions of self- and open pollination, respectively. In 1991 we found that the sterility of the mutant, 2P1, could also be maintained and restored. The varieties which could maintain the sterility of 2P1 mutant are: (1) maintainer line of a WA-type CMS line including Er Jui-aiB, 7017B, and Zhen Shang 97B. Using Er Jui-aiB, a male sterile of the mutant 2P1 was stably maintained for five generations (until July 1995); (2) the maintainer of the CMS line, Tien Ye A; and (3) 162–5, the maintainer of somaclonal ms mutant, 54257. The varieties which could restore the fertility of 2P1 mutant are 3550, Wai Xian 25, 114 Hui, and Ce222; but $IR_{24}$ and $IR_{50}$, which are good, well-known restorer lines of WA-type CMS line, only half restored the fertility of the 2P1 mutant (the frequency of seed setting in 2P1/$IR_{24}$ or $IR_{50}$ $F_1$ was around 50–60%).

For the mutant 54257, more detailed research on inheritance, breeding, biochemistry, and mtDNA analysis, was done and the results are discussed in the next section.

The two mutants Q925 and 3936 were derived from the $R_1$ generation in young panicle cultures of Qing Er-ai and 3550, respectively. The maintainers of the mutant Q925 are the established maintainer of WA-type CMS (Er Jui-aiB, Zhen Shang 97B); moreover, $IR_{28}$, $IR_{26389}$, and Nan Te could also maintain their sterility. It is well known that the IR line of rice is an accepted restorer in the WA-type CMS line. It is interesting that two of the IR varieties can maintain sterility in the mutant Q925. Using this result, we are now using more IR lines and other different varieties to make more test crosses. Because the test generations in these two mutants are only primary, we must wait to obtain more results from the further BC generations.

## 3 A Male Sterile Line, 54257/162–5, from Somaclones

We developed several ms lines from the two ms mutants, 54257 and 2P1 (Table 4). In 54257, four ms lines were established whose maintainer lines were Zhen Shang 97B, Er Jui-aiB, Gui Chao 2#, and 162–5. Here only the mutant 54257 and its ms line 54257/162–5 are described.

### 3.1 Genic-Cytoplasm Mutant, 54257, from Somaclone

In spring 1986, from 57 lines of the $R_2$ generation derived from young panicle culture of $IR_{54}$, one (code 54257) was found to segregate in fertility. Out of 147 plants of the line 54257, eight were sterile (Fig. 12). The pollen grains of the ms plants were not stained by 1% I-KI solution. The seed setting frequency under conditions of free pollination was 0.25%. After the fall of 1986, the ms plants (female) were crossed with 13 different varieties as male parents, and the frequency of seed setting of the mutant after pollination was from 8.8 to 49.5%, showing that 54257 was male sterile.

The fertility of the mutant could be maintained and restored by some varieties, showing that the mutant was genic-cytoplasmic male sterile (Ling et al. 1987b, 1991a). The varieties which could maintain the ms mutant 54257 were Er Jui-aiB; Zhen Shang 97B; Gui Chao 2#, and 162–5 (Table 4). $IR_{36}$, $IR_{54}$, $IR_{24}$, Shuang Er-zhan, Wai Xian 25# Ce 64, Ce 222, etc. could restore its fertility (Table 4).

**Fig. 12.** The ms mutant, 54257 (*left*) and its fertile sister plant (*right*). (Ling 1991)

## 3.2 The Male Sterile Line 54257/162–5

The maintainer 162–5 is a somaclone which was derived from $IR_{52}$. After continuous backcrossing for more than ten generations (only eight generations are shown in Table 6), 54257/162–5 became a stable male sterile line (Fig. 13).

**Fig. 13.** The ms line 515A population 100% male sterile. The paper bags cover the panicles

### 3.2.1 *The Sterile Characters of 54257/162–5 (515A)*

During all of the more than ten BC generations, no or few seeds (less than 10%) were set under conditions of self and open pollination, respectively. For population sterility, the frequency of ms plants of each generation was 100% (Fig. 13, besides the primary generation $B_2F_1$, 90.1% Table 6), showing that the sterility of the ms line 515A was stable; but the pollen sterility in the ms line, 515A, was very special. During all the more than ten BC generations from $F_1$, the pollen from around 30% plant of the population was not stained by I-KI (Fig. 10); 20% stained well (Fig. 14), and 50% was half-stained (Fig. 15). In the case of full staining, although the pollen was stained by I-KI, all had no ability to fertilize, and no seeds were set (Table 6). This character was quite different from many CMS lines of rice, including WA-type cytoplasm.

### 3.2.2 *Differences in mtDNA Between 515A and the WA-Type CMS Line*

Mitochondrial DNA of the CMS line 515A and WA-type control material of the ms line Zhen Sheng 97A and the maintainer line Zhen Shang 97B (CMS, A, and B, respectively in the following text and Figs. 16–18) were isolated. The mtDNA of CMS, A, and B were digested by Bam HI. Following electrophoresis through 0.8% agarose gel, DNA was transferred to hybond-N nylon membranes. The ms

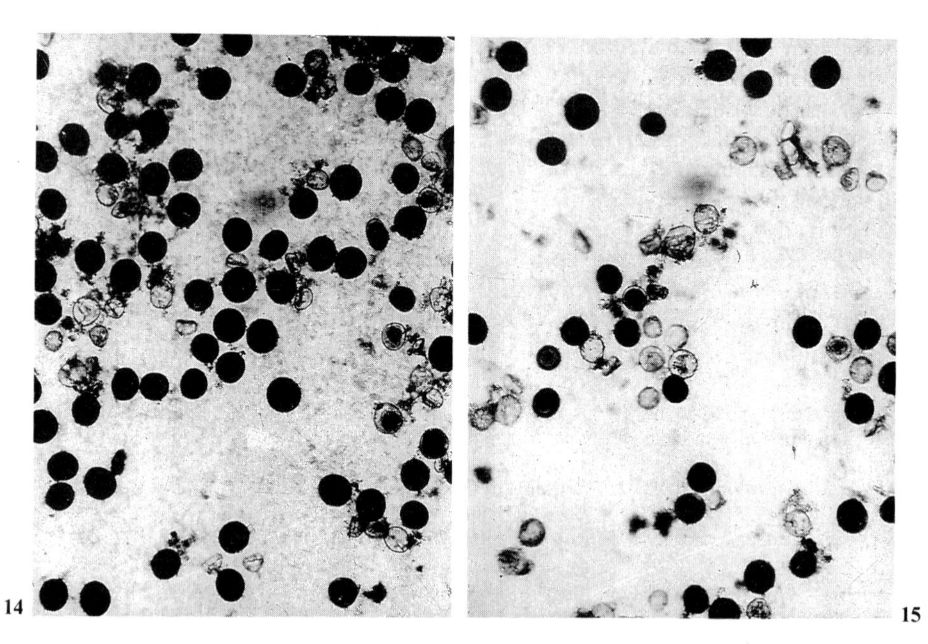


**Figs. 14, 15.** The abortive pollen of 515A were full- (Fig. 14) and half- (Fig. 15) stained by I-KI, (cf. Fig. 10) (Ling et al. 1991b)

gene *cox1* and *atp6* and the total mtDNA from A (the cms line of WA cytoplasm as control) were used as the radioactive probe with ($\alpha$-$^{32}$P) dATP, then hybridization with the radio-probe was carried out (Figs. 16–18).

From these three probes, some bands could be found that were similar between CMS and A: for the *cox*1 probe, at least three bands were similar, one in 10 kb and the other two in around 20 kb (Fig. 16), although there were some clear differences in the intensity of the bands.

*1. With the cox1 Probe.* At least three major differences could be found between CMS, A, and B (Fig. 17). (1) At ca. 14 kb, there is a band in both A and B, which was stronger in maintainer B, but not in CMS; (2) at ca. 20 kb, there are two strong bands in A, weak in CMS, and almost absent in B; (3) at the highest molecular weight, there is a very strong band in CMS, which is absent in both A and B.

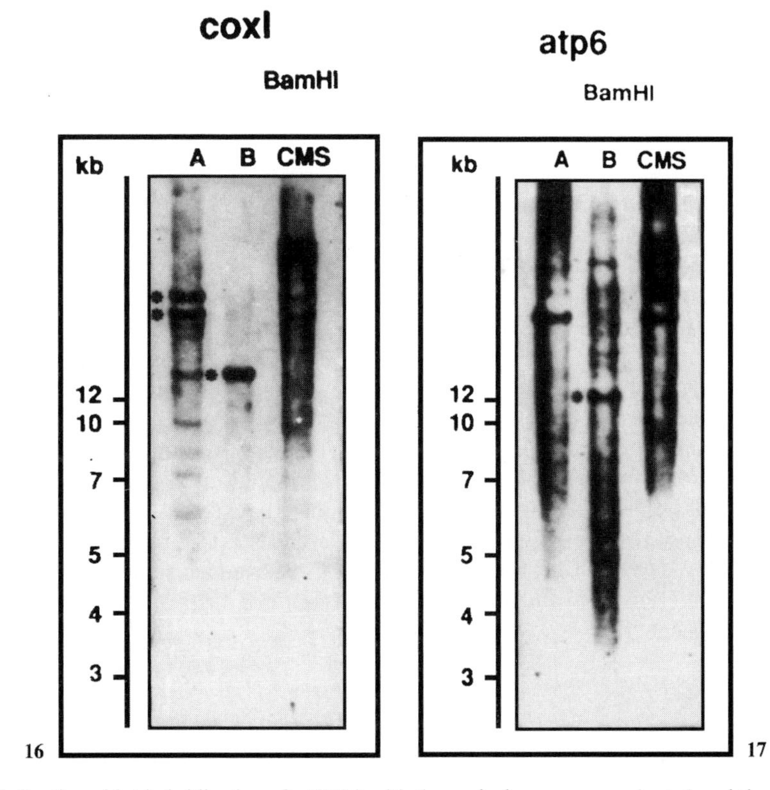


**Figs. 16–18.** Southern blot hybridization of mtDNA with the probed ms-genes *cox1*, *atp6*, and the total mt-DNA from 515A, respectively. The mtDNA digested with BamH1. Lanes: *A* Zhen Shang 97A; *B* Zhen Shang 97B; *CMS* 515A

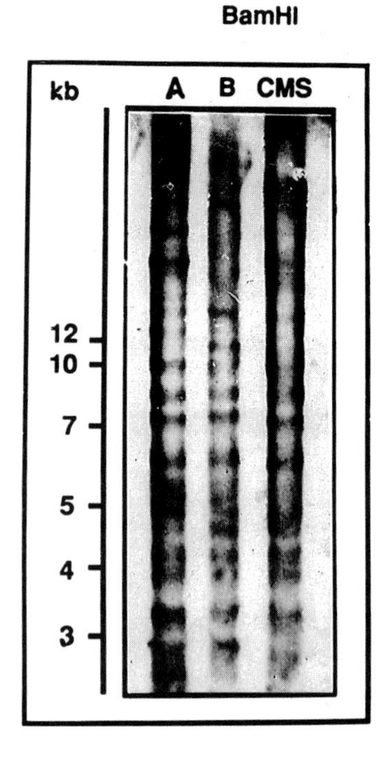


Fig. 18

2. *With the atp6 Probe.* Three differences could be found (Fig. 16): (1) At 12 kb, there is a band in B, which is absent in A and CMS; (2) at ca. 20 kb there is a band in A and CMS, but in B it is very weak; (3) at higher molecular weights (more than 20 kb), in CMS there are two stronger bands but in A and B they are very weak (almost absent).

3. *With the Probe of Total mtDNA of the A Line.* In this probe the differences are quite clear (Fig. 18). (1) All of the bands in A are stronger than those in CMS. (2) The number of bands in CMS is obviously less than in A. Compared to A, CMS has at least four bands less, each in the following fragment: below 3 kb, between 5 and 7 kb, between 7 and 10 kb, and higher than 12 kb.

These results showed that some differences exist in mtDNA between 515A and WA-type CMS. Thus 515A, which was derived from in vitro culture of $IR_{54}$, is a different cms line from the WA-type cms line on the mtDNA level of cytoplasm.

### *3.2.3 Comparison of the Relationship of Restoring and Maintaining (RRM) Between the 515A and WA-CMS Lines*

To know whether there are any differences in the RRM of 515A from the WA-CMS line, several test crosses (TC) were conducted. The varieties of male parents which had proved good restorers of the WA-CMS line and were used for these test crossings were 114Hui, Ce64, $IR_{24}$, 3550, Ce222, and Fun Gui6#. Among these six varieties, only three (114Hui, Ce64, and Ce222) could fully restore the fertility of 515A; two ($IR_{24}$ and 3550) could only half (incompletely) restore, and the other (Fen Gui 6 #), which is a good restorer of the WA-CMS line, not only could not restore fertility but also maintained sterility in 515A. On comparing these results in the RRM of the WA-CMS line with these six TC varieties, the following three relationships could be found.

1. The ms lines 515A and WA-CMS could be restored by some of the same varieties, in other words, some restorers are common to these two CMS lines. 114 Hui and Ce 64 could restore the fertility of both 515A and WA-CMS lines (Zhen Shang 97A and 7017 A). In the population of these TC $F_1$s, no sterile plant was found, pollen was well stained, and the frequency of seed setting was more than 80% (Table 7).
2. Some differences existing at the restoration level between 515A and WA-CMS, $IR_{24}$ and 3550, are the best restorer lines of WA-CMS. In $F_1$ (WA-CMS × these two varieties), no sterile plant was observed, pollen was full stained and the frequency of seed setting was more than 88% (Table 7). The fertility of (515A × $IR_{24}$ or 3550) $F_1$, however, was only half restored. In the population of these two TC $F_1$, around 40% or half sterile plants were found, the pollen, about 10–20%, was not stained by I-KI solution, and the frequency of seed setting was only 37.2–53.5% (Table 7).
3. The RRM was completely opposite between 515A and WA-CMS:
   RRM of the test variety, Fung Gui 6 #, to 515A, and the WA-CMS line was complete opposite. Fung Gui 6 # could fully restore fertility in the WA-CMS line, but in 515A, not only could not restore fertility but even maintained sterility (Table 7). Moreover, Ce222 could restore fertility in 515A, in which both the frequency of pollen staining and seed setting were very high (Table 7), but for Tian ZhenA, all the 15 $F_1$ test plants were sterile (pollen not stained and the frequency of seed setting only 8%, Table 7).

These three points clearly demonstrated the similarity and differences in RRM between 515A and WA-CMS. These similarities and differences in RRM of breeding research correspond to those of mtDNA in these two cms lines.

RRM is the most important basis for classifying CMS lines of rice (Lee 1980; Kaul 1988). According to the different RRM, Lee classified the CMS line of indica rice in China into three different cytoplasmic types: wild rice abortive (WA), Hong Lian, and Dian-I type (Lee 1980). Any CMS line of rice has its own special RRM and is different from the others. The RRM usually was used to identify the source of cytoplasm in a CMS line of rice.

If we consider the results concerning mtDNA and RRM in both 515A and WA-CMS, it is interesting to find that for these two very important and quite different-level characters, 515A and WA-CMS both have some differences on one hand and similarities on the other. Several different bands in mtDNA (Figs. 16–18) correspond to opposite and different levels in RRM (Table 7) between 515A and WA-CMS lines. On the other hand, the common restorer in 515A and

**Table 7.** Comparison of restoring effect of some restorer between 54257/162–5 and some other ms lines

| Fertile type of test hybrid | Combination and generation ($F_1$) | Population fertility | | | Frequency of pollen | | | Frequency of seed |
|---|---|---|---|---|---|---|---|---|
| | | No. of plants | | | Staining (%) | | | Setting (%) |
| | | Total | ms | ms (%) | 0–10 | 11–50 | > 50 | |
| Restored by the common restorer line | 515A/114 Hui | 31 | 0 | 0 | 0 | 0 | 100 | 79.4 |
| | 7017A/114 Hui | 19 | 0 | 0 | 0 | 0 | 100 | 82.8 |
| | 515A/Ce64 (1991) | 14 | 0 | 0 | 0 | 0 | 100 | 67.2 |
| | 515A/Ce64 (1992) | 14 | 0 | 0 | 0 | 0 | 100 | 74.9 |
| | 7017A/Ce64 | 16 | 0 | 0 | 0 | 0 | 100 | 76.9 |
| Difference in restoration levels | 515A/IR24 | 10 | 4 | 40 | 10 | 30 | 60 | 53.5 |
| | ZS97A/IR24 | 43 | 0 | 0 | 0 | 0 | 100 | 90.0 |
| | 515A/3550 | 15 | 6 | 40 | 20 | 20 | 60 | 37.2 |
| | ZS97A/3550 | 22 | 0 | 0 | 0 | 0 | 100 | 88.1 |
| Opposite in the RRM | 515A/Ce222 (1991) | 24 | 3 | 13 | 4 | 8 | 88 | 80.3 |
| | 515A/Ce222 (1992) | 21 | 2 | 9.5 | 0 | 42 | 58 | 74.5 |
| | TZA/Ce 222 | 15 | 15 | 100 | 100 | 0 | 0 | 8.0 |
| | 515A/FengGui6# | 16 | 16 | 100 | 100 | 0 | 0 | 7.6 |
| | ZS97A/FengGui6 | 10 | 0 | 0 | 0 | 0 | 100 | 59.1 |

TZA = Tian ZhenA, belong to WA – MS cytoplasm. ZS 97A = Zhen Shang 97A.

WA-CMS (Table 7) corresponds to similar bands of mtDNA (Figs. 16–18) in these two kinds of ms lines. The opposite/different level in RRM corresponds to the difference in the bands. This is a good example that inherent/breeding research of RRM on the individual and population level corresponds to mtDNA research on the molecular level in the same CMS line. From these results, the following conclusions can be drawn:

1. The results of the data from heredity/breeding research on RRM cms line can correspond to those from mtDNA at the molecular level.
2. The differences in RRM in rice breeding research corespond to the distinction in cytoplasm (mtDNA) of a cms line, so that using RRM as the standard to identity the cytoplasm source of a cms line in hybrid rice breeding is correct.
3. The cytoplasm of 515A, which was derived from the mutant 54257 in vitro culture, is partially different from WA-type cytoplasm of the cms line.
4. From our experiment, in vitro culture can induce a new male sterile mutant in which the cytoplasm is different from that in the WA-type CMS line. It is recognized that the search for different sources of cytoplasm for male sterility is very important (Virmani et al. 1986), especially for hybrid rice breeders in China (Ling 1991).

### 3.3 Effect of the Maintainer 162–5 on the Special Expressions of Pollen Staining in the ms Line 515A

From Table 6, it is clear that, although the pollen of the ms line 515A was sterile (even when fully stained), its pollen staining was segregated in each generation. The pollen from different ms plants was usually expressed as non-, half-, and full-staining (Table 6). In order to understand the reason for this phenomenon, special test crosses in which the female and male parent plants were fixed were designed (Table 8). From these series of test crossings, we may understand the effect of male or female parent on the "segregation" of pollen staining in the 515A.

**Table 8.** Design to fix plants of male/female parents for backcrossing

| Female parents (54257/162–5) | Male parents 162–5 | | |
|---|---|---|---|
| | 1 | 2 | 3 |
| A. Pollen non-stained | A × 1 | A × 2 | A × 3 |
| B. Pollen half-stained | B × 1 | B × 2 | B × 3 |
| C. Pollen full-stained | C × 1 | C × 2 | C × 3 |

A, B, C and 1, 2, and 3 were the code of the fixed plant as female and male parent, respectively.
Non-stained = Frequency of pollen staining was 0–10%.
Half-stained = Frequency of pollen staining was 11–50%.
Full-stained = Frequency of pollen staining was more than 50%.

Forty one combinations were backcrossed, and the results of all combinations were the same as shown in Table 9 (only nine combinations are listed). The backcrossing $F_1$ in all combinations did not set any seed and were completely sterile. The frequency of seed setting was 0% and less than 7.6% under conditions of controlled self- and open pollination, respectively. This result was the same as that from all the more than ten generations of the ms line (Table 6). The sterile type of pollen staining in all combinations, however, varied. When plants of the female parent, in which the pollen was not stained (0%), were selected and backcrossed to the male parents (combination A in Table 9), the pollen staining of the BC hybrid $F_1$s was not expressed as for the female parent (no staining), but still showed three types: non-, half-, and full-stained. For the combinations B, in which the female was selected half-staining plants and backcrossed to the three fixed male parents, the pollen staining of the BC hybrids in the three combinations did not express as their female parent (half-stained), but also showed three types of staining (non-, half-, and full staining; Table 9). As in the result of the combinations A and B, the type of pollen staining in the BC hybrid $F_1$ from combination C, in which the female parent belonged to full staining, also expressed the three types of pollen staining (non-, half- and full staining), unlike their female parent (full staining). Thus in all of the BC $F_1$, including A, B, and C combinations, there was no seed setting, but pollen sterility demonstrated three different staining types: non-, half- and full staining, no matter to which type of pollen sterility the female belonged.

The results in Table 9 also show that when the same male parent was used for backcrossing (A × 1, B × 1, and C × 1 in Table 9), the pollen staining of BC $F_1$ in different combinations was still expressed variously (non-, half- and full staining),

**Table 9.** Effect of 162–5 on pollen fertility of the ms line 54257/162–5

| Type of pollen sterility in the female parent[a] | Code of combination | Type of pollen sterility in TCF1[b] | | | Frequency of seed setting (%) | |
|---|---|---|---|---|---|---|
| | | Non- (%) | Half- (11–50%) | Full- (>50%) | Open pollination | Self-pollination |
| Pollen non- | A × 1 | 50 | 50 | 0 | 0.5 | 0 |
| stained | A × 2 | 25 | 37.5 | 37.5 | 7.6 | 0 |
| (0%) | A × 3 | 0 | 75 | 25 | 2.2 | 0 |
| Pollen half- | B × 1 | 33.3 | 50 | 16.7 | 4.8 | 0 |
| stained | B × 2 | 20 | 60 | 20 | 2.9 | 0 |
| (11–50%) | B × 3 | 25 | 25 | 50 | 1.9 | 0 |
| Pollen fully | C × 1 | 66.7 | 0 | 33.3 | 3.3 | 0 |
| stained | C × 2 | 33.3 | 66.7 | 0 | 4.4 | 0 |
| (>51%) | C × 3 | 20 | 60 | 20 | 2.7 | 0 |

[a] The fixed female parents were 54257/162–5, $B_5F_1$.
[b] The figure in the table is a percentage:

$$\% = \frac{\text{the plant number of observation in each item}}{\text{total } F_1 \text{ plants in each combination}} \times 100.$$

but no seeds were set. The results of the TC hybrid with the other two fixed male parents 2 and 3 (corresponding to A × 2/3, B × 2/3, and C × 2/3) were the same (Table 9). These results demonstrated that pollen sterility with non-, half- and full staining expression is not segregated, but is a special inheritance character in the 515A line, which is stable in heredity.

The male plants (Nos. 1, 2, and 3, in Tables 8 and 9) used for test crossing developed into three plant families in the next generation. Like the population of the parent, no segregation and/or differences in or among the three families was observed (data not listed), showing that the male parent, 162–5, was stable and pure in inheritance.

Pollen sterility, whether staining or not, is related to the pollen abortive stage (developmental stage). If the pollen aborts before the uninuclear stage, it will be non-staining sterile. If the abortion of pollen occurs after the two-nuclear stage of the microspore, the pollen will be full-staining sterile, and if it occurs in both stages, half-staining sterile. The expression of pollen abortion in most CMS lines was usually uniform (either nonstained abortion or stained abortion). That the expression of the abortive pollen in 515A was not uniform showed that the stage of pollen to become abortive was not synchronous. A similar phenomenon was observed in the Gang-type CMS line. Most CMS lines in Gang-type cytoplasm were non-staining abortive pollen, but the pollen in the CMS line, Qing Shan-JinA was both non-staining and staining (Zhu 1979).

### 3.4 Effect of the Maintainer, 162–5, on the Sterility of 515A and the Other CMS Line

In order to understand whether the characters of 515A, which was no seed set but pollen sterility of different types, were caused by the maintainer, 162–5, or not, the following test crosses were carried out. The maintainer, 162–5, as the male parent, was crossed with five CMS lines: Zhen Shang 97A, 7017A, Tian YeA, Bao YuanA, and Hong LianA. The results in these test crosses are listed in Table 10. Among all the Tc $F_1$s, only (Hong LianA/162–5) $F_1$ was fertile, and here the pollen was full-stained and seeds were well set (Table 10). Moreover, the frequency of seed setting was 0 (self-pollination) or 0.15–2.3% (open pollination). The following points: can be made about pollen sterility.

1. In all four sterile combinations (Nos. 1–4 in Table 10). We found no combination in which the pollen was only non-staining sterile. In other words, the pollen in all these four combinations are either full-staining sterile, half-staining or no-staining sterile.
2. It is well known that pollen in the two WA-cms lines (7017A and Zhen Shang 97A) is non-staining sterile. When the pollen were stained by I-KI, all of them were non-stained, but when these ms lines were crossed with 162–5, the pollen from all of the Tc $F_1$ expressed different types: non-, half- and full-staining sterile. From this result, it can be concluded that 162–5 is the major factor causing the pollen in the cms lines (including 515A 7017A/162–5, Zhen Shang 97A/162–5, etc.) to become staining sterile.

**Table 10.** Fertility of $TCF_1$ of other ms lines/162–5[a]

| Combination and generation | Population fertility | | | Pollen fertility | | | Seed setting (%) | |
|---|---|---|---|---|---|---|---|---|
| | No. of plants | | | Pollen Staining (%) | | | Open pollination | Self-pollination |
| | Total | ms | ms (%) | 0–10 | 11–50 | > 50 | | |
| 1 7017A/162–5$F_1$ | 48 | 48 | 100 | 18.8 | 62.5 | 18.7 | 2.4 | 0 |
| 1 Same $B_1F_1$ | 41 | 41 | 100 | 29.3 | 34.1 | 36.6 | 1.1 | 0 |
| 2 ZS97A/162–5$F_1$ | 30 | 30 | 100 | 63.4 | 23.3 | 13.3 | 0.6 | 0 |
| 3 TianA/162–5$F_1$ | 29 | 29 | 100 | 93.1 | 6.9 | 0 | 0.2 | 0 |
| 4 Bao A/162–5$F_1$ | 13 | 3 | 100 | 0 | 0 | 100 | | |
| 5 HongA/162–5$F_1$ | 13 | 0 | 0 | 0 | 0 | 100 | | |

ms(%) = No. of ms/total no. of plants.
[a]ZS97 A = Zhen Shang 97A; Tian A = Tianye A; Bao A = Baoyuan A; Hong A = Hong Lian A.

3. For the cms line Hong LianA, 162–5 did not maintain sterility, but restored fertility, showing that the cytoplasm of Hong LianA is quite different from that of WA-cms.

## 4 Origin of the Maintainer, Somaclone 162–5

Tests with 18 $R_2$ plant families derived from young panicle culture of $IR_{52}$ were conducted in 1986. Among them, one coded as 162, of which 40 plants were tested, was segregated according to the following characters: short (only 40 cm) or higher in plant height, earlier or late in mature stage, etc. The short/earlier plant was selected, 22 plants of which were developed in the $R_3$ generation. At $R_3$, tall in plant height (5 plants) and short (17 plants) were again segregated. From the 17 short individuals, late mature with long seed shape and early mature with widen seed shape, were segregated. The plant selected for short height earlier and wide seed shape was 162–5. After the $R_4$ generation, the somaclone was uniform and no segregation was observed where the populations of the somaclone were 65 to 1500 plants. Compared with the parent variety, $IR_{52}$, 162–5 was earlier in maturing stage, shorter in plant height, wider/shorter in seed shape, and darker in leaf colour. The fertility of 162–5 was normal and the frequency of seed setting more than 80%. Thus 162–5 was of heterozygous origin and, after selecting continuously for three generations, the somaclone became stable.

Compared with the donor variety $IR_{52}$, however, almost all of the major characters in 162–5 were changed. Additionally, the most important difference in the somaclone was the change in the relationship of R and M with CMS lines. It is well known that $IR_{52}$ is a good restorer line of the WA-type CMS line, but its somaclone, 162–5, became a maintainer line of WA-CMS. This result showed that in vitro culture not only can cause variation or mutation in morphology or physiological appearance, but also in the characters of RRM. This is a typical sample in which in vitro culture made a restorer, $IR_{52}$, became a maintainer, 162–5. Until now, various mutants have been reported from in vitro culture (Bajaj 1990; Sun et al. 1990), but a restorer becoming a maintainer by in vitro culture has never been observed. By $Co^{60}$ gamma radiation treatment, Lin et al. (1979) reported that two mutants which could maintain the sterility of the WA-CMS line were induced from $IR_{26}$. $IR_{26}$ is an excellent restorer of the CMS line. The case of 162–5 was similar to this result, showing that for inducing new genetic germplasm, somaclonal variations have great potential abilities.

## 5 Summary and Conclusions

Sterility is common in mutants in somaclones of rice; this include chromosomal and gene mutants. Although the frequency of male sterile mutants from somaclones in rice was not high, we found this mutant each year in our experiments in the years 1984–1994. Among these, three kinds of ms mutants were found: pollen-free, pollen-abortive, and antherless. All the $F_1$ hybrids of test crosses from pollen-free ms mutants were fertile and their sterility could not be maintained. This showed that pollen-free ms mutants belonged to the nuclear gene-controlled group. The ms mutant antherless could also not be maintained, and these two kinds of ms mutants were not used for hybrid rice production. Only the pollen-abortive ms mutant could be both maintained and restored by some varieties. Since 1984, we have found six pollen-abortive ms mutants which have been maintained and restored (Table 4). Only one mutant, 54257, was continuously maintained for a long time, and this was controlled by a genic-cytoplasm gene. Two mutants, Q925 and 3936, which were found recently, are still being tested.

From the mutant 54257, a new sterile line, 515A, derived from continuous backcrossing with 162–6 as the male parent (54257/162–5), was developed. No or very few seeds (less than 10%) were set under conditions of self and open pollination, respectively, and in each generation, the population sterility of the ms line was 100%. The pollen from ca. 30% of the plants of the population in each generation were not stained by I-KI solution; 20% stained well, and 50% were half-stained, although all the well- and half-stained pollen were sterile. A comparative study of Southern blot analysis of mtDNA and the relationship of restoring/maintaining (RRM) between 515A and the CMS line with WA-type cytoplasm (Zhen Shang 97A) was conducted. For both Southern blot analysis of mtDNA and RRM between these two kinds of CMS lines, there were some common points (same bands/RRM), but at the some time some clear differences (different bands/RRM) could be found. The results from molecular and breeding levels corresponded well. From these experiments, it can be concluded that from somaclones, not only ms mutants but also ms mutants controlled by genic-cytoplasmic genes can be induced, from which a male sterile line could be developed.

## References

Aarts MGM,Drikse WG, Stiekema WJ Pereira A (1993) Transposon tagging of a male sterility gene in *Arabidopsis*. Nature 363: 715–717

Akamatsu T, Todhita M Ohawa Y Shiga T (1988) Introduction of cytoplasmic male sterility (ogura cytoplasm) into cabbage (*Brassica oleracea* L.) by protoplast fusion. Jpn J Plant Breed 38: 14–15

Bajaj YPS (ed) (1990) Biotechnology in agriculture and forestry, vol 11. Somaclonal variation in crop improvement I. Springer, Berlin Heidelberg New York

Fu Xiang-quan, Gong Shoa-wen (1994) The great achievement of hybrid rice in thirty years and its further developing strategies in China. Hybrid Rice 1994, 3–4: 17–21

Kaul K (1988) Male sterility in higher plants. Springer Berlin Heidelberg New York, pp 97–176

Lee CB (1980) Primary review on the classification of male sterile line in China. Crop Sci, China 6(1): 17–26

Lin YM, Zhou X, Wang GY (1979) Radiation genetics on male sterile and three lines of rice. Agric Technol Sci Huna 6: 1–6

Ling (1987) Aquintuple reciprocal translocation produced by somaclonal variation in rice. Cereal Res Commun 15(1): 5–12

Ling DH (1991) Male sterile mutant from rice somaclones In: Bajaj YPS, (ed) Biotechnology in agriculture and forestry vol 14, Rice. Springer, Berlin Heidelberg New York, pp 347–367

Ling DH, Wang XH, Chen MF (1981) Cytogenetical study on homologous asyndetic triploid derived from anther culture in indica rice. Acta Genet Sin 8: 262–268

Ling DH, Ma ZR, Ahen WY, Chen FM (1987a) Chromosomal variation of regenerated plants from somatic cell culture in indica rice. Acta Genet Sin 14: 249–254

Ling DH, Ma ZR, Chen WY, Chen MF (1987b) Male sterile mutant from somatic cell culture of rice. Theor Appl Genet 75: 127–131

Ling DH, Chen MF, Chen WY, Ma ZR (1988) Desynaptic mutant from somatic cell culture in rice. Acta Genet Sin 15: 86–88

Ling DH, Ma ZR, Chen MF, Chen WY (1991a) Female mutant from somaclones in somatic cell culture of indica rice. Acta Genet Sin 18: 446–451

Ling DH, Ma ZR, Chen MF, Chen WY (1991b) Types of male sterile mutant in somaclones from somatic cell culture of indica rice. Acta Genet Sin 18: 132–139

Mariani C, Beuckeleer MD, Truettner J, Leemans J, Goldberg BR (1990) Induction of male sterility in plants by a chimeric ribonuclease gene. Nature 347: 737–741

Sakai T, Imamura J (1988) Transfer of *Raphanus sativa* CMS cytoplasm to *Brassica napus* via protoplast fusion. Jpn J Plant Breed 38: 12–13

Sun ZX, Zheng KL (1990) Somaclonal variation in rice In: Bajaj YPS (ed.) Biotechnology in agricultural and forestry vol. 11. Somaclonal variation in crop improvement I. Springer, Berlin Heidelberg New York, pp 288–325

Virmani SS, Govinda RK, Coal B, Dakmacio R, Auria PA (1986) Current knowledge of and outlook on cytoplasmic genetic male sterility and fertility restoration. In: Rice genetics. IRRI, Manila, pp 633–647

Yuan LP, Mao CX (1991) Hybrid rice in China—techniques and production. In: Bajaj YPS (ed) Biotechnology in agriculture and forestry, vol 14: Rice. Springer, Berlin Heidelberg New York, pp 129–148

Zhou YG (1979) Study on the different cytoplamic types of male sterile in rice. Crop Sci Sin 5: 29–38

# I.3 Release of the Rice Variety Dama Developed by Haploid Somaclone Breeding

L.E. HESZKY[1], I. SIMON-KISS[2], and D.Q. BINH[1]

## 1 Introduction

In May 1992, on the basis of data from nationwide experiments lasting 3 years, the National Council for Variety Qualification registered a new rice variety in Hungary, improved by the use of the Haploid Somaclone Method (HSM) with the name Dama (Heszky and Simon-Kiss 1992). This new rice variety is the product of more than 10 years of research and cooperation between the Rice Breeding Section (Irrigation Research Institute of Szarvas, Hungary) and the Department of Genetics and Plant Breeding (University of Agricultural Sciences, Gödöllö, Hungary).

In this chapter the Haploid Somaclone Method, the main steps of the procedure, and the rice variety produced by it are described.

A short history of the new rice variety Dama of haploid somaclone origin is as follows:

| | |
|---|---|
| Setting up a new hypothesis (Ploidy-dependent somaclonal variation) | 1983 |
| Proving the hypothesis by a new approach (Haploid somaclone) | 1984–85 |
| Elaboration of a new method (Haploid somaclone breeding HSM) | 1985 |
| Application of HSM in rice breeding | 1985–89 |
| Official test of developmental varieties | 1989–91 |
| Registration of a new variety (Dama) of HSM origin | 1992 |

## 2 Somaclonal Breeding

### 2.1 New Approaches (Doubled Haploid Somaclone)

Somaclonal variants for qualitative and quantitative characters have now been reported in more than 60 species of higher plants (see Bajaj 1990). Considerable

[1] Department of Genetics and Plant Breeding, University of Agricultural Sciences, H-2103 Gödöllö, Hungary
[2] Rice Breeding Section, Irrigation Research Institute, H-5541 Szarvas, Hungary

Biotechnology in Agriculture and Forestry, Vol. 36
Somaclonal Variation in Crop Improvement II (ed. by Y.P.S. Bajaj)

work has been done on somaclonal variation in rice (Sun and Zheng 1990; Sun et al. 1991). Variation was also dramatic in rice somaclones from a true-breeding doubled haploid derivative of the cvs. Norin (Oono 1981), Calsose 76 (Schaeffer 1982), and various Russian (Kucherenko 1991), and Chinese cultivars (Sun et al. 1991).

Two essential prerequisites in plant breeding are the presence of sufficient genetic variation and the availability of efficient selection procedures (Ceulemans et al. 1986). In this respect, we assumed that at cellular level many more genetic changes occur than can manifest themselves phenotypically after plant regeneration.

Our theory (Li et al. 1986; Heszky et al. 1989) was that the phenotypic manifestation of molecular and chromosomal changes (somaclonal variation) depends on the origin and the ploidy level of the initial explants and primary callus. This correlation means that the higher the number of DNA copies in the genome of cultured cells, the less probable the phenotypic manifestation of a given genetic change. Consequently, the variation of somaclones can be increased by reducing the ploidy level of initial explants.

The introduction of the new term haploid somaclone necessitates revising the relevant terminology. Such a terminology is necessary and can help to define the concepts and origins of the variation (Fig. 1).

*Somaclone (Conventional).* Diploid (2n) plants regenerated from somatic tissue cultures of diploid plants of zygotic origin. Variation among the somaclones originates from the genetic instability of cultured diploid somatic cells (Fig. 1).

*Doubled Haploid Somaclone.* Diploid (2n) plants regenerated from cultured somatic tissue of pollen haploid plants (n) of androgenic origin. Variation among doubled haploid somaclones originates from the genetic instability of cultured haploid somatic cells (Fig. 1).

The results of Ramulu et al. (1986) in potato supported our theory. Their data suggested that the explant source and the initial ploidy level of the genotype play an important role in influencing the degree of somaclonal variation.

## 2.2 Production of Somaclones

The Hungarian rice cultivars Nucleoryza and Carolina were the plant material in our investigations. Seeds (A), plumule meristems (B) of diploid and immature inflorescences (C) of haploid plants were used as explants. To produce homogenous calli with high morphogenic potential, callus induction and plant regeneration from anthers, plumule meristem, and haploid immature inflorescence-derived callus were studied (Heszky and Pauk 1975; Heszky et al. 1986; Li and Heszky 1986; Binh and Heszky 1990). Methods used for the preparation of various explants, callus induction, subculture, and plant regeneration have been described earlier (Heszky et al. 1989, 1991; Binh et al. 1992).

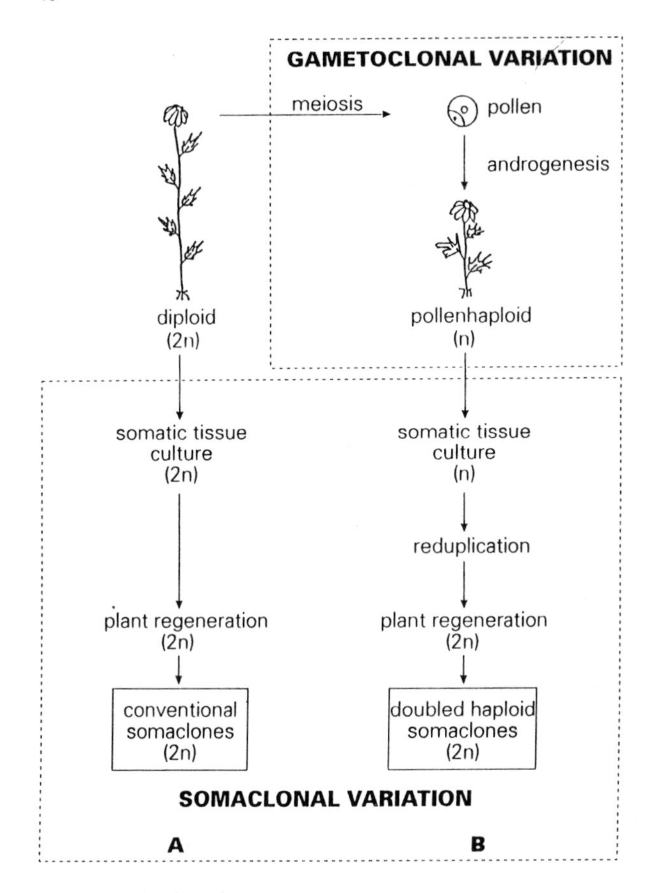


**Fig. 1.** Production of somaclones by **A** conventional and **B** haploid method

## 2.3 Production of Somaclones (2n) of Haploid Origin

Shoots bearing panicles (at the uninucleate pollen stage) were cut and pretreated for 7 days at 10 °C before further isolation. Shoots were surface-sterilized in 0.2% $HgCl_2$ for 10 min and rinsed three times with sterilized water. Whole panicle was excised aseptically, and anthers inoculated on medium in petri dishes.

Anther calli were obtained by culturing anthers on $N_6$ medium containing 2 mg/l 2,4-D and 3% sucrose. The pH of the medium was adjusted to 5.6–5.8 before adding agar (0.8%), and then sterilized for 15 min at 121 °C.

Regeneration medium was the same as that used for callus induction and subculture, except that 2,4-D was substituted for kinetin (10 mg/l) and the concentration of sucrose was increased to 4%. The calli were subcultured three times at 30-day intervals. Plants of 1–4 cm in height were removed individually from the callus mass and transferred to a medium containing MS mineral salts, 2 mg/l kinetin, 0.2 mg/l indole-3-acetic acid (IAA), 2% sucrose, and 0.6% agar. When the roots were well developed, plantlets were placed in commercial potting soil, acclimatized in mist chambers for 2 weeks, and grown to maturity in a greenhouse.

Immature inflorescences were excised from pollen haploid plants maintained and propagated vegetatively in a greenhouse. The stem was selected prior to emergence of the flag leaf, and was cut below the peduncular node and surface-sterilized in the same way as the seeds. The whole immature inflorescence, 3–10 mm in length, was isolated.

After three passages of the primary haploid callus, nearly 100% of regenerants were diploid (Li et al. 1986; Binh and Heszky 1990; Heszky et al. 1991). These have been termed doubled haploid somaclones, referring to the gamete (n) origin and to distinguish them from the somaclones of normal (zygote) 2n origin.

## 2.4 Field Performance

In the early 1980s 6500 somaclones from somatic cell cultures of haploid and diploid lines were produced. Selfed seeds ($SC_2$) were harvested from individual regenerants ($SC_1$) grown in a greenhouse between 1983 and 84 in the Research Center for Agrobotany, Tápiószele. The second and third generations ($SC_2$ and $SC_3$) were sown in rice field trials of the Research Institute of Irrigation (Szarvas) in 1985–86. Earliness (number of days between sowing and flowering) and tillering ability (number of tillers per plant) were examined under upland conditions. Yield components such as panicle length, grain number/panicle, and number of branching/panicle were observed in lowland field trials. The data were statistically evaluated and compared by F- and t-tests.

The results of comparative studies of different somaclones in the field (1986–1988) support our hypothesis (Table 1). As compared to somaclones produced in the conventional way, somaclones of haploid origin showed greater

**Table 1.** Distribution of earliness (heading date), tillering capacity, and deviation (D), coefficient of variance (CV) of several yield components in somaclones ($SC_3$) of different origin

| | Origin of somaclones | | | | | |
|---|---|---|---|---|---|---|
| | | Inflorescence n | | Seed 2n | | Meristem 2n |
| Heading date (%) | | | | | | |
| early | | 25 | | — | | — |
| medium | | 25 | | 50 | | 11 |
| late | | 50 | | 50 | | 89 |
| Tillering (%) | | | | | | |
| low | | 25 | | 100 | | — |
| medium | | 50 | | — | | 100 |
| high | | 25 | | — | | — |
| Yield components | D | CV% | D | CV% | D | CV% |
| panicle length | 1.16 | 8.44 | 0.90 | 6.19 | 0.93 | 6.40 |
| No. of branching/panicle | 1.11 | 18.50 | 0.64 | 7.62 | 1.00 | 12.94 |
| No. of grains/panicle | 11.95 | 41.89 | 6.20 | 12.44 | 7.81 | 14.69 |

variability in the majority of agronomic, morphological, and phenological properties (Heszky et al. 1989, 1991).

## 2.5 The New Method (Haploid Somaclone Breeding)

From the results, we drew the conclusion that somaclonal variability manifesting on the plant level can be increased by reducing the ploidy level of in vitro cultures. Individuals having plant improvement value occur with higher probability among these types of somaclones. A new method, named the Haploid Somaclone Method (HSM), was developed to facilitate practical application (Fig. 2). The breeding scheme of this method consists of the following steps:

1. *Reduction of ploidy level.* Haploid plants are produced from the desired breeding material (hybrid, line, etc.) by anther (androgenesis) or ovule (gynogenesis) cultures.

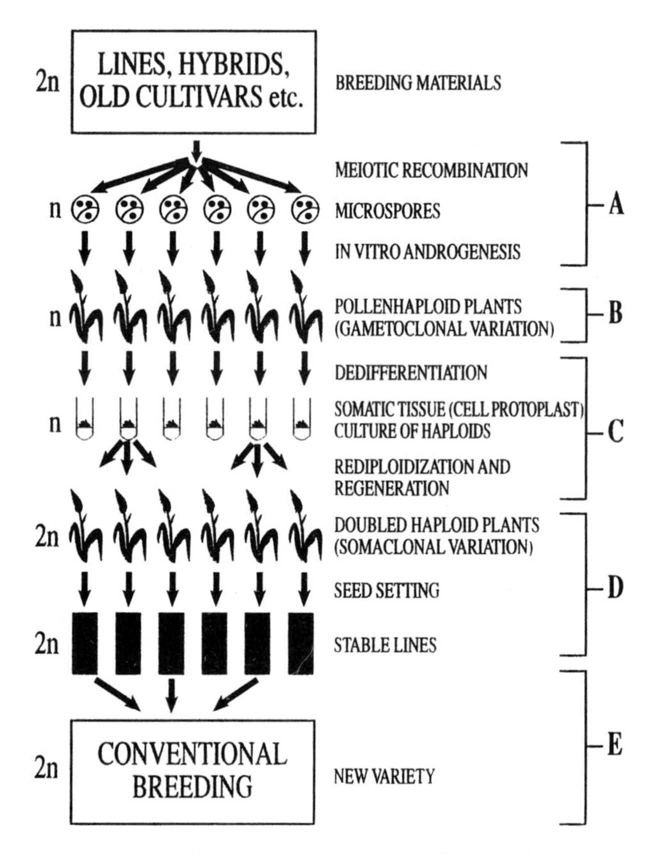


**Fig. 2.** Method of Haploid Somaclonal Breeding: **A** reduction of ploidy level; **B** maintenance of plants on reduced ploidy level; **C** production of doubled haploid somaclones; **D** field test of DH somaclones; and **E** conventional breeding of DH somaclones

2. *Maintenance and propagation of somatic tissue on reduced ploidy level.* Haploid plants are grown, propagated, and maintained in vitro and/or in the greenhouse, depending on the given species.
3. *Production of somaclones from somatic tissue at reduced ploidy level.* Callus is induced from the somatic tissue of haploid plants (flower, meristem, leaf, etc.), and after several passages diploid plants (doubled haploid) are regenerated. Genetic changes taking place at the haploid cell level become homozygous during rediploidization and manifest phenotypically in the regenerated doubled haploid somaclone plants (DH somaclone).
4. *Field test of doubled haploid somaclones.* Seeds are obtained from diploid regenerants ($SC_1$). The variability among DH somaclone lines of successive generations can be regarded as a genetically changed new character in homozygous form, so that further segregation cannot be expected.
5. *Breeding of DH somaclones.* Doubled haploid somaclones mean diploid (2n) plants regenerated from somatic tissue cultures of haploid plants (n) of androgenic or gynogenic origin. Variation among DH somaclones originated from the genetic instability of cultured haploid somatic cells. These lines can then be used directly to produce a new variety by conventional breeding procedures.

## 2.6 Application of the HSM in Rice Breeding

In order to test the applicability of HSM in breeding, DH somaclones of haploid origin were produced from the varieties Nucleoryza, Oryzella, and Karolyna at the Gödöllö University of Agricultural Sciences. Breeding work with DH somaclones was carried out at the Irrigation Research Institute in Szarvas. The improvement was aimed at selecting new individuals of earlier maturation, better seed profile, and increased blast resistance, with a yielding capacity and agronomical traits identical with those of the old varieties. As a results of several years selection work, we succeeded in producing two developmental varieties which were examined by the official test (Heszky et al. 1989, 1991).

### *2.6.1 New Variety Dama*

Based upon the results of nationwide field tests lasting 3 years (1989–91), the National Council for Variety Qualification registered the HSC-2 candidate by the name Dama in 1992. Of all the cultivars approved so far and cultivated in Hungary at present, this variety is the most resistant to Piricularia, and has the best seed profile and cooking quality (Heszky et al. 1992).

The variety Dama exhibits improved performance in some agronomic traits compared to the old source cultivar Karolina (Table 2).

The brief description of the variety Dama is as follows. A medium-early variety. Its duration is 140 days. On the average of the 3-year official test it was superior to the standards (Table 3). Plant height is 80–85 cm. It has very good resistance to lodging and piriculariosis. It is easy to thresh, and has a good

**Table 2.** Differences in traits of DH somaclone variety Dama compared to the source cultivar Karolina

| Varietal trait | Dama |
|---|---|
| Yield capacity (%) | +8–10 |
| Duration (days) | +6–10 |
| Plant height (cm) | +11.0 |
| Thousand-grain mass (g) | +0.3 |
| Grain (l:w) | +0.2 |
| Milling yield (%) | +2.0 |
| Amylose (%) | −1.6 |
| Field resistance to blast | + |
| Cold tolerance | ~ |
| Origin | Karolina |

**Table 3.** Data[a] from the National Rice Variety field test

| Variety | Paddy yield | | Cargo yield | | Duration | Plant height | Resistance to blast |
|---|---|---|---|---|---|---|---|
| | (t/ha) | (%) | (t/ha) | (%) | (days) | (cm) | |
| M-225 St. | 4.37 | 103 | 3.59 | 104 | 137 | 67 | MS |
| Dama (HSC-2) | 4.36 | 102.8 | 3.57 | 103.5 | 140 | 88 | R |
| Köröstaj | 4.31 | 101.6 | 3.43 | 99.4 | 139 | 67 | MS |
| Ringola | 4.23 | 99.7 | 3.45 | 100 | 135 | 74 | MR |
| ÖKI-3 | 4.23 | 99.7 | 3.44 | 99.7 | 132 | 72 | MR |
| HSC-1 | 4.18 | 98.5 | 3.46 | 100.2 | 133 | 74 | MR |
| Olympia 72 | 4.13 | 97.4 | 3.33 | 96.5 | 140 | 70 | S |
| Karmina | 4.1 | 96.7 | 3.31 | 95.9 | 137 | 74 | MR |
| Mean | 4.24 | 100 | 3.45 | 100 | 136 | 73.25 | — |
| LSD(0.05) | Ns | 8.7 | Ns | 9.1 | — | — | — |

[a] Mean of 3 years 1989–1991 and five locations.

**Table 4.** Data[a] from the National Rice Variety field test

| Variety | Panicle length | Spikelet fertility | Paddy seed 1000-grain mass | Cargo seed | | | Milled head rice |
|---|---|---|---|---|---|---|---|
| | | | | Length | Profile | 1000-grain | |
| | (cm) | (%) | (g) | (mm) | value | mass (g) | (%) |
| M-225 St. | 12.1 | 90 | 27.9 | 5.4 | 1.92 | 23.1 | 67.1 |
| Dama (HSC-2) | 20 | 89 | 30.8 | 7.2 | 2.98 | 25.1 | 63.6 |
| Köröstaj | 14.6 | 87 | 29.3 | 6.1 | 2.42 | 20.6 | 63.6 |
| Ringola | 17.6 | 90 | 33.5 | 7.4 | 2.95 | 27.4 | 65.5 |
| ÖKI-3 | 16.9 | 92 | 33.8 | 7.3 | 2.81 | 27.5 | 65.8 |
| HSC-1 | 16.5 | 91 | 31.1 | 7.2 | 2.89 | 25.4 | 66.3 |
| Olympia 72 | 14.3 | 88 | 29 | 6.6 | 2.58 | 24.5 | 64.4 |
| Karmina | 17.1 | 92 | 30.7 | 7 | 2.84 | 24.6 | 64.6 |

[a] Mean of 3 years 1989–91 and five locations.

resistance to shattering. The grain's mass of paddy seed is 30.8 g. It is easy to husk, the average milling yield is 64%, the whole grain yield 52%. The length of cargo seed is 7.2 mm (Table 4). Its profile value is 3 (the narrowest of all varieties). Its grain is glassy and of excellent cooking quality (amylose content 20%; undergoes a 2.3-fold increase in volume). Its potential genetic yield is 6–7 tons/hectare.

The seed production started on 200 ha in 1993. The yield achieved was high 5.2 t/ha as contrasted with 3 t/ha, the average yield in Hungary.

## 3 Summary and Conclusions

Somaclonal variation has been reported in a large number of species, however, only a few varieties of somaclone origin have been registered in the world. This contradiction is mainly due to the origin and nature of variation, including the advantages and disadvantages in plant breeding. Thus, in order to widen the application possibilities of somaclonal variation in plant breeding, a new hypothesis "ploidy-dependent somaclonal variation" was set up in 1983, and proved experimentally by a new approach "haploid somaclone" between 1984 and 1985.

From the results we drew the conclusion that the manifestation of somaclonal variability on plant level can be increased by reducing the ploidy level of in vitro cultures. A new method named *Haploid Somaclone Breeding* (HSM) was developed in 1985 to facilitate the practical application and was applied in rice breeding between 1985 and 1989. As a result of the application of HSM in rice breeding, two somaclonal varieties were developed. One of the candidates was registered by the name Dama in 1992.

The registration of the new variety Dama proved the applicability of HSM in breeding. The HSM method can be used to obtain new variation in one or more characters in existing cultivars or breeding materials, which may lead to their improved performance or a new variety.

Dama, as a somaclone variety of haploid origin, can be considered a world novelty. In Hungary, it is the first qualified plant variety improved by the use of a method of plant biotechnology. In our opinion, the Haploid Somaclone Method of improvement appears to be suitable for successful use by breeders of other plant species as well.

## References

Bajaj YPS (ed) (1990) Biotechnology in agriculture and forestry, vol 11. Somaclonal variation in crop improvement I. Springer, Berlin Heidelberg New York

Binh DQ, Heszky LE (1990) Restoration of the regeneration potential of long-term cell culture in rice (*Oryza sativa* L.) by salt pretreatment. J Plant Physiol 136: 336–340

Binh DQ, Heszky LE, Gyulai G, Csillag A (1992) Plant regeneration of NaCl-pretreated cells from long-term suspension culture of rice (*Oryza sativa* L.) in high saline conditions. Plant Cell Tissue Organ Cult 29: 75–82

Bulk RW (1991) Application of cell and tissue culture and in vitro selection for disease resistance breeding—a review. Euphytica 56: 269–285

Ceulemans E, Lefebvre M, Mandon D, Vermoote D (1986) A breeder's point of view about new breeding tools. In: Semal J (ed) Somaclonal variations and crop improvement. Martinus Nijhoff, Dordrecht, pp 100–107

Heszky LE, Pauk J (1975) Induction of haploid rice plants of different origin in anther culture. Riso 24: 197–204

Heszky LE, Simon-Kiss I (1992) Dama, the first plant variety of biotechnology origin in Hungary registered in 1992. Hung Agric Res 1: 30–33

Heszky LE, Li Su Nam, Horváth Zs (1986) Rice tissue culture and application to breeding. II Factors affecting the plant regeneration during subculture of diploid and haploid callus. Cereal Res Commun 14: 289–296

Heszky LE, Li Su Nam, Simon-Kiss I, Kiss E, Lökös K, Gyulai G, Kiss J (1989) Organ specific and ploidy-dependent somaclonal variation in rice—new tool in plant breeding. Acta Biol 40: 381–394

Heszky LE, Li Su Nam, Simon-Kiss I, Kiss E, Lökös K, Binh DQ (1991) In vitro studies on rice in Hungary. In: Bajaj YPS (ed) Biotechnology in agriculture and forestry, vol 14. Rice. Springer, Berlin Heidelberg New York, pp 619–637

Heszky LE, Simon-Kiss I, Binh DQ, Kiss E, Kiss J, Gyulai G (1992) New plant varieties developed by conventional and haploid somaclone method. Proc Int Symp Biotech. Assiut, Egypt (Nov 21–23, 1992), pp 139–146

Kucherenko LA (1991) Rice improvement through tissue culture in the USSR. In: Bajaj YPS (ed) Biotechnology in agriculture and forestry, vol 14. Rice. Springer, Berlin Heidelberg New York, pp 575–590

Li Su Nam, Heszky LE (1986) Rice tissue culture and application to breeding. I. Induction of high totipotent haploid and diploid callus from the different genotypes of rice (*Oryza sativa* L.). Cereal Res Commun 14: 197–203

Li Su Nam, Heszky LE, Simon-Kiss I, Horváth Zs (1986) Production and applicability of doubled haploid somaclones in rice. Oryza 23: 229–234

Oono K (1981) In vitro methods applied to rice. In: Thorpe TA (ed) Plant tissue culture. Academic Press, New York, pp 273–298

Ramulu KS, Dijkhuis R, Rovert S, Bokelmann GS, Groot V (1986) Variation in phenotype and chromosome number of plants regenerated from protoplasts of dihaploid and tetraploid potato. Plant Breed 97: 119–128

Schaeffer CW (1982) Recovery of heritable variability in anther-derived doubled-haploid rice. Crop Sci 22: 1160–1164

Semal J (ed) (1986) Somaclonal variations and crop improvement. Martinus Nijhoff, Dordrecht, pp 14–188

Sun Z, Sun L, Shu L (1991) Utilization of somaclonal variation in rice breeding In: Bajaj YPS (ed) Biotechnology in agriculture and forestry, vol 14. Rice. Springer, Berlin Heidelberg New York, pp 328–346

Sun Z-X, Zheng K-L (1990) Somaclonal variation in rice. In: Bajaj YPS (ed) Biotechnology in agriculture and Forestry, vol 11. Somaclonal variation in crop improvement I. Springer, Berlin Heidelberg New York, pp 288–325

# I.4 Somaclonal Variation for *Fusarium* Tolerance in Maize

H.S. Zhou[1] and Y.H. Deng[1]

## 1 Introduction

### 1.1 Botany, Importance, and Distribution of Maize

Maize (*Zea mays* L.) is a cross-pollinating species, with the male (tassel) flowers at the top and female (ear) flowers at lower nodes on the stalk. The grains, developed in ears or cobs, are the part mainly harvested. There are a number of grain types distinguished by differences in chemical compounds deposited or stored in the kernel: flint corn, dent corn, floury corn, waxy corn, sweetcorn, popcorn, sweet floury corn, and pod corn. Maize provides nutrients for human and animals, and serves as a basic raw materials for industry such as in starch, oil, protein, alcoholic beverages, food sweeteners, seasonings, fuel, etc. The green or dried plant has been used with much success as feed for livestock. Maize is one of the world's leading cereal crops with 129 million hectares of havesting area in the world, producing 478 million tons in 1991 (FAO 1991).

### 1.2 Available Genetic Variability and Breeding Objectives

Maize genetic variability is found mostly in its center of origin. The largest collection of maize germplasm is at the International Maize and Wheat Improvement Center (CIMMYT). About 12 500 entries are maintained and made available to breeders all over the world (Breth 1986). Other countries also maintain numerous collections. The genetic variability available for the commercial production of temperate maize is extremely limited, however, because only a few maize cultivar races are adapted to the short growing season of the northern latitude. Most of the genetic diversity in maize is discovered in tropical photoperiod-sensitive germplasm from South and Central America. This maize is difficult to work with in temperate regions. Most of the temperate region maize is selected from a few populations. For example, most of the US inbreds were selected from Red Yellow Dent and Lancaster Sure Crop. This narrows the maize genetic variability and broadens the way for the proliferation of diseases and

[1] Institute of Crop Breeding and Cultivation, Chinese Academy of Agricultural Sciences, Beijing, 100081, China

Biotechnology in Agriculture and Forestry, Vol. 36
Somaclonal Variation in Crop Improvement II (ed. by Y.P.S. Bajaj)

dpests. As indicated in a study by the Committee on Genetic Vulnerability of Major Crops (1972) of the National Academy of Sciences, corn has undergone a gradual, but continual decrease in genetic diversity over the past 50 years. The decrease in genetic diversity has been accompanied by an increase in genetic vulnerability. As the genetic base of corn germplasm used commercially is diminished, the risk of economic loss of corn due to diseases, insects, or unusual stress conditions increases correspondingly.

Breeding objectives for maize include: (1) increased kernel yield, oil and protein content, biological yield, adaptation (to cold, heat, salt, drought, soil fertility, wind), resistance to diseases and pests (leaf blights, downy mildew, rust, rootworms, earworms, borers, ear and root rots, etc.), resistance to herbicides and insecticides; (2) improved quality (protein quality, popoing quality, milling quality, etc.); (3) production of male sterile lines in order to save labor in hybrid seed production. Most breeders approach these objectives by working at the population or the inbred level. Sometimes they incorporate wild species into maize to increase the resistance and to improve quality, but this is both energy- and time-consuming.

### 1.3 Need to Induce Somaclonal Variation for *Fusarium* Tolerance in Maize

The fungus *Fusarium* can cause stalk rot, root rot, and ear and kernel rot. *F. graminearum* and *F. moniliforme* are pathogenic on maize stalk and root. *Fusarium* kernel or ear rot is the most widespread disease of corn ears. Losses result from reduced ear weight, poor grain quality, and mycotoxins that may contaminate feeds and food. It is caused by *F. moniliforme* and the closely related *F. moniliforme* var. *subglutinans*. Most corn inbreds are sensitive to *F. moniliforme* but greater susceptibility occurs in high-lysine, brown midrib, cms-T male sterile maize and sweetcorn (Ullstrup 1971; Warmke and Schenck 1971; Nicholson et al. 1976; Ooka and Kommedahl 1977; Tomov and Ivanova 1990).

When an elite inbred is sensitive to *Fusarum* sp., a traditional improvement method is to cross it with a resistant resource and then backcross with the inbred for at least six generations. This often changes the elite line with some disadvantages. Although this method does work well sometimes, it is always a time-consuming procedure which broadens the gap between breeding and the need for production, and normally there are limited resistant resources and useful elite genetic variability in temperate-region maize. Somaclonal variation may be the potential variability for resistance breeding in maize. Resistant mutations could be selected from the somaclonal variations without changing other characters of the elite inbred.

## 2 Brief Review of *Fusarium* Studies on Maize

Analyses of resistance of maize to ear-rotting pathogens were carried out by many researchers (Odimeh and Manninger 1984; Huang 1990; Reid et al. 1992;

Zummo and Scott. 1992). Mesterhazy (1989) concluded that resistant genotypes can be developed for maize and wheat as significant differences in disease susceptibility to *Fusarium* sp. exist among commercial cultivars, inbred lines and hybrids. In 15 hybrids from a diallel set of crosses involving 2 resistant, 2 moderately resistant, and 2 susceptible lines, resistance was studied under conditions provoking natural infection and after inoculation with *F. moniliforme*, and general combining ability (GCA) + special combining ability (SCA) effects for resistance were estimated. Both effects were involved, the former predominating. The genetic system conditioning resistance under natural infection appeared to differ from that under artificial infection (Nikonorenkov and Ivashchenke 1989). A study of the hybrids resulting from a diallel cross involving seven early inbred lines showed that resistance to *F. graminearum* and *F. moniliforme* was controlled by dominant genes which were strongly expressed when the hybrids were artificially infected. In inoculated plants the ratio of dominant to recessive genes expressed was 3:1 while in naturally infected plants the ratio was 2:1. Genetic analysis showed the presence of overdominance for lodging resistance (Cosmin et al. 1988). Popa et al. (1988) studied the response to *F. graminearum* under artificial infection and found that additive, epistatic, and environmental effects were significant for ear infection, while additive and environment effects were significant for stem infection. Leon (1989) modified ear-to-row selection scheme to improve resistance to *F. moniliforme* ear and stalk rots and five agronomic traits in eight tropical maize gene pools. Six pools were selected for stalk-rot resistance (SRR) and two for ear rot resistance (ERR). Averaged across gene pools, progress from selection per cycle was significant for all traits. Yield improvement was higher in pools selected for SRR than in those selected for ERR.

Two important types of toxins, zearalenone and trichothecenes, are produced by the *Fusarium* sp. mainly *G. zaea*, that produce ear rot in corn. Zearalenone is an estrogenic compound causing infertility and other reproductive problems in swine. Zearalenone can be assayed by thin-layer chromatography or by high-performance liquid chromatography (Sharman et al. 1991; Paster et al. 1991). The trichothecenes are a group of about 40 toxins, one of which was first discovered in a culture of the fungus *Trichothecium roseum*. Some of the trichothecenes are extremely toxic when ingested. The principal trichothecenes found in corn are primarily deoxynivalenol (DON), and less frequently T-2 toxin, diacetoxyscirpanol, and nirvalenol (Mirocha et al. 1980). A specific toxin has not been identified, but *F. moniliforme* and other *Fusarium* sp. produce a toxin known as moniliformin that is toxic to ducklings (Rabie et al. 1982), to baby chickens (Vesonder et al. 1976), and to plants (Cole et al. 1973). Pathre and Mirocha (1978) studied the deoxynivalenol from culture of *Fusarium* species. Mycotoxins of *Fusarium* were reviewed in detail by Wyllie and Morehouse (1978). Cordero found (1992) induction of PR protein in germinating maize infected with the fungus *F. moniliforme*. The two major acid-soluble proteins had molecular weights of 23 and 24 kDa. Casacuberta (1991) selected a clone containing an open reading frame encoding a protein homologous to PR protein from tomato and tobacco (PR-1 group). DNA blot hybridization indicated that this PR-like protein was encoded by a single-copy gene in maize.

The use of tissue and cell culture in breeding for resistance to *Fusarium* in tomato, potato, lucerne, and barley was reviewed by Mezentseva (1990). Schutze (1988) gave an example in vitro selection resistance to *F. oxysporum* in tomatoes. Ahmed et al. (1991) selected wheat somaclonal variation for *Fusarium* resistance using the double-layer culture technique. $R_2$ seedling populations from self-fertilized $R_1$ plants of four varieties were tested for *Fusarium* resistance by artificial infections in the greenhouse, and 3% of the regenerated $R_2$ plants have been found to be more resistant than the original cultivars.

We use calli of some elite corn inbreds susceptible to *F. moniliforme* ear rot (FER) for screening somaclonal variation resistant to FER in the hope that the improved lines can be immediately applied in corn-breeding projects. The primary results were reported by Zhou (1992). Therefore, in vitro selection of somaclonal variants insensitive to toxic metabolites produced by *Fusarium* spp. seems to be a viable approach to obtain resistant or tolerant plants.

## 3 Somaclonal Variation for *Fusarium moniliforme* Tolerance in Maize

Extensive studies have been conducted on somaclonal variation in maize (see Bajaj 1990, 1994) for tolerance to *Helminthosporium maydis* (Gengenbach et al. 1977), salt (Dupotto et al. 1994), herbicides (Somers and Anderson 1994), insecticides (Kuehnle and Earle 1994), and for increased amino acids (Gengenbach and Diedrick 1994). The work on tolerance to *Fusarium* is discussed here.

### 3.1 Toxicity of Exotoxin Extractive from Cultures of *F. moniliforme*

Fifteen pathotypes of *F. moniliforme* isolated from infected kernels in Beijing and Hainan were cultured at 25 °C on corn kernel medium. The cultures were tested by thin layer chromatography (TLC sensitivity ≈ 1 ng) to detect whether they produced moniliformin, a myco- and phytotoxin found in metabolites of *F. moniliforme*. The crude exotoxins of the fungus were extracted first by water and then by ethylacetate. After vacuum distillation of the extractive, two kinds of exotoxin extractives, named water-soluble exotoxin (WST) and oil-soluble exotoxin (OST), were obtained. Their toxic effects were tested according to their ability to inhibit the germination of corn kernels. The TLC result showed that 15 pathotypes were all free from producing monilifomin during their life cycle. Other faculties at the Beijing Agricultural University had found no moniliformin in other lines collected in Beijing (unpubl. data). This indicated that moniliformin is not necessary for the infection of *F. moniliforme*. We therefore changed the original plan of using moniliformin as selective agent for screening somaclonal variations resistant to *F. moniliforme*.

In Fig. 1a, serious inhibition to seed germination was observed when corn kernels were treated with 4% WST in sterile water, while the 6% OST (Fig. 1b)

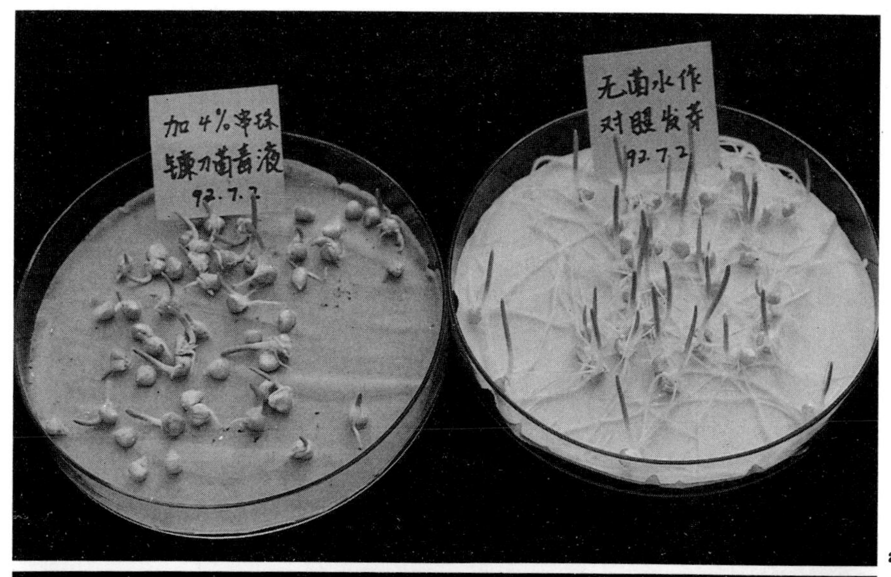

a

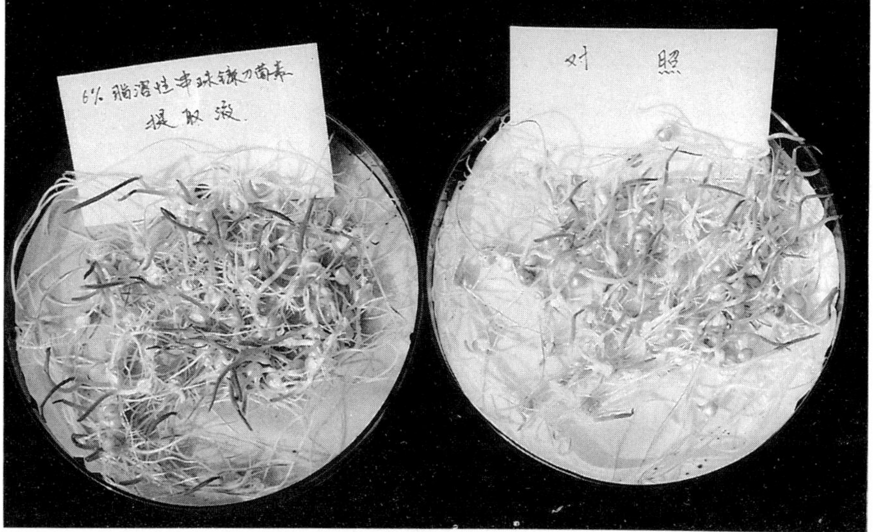

b

**Fig. 1. a** Tests of WST on germination of corn kernels by Y.H. Deng in 1993 (unpubl.). Seeds in *left* Petri dish treated with 4% WST, *right* with sterile water as control. **b** Tests of OST on germination of corn kernels by Y.H. Deng in 1993. Seeds in *left* Petri dish treated with 6% OST, *right* with sterile water as control

had little toxic effect on germination. Tests with calli also showed the same trend. Toxic effects of WST were mainly expressed as seriously inhibiting the growth of radicle. Some negative effects were also found on plumule development. On this basis, we employed the WST as the selective agent for screening the resistant calli.

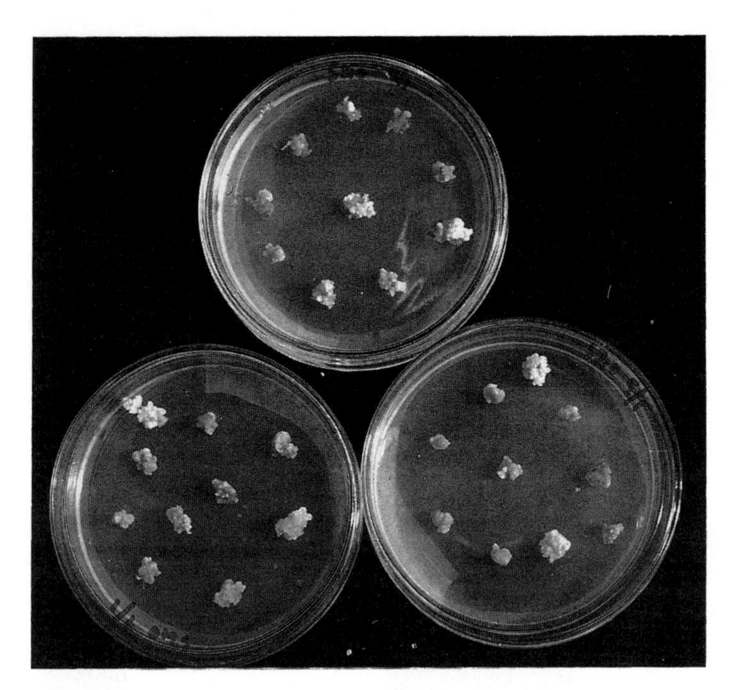

**Fig. 2.** Corn calli on medium with 0.08% WST grow well

The correlation between the survival rate of calli and WST concentration in the medium is shown in Table 1. The concentration of WST from 0.05 to 0.12% results in a lethal inhibition to callus growth.

### 3.2 Screening for Resistant Variants

Calli were obtained from three elite corn inbreds but susceptible to FER. These were chosen from 12 tested inbreds since their calli can regenerate plants. The callus was initiated from immature embryos on $N_6$ medium modified by adding 2 mg/1 NAA (α-naphthyl acetic acid), 2 mg/l 2, 4-D (2, 4-dichlorophenoxy acetic

**Table 1.** Survival rate of corn inbred calli on medium with different concentration of WST by Y.H. Deng in 1993

| Inbreds | Concentration of WST (%) | | | |
|---|---|---|---|---|
| | 0 | 0.05 | 0.08 | 0.12 |
| 3358 | 100 | 48 | 4.0 | 0 |
| P9–10 | 96 | 36 | 4.0 | 0 |
| Syn 31 | 94 | 38 | 6.0 | 0 |

acid) and 1.6 g/l proline in the medium. Immature embryos were inoculated at approximately 1.5 mm in length (11 to 13 days after pollination) and cultured at 25 °C in the dark for about 3 weeks, then the yellow granular calli were transferred onto maintenance media for propagation.

Pieces of calli (size about 2 mm) were placed ten in each Petri dish containing toxic media; additional mutagenic agents were avoided. Three inbred calli were transferred onto toxic medium containing 0.05% WST for the first round of screening. More than half the calli showed no further growth, became brownish and died after about 3 weeks. It was impossible to revive them on a nontoxic medium. In order to obtain reproducible results, calli of nearly identical size (2 mm ± ) were used for selective procedure. After 3 or 4 weeks of selection, about 40% of the inoculated calli survived and began to expand (Fig. 2).

These calli were transferred on 0.08% WST media, and showed a seven fold rate of survival in contrast to the calli without previous selection. Spontaneous mutations or physiological adaptations might have occurred during the first selective cycle. After propagation of the survived calli on 0.08% toxic media, the third round of selection was initiated with toxic media with 0.12% WST, which was lethal to calli of corn inbreds. The survival rate of the calli was still substantial as a result of the two rounds of preselections (Table 2). This result agrees in general with our early work using other maize inbreds in 1990–1991 (Table 3).

**Table 2.** Results of screening corn inbred calli resistant to WST in 1993

| Inbred | 0.05% WST | | | 0.08% WST | | | 0.12% WST | | |
|---|---|---|---|---|---|---|---|---|---|
| | No. of calli | No. of surv. | Surv. rate (%) | No. of calli | No. of surv. | Surv. rate (%) | No. of calli | No. of surv. | Surv. rate (%) |
| 3358 | 100 | 37 | 37 | 80 | 25 | 31.5 | 80 | 16 | 20 |
| P9–10 | 100 | 33 | 33 | 70 | 18 | 25.7 | 80 | 9 | 11.3 |
| Syn31 | 100 | 29 | 29 | 70 | 20 | 28.6 | 80 | 11 | 13.8 |

**Table 3.** Selective process of somaclonal variations resistant to toxin of *F. moniliforme* in 1990–1991

| Material | | Selective process (toxin %) 0.01 → 0.01 → 0.05 → 0.05 → 0.1 → 0.1 | | | | | |
|---|---|---|---|---|---|---|---|
| Mo 17 | No. of calli | 545 | 20 | 30 | 38 | 50 | 37 |
| | No. of surv. | 6 | 9 | 12 | 27 | 30 | 30 |
| | Rate of surv. | 1.1 | 45.0 | 40.0 | 71.05 | 60.0 | 81.0 |
| Zhong 017 | No. of calli | 690 | 98 | 164 | 159 | 201 | 200 |
| | No. of surv. | 34 | 54 | 82 | 142 | 126 | 182 |
| | Rate of surv. | 4.9 | 55.1 | 50.0 | 89.3 | 62.7 | 91.0 |
| CA 11 | No. of calli | 554 | 12 | 20 | 20 | 32 | 31 |
| | No. of surv. | 5 | 7 | 7 | 12 | 13 | 23 |
| | Rate of surv. | 0.9 | 58.3 | 35.0 | 60 | 40.0 | 74.2 |

Calli which could be maintained on 0.12% toxic medium were tentatively called resistant calli.

### 3.3 Plant Regeneration

The resistant calli were transferred onto regeneration medium (a modified MS medium, Murashige and Skoog 1962) with 2 mg/l 6-BA and without auxins. The cultures were illuminated 14 h a day by fluorescent lights of about 2500 lx.

The plantlets (Fig. 3) regenerated from the resistant calli were transferred onto a nursery medium with 2 mg/l IAA and 2 mg/l paclobutrazol to enhance the development of the root systems before they were transplanted into soil. The nursery medium could greatly improve the survival rate of transplantation of plants. The effect of paclobutrazol should be emphasized here for its significant effect on root development.

The $R_0$ plants were not tested for the resistance to *F. moniliforme* due to the weaker vigor and lower setting rate (see Fig. 4). The progeny plants ($R_1$) of $R_0$ and the calli induced from $R_1$ embryos will be tested for the resistance to FER and WST.

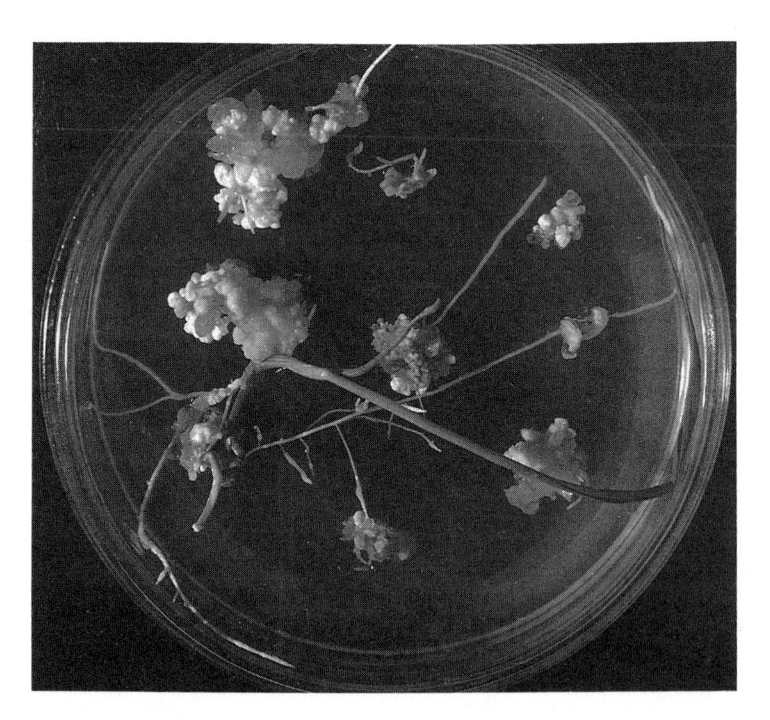

**Fig. 3.** Differentiation of the resistant calli after three-cycle screening on toxic medium

**Fig. 4.** Regenerated plants from resistant calli of Zhong 017 growing in the greenhouse showed weaker vigor. No seeds were harvested from the regenerated Zhong 017 plants by H.S. Zhou in 1991 (unpubl.)

## 4 Summary

All 15 pathotypes of *Fusarium moniliforme* used produced moniliformin during their life cycle. This indicated that moniliformin was not necessary for the infection of *F. moniliforme*. However serious inhibition to seed germination was observed when corn kernels were treated with 4% water-soluble exotoxin (WST), while the 6% oil-soluble exotoxin (OST) had little toxic effect on germination. Tests with calli also showed the same trend. Calli of six maize genotypes were screened on medium containing toxin from low to high concentrations. Resistant calli were obtained after several selections. The regenerated plants were recovered from two genotypes; but the $R_0$ plants were not tested for resistance to *F. moniliforme* due to weaker vigor and lower seed-setting rate.

## 5 Protocol

On the basis of our primary study, a protocol for screening somaclonal variants resistant to *F. moniliforme* was outlined as follows. It may be used as reference in experiments with other *Fusarium* species or other pathogens whose interactions with their host are not clear, and specifically acting toxins are not found. The gradual screening procedure in our work shows an advantage over the one-step procedure which is conducted at the same toxic level. This method can greatly increase the toxic level in medium during the screening procedure and produce more resistant calli.

1. Extract crude exotoxin of the fungus and test the toxicity of the toxin by bioassay to determine whether it can be used as a selective agent or not.
2. Initiate Type II calli from immature embryos and proliferate them to a large quantity.
3. Determine the tolerant scope of calli on toxic medium so as to determine the toxin concentration used in the whole screening procedure.
4. Screen the calli massively on low concentration toxic medium and then transfer the surviving gradually to high-concentration toxin medium to stabilize the resistant variation.
5. Regenerate plants from the resistant calli, conduct the resistance test on their progeny to determine the pattern of the resistant trait.

## References


Ahmed KZ, Mesterhazy A, Sagi F (1991) In vitro techniques for selecting wheat (*Triticum aestivum* L.) for *Fusarium* resistance. I. Double-layer culture technique. Euphytica 57: 251–257

Bajaj YPS (1990) Biotechnology in agriculture and forestry, vol 11. Somaclonal variation in crop improvement L. Springer, Berlin Heidelberg New York

Bajaj YPS (1994) Biotechnology in agriculture and forestry, vol 25. Maize. Springer, Berlin Heidelberg New York

Breth SA (1986) Mainstreams of CIMMYT research: a retrospective. CIMMYT, Mexico, DF, p 12

Casacuberta JM (1991) A gene coding for a basic pathogenesis-related (PR-like) protein from *Zea mays*, molecular cloning and induction by a fungus (*Fusarium moniliforme*) germinating maize seeds. Plant Mol Biol 16: 4, 527–536

Cole RJ, Kirksey JW, Cutten HG, Doubnik BL, Peckham JC (1973) Toxin from *Fusarium moniliforme* effects on plants and animals. Sciences 179: 1324–1326

Committee on Genetic Vulnerability of Major Crops (1972) Genetic vulnerability of major crops. NAS, Washington

Cordero MJ (1992) Induction of PR proteins in germinating maize seeds infected with the fungus *Fusarium moniliforme*. Physiol Mol Plant Pathol. 41(3), pp 189–200

Cosmin O, Craiciu D, Sarca T, Bica N, Ciocazanu I, Restea T (1988) The inheritance of resistance to stalk lodging and ear rot, caused by *Fasarium graminearum* Schw. and *Fasarium moniliforme* Sheld. in maize and its importance for breeding programmes. Probl Genet Teoret Apl 20(2): 75–107

Duncan DR, Williams ME, Zehr, BE, Widholm JM (1985) The production of callus capable of plant regeneration from immature embryos of numerous *Zea mays* genotypes. Plants 165: 322–332

Evans DA, Sharp WR (1986) Applications of somaclonal variation. Biotechnol Adv 4: 528–532

FAO (1991) Production yearbook. FAO, Rome

Gengenbach BG, Diedrick TJ (1994) Alterations in the biosynthesis of lysine, threonine, and methionine by tissue culture approaches. In: Bajaj YPS (ed) Biotechnology in agriculture and forestry, vol 25. Maize. Springer, Berlin Heidelberg New York, pp 587–603

Gengenbach BG, Green GE, Donovan CM (1977) Inheritance of selected pathotoxin resistance in maize plants regenerated from cell cultures. Proc Natl Acad Sci USA 74: 5113–5117

Huang CL (1990) Studies on some problems in identifying resistance of opaque-2 maize to *Fusarium moniliforme* ear rot. Sci Agric Sin 23: 5, 12–20

Kuehnle AR, Earle ED (1994) Insecticide resistant maize plants regenerated in vitro. In: Bajaj YPS (ed) Biotechnology in agriculture and forestry, vol 25. Maize. Springer, Berlin Heidelberg New York, pp 276–292

Leon C de (1989) Improvement of resistance to ear and stalk rots and agronomic traits in tropical maize gene pools. Crop Sci 29: 12–17

Liang GQ, Dai JR, Xie YJ (1988) Selection of C-cytoplasm maize mutants resistant to *Helminthosporium maydis* race C pathotoxin. Genome 30: 464
Lupotto E, Locatelli F, Lusardi MC (1994) In vitro selection for salt tolerance in maize. In: Bajaj YPS (ed) Biotechnology in agriculture and forestry, vol 25. Maize. Springer, Berlin Heidelberg New York, pp 314–330
Mesterhazy A (1989) Progress in breeding of wheat and corn genotypes not susceptible to infection by fusaria. In: Chelkowski J (ed) *Fusarium* mycotoxins, taxonomy and pathogenicity vol 2. Topics in secondary metabolism. Elsevier, Amsterdam, pp 357–386
Mezentseva OY (1990) Use of tissue and cell cultures in breeding for resistance to phytopathogens. Selek Semeno Moskva 4: 59–62
Mirocha CJ, Pathre SV, Christensen EM (1980) Mycotoxins. In: Pomeranz Y (ed) Advances in cereal science and technology vol 3. Am Assoc Cereal Chem St Paul, pp 159–225
Murashige T, Skoog F (1962) A revised medium for rapid growth and bio assays with tobacco tissue cultures. Physiol Plant 15: 473–497
Nicholson RL, Bauman LF, Warren HF (1976) Association of *Fusarium moniliforme* with brown midrib maize. Plant Dis Rep 60: 908–910
Nikonorenkov VA, Ivashchenko VG (1989) Analysis of the resistance of maize to stem rots based on different methodological approaches. Nauchno Tekh Byull Vses Ordena Lenina Ordena Druzhby Narodov Nauchro Issled Skogo Inst Rastenie Vod Imeri N I Vavilova 189: 19–22
Odimeh M, Manninger I (1984) Diallel analysis of resistance of maize to ear-rotting pathogens. Acta Agron Acad Sci Hung 33: 3/4 441–444
Ooka JJ, Kommedahl T (1977) Kernels infected with *Fusarium moniliforme* in corn cultivars with opaque-2 endosperm or male-sterile cytoplasm. Plant Dis Rep 61: 161–165
Paster N, Blumenthal-Yonassi J, Barki-Golan R, Menasherov M (1991) Production of zearalenone in vitro and in corn grains stored under modified atmospheres. Int J Food Microbiol 12(2/3): 157–165
Pathre SV, Mirocha CJ (1978) Analysis of deoxynivalenol from culture of *Fusarium* species. Appl Environ Microbiol 5: 992–994
Popa M, Cristtea M, Dutu H (1988) Experimental results concerning the inheritance of response to attack by *Fusarium graminearum* in maize. Aaalele Inst Cercetari Pentru Cereale Plant Tehn Fundulea 56: 45–55
Rabie CJ, Marasas WFO, Thiel PG, Lubben A, Vleggaar R (1982) Moniliformin production and toxicity of different *Fusarium* species from South Africa. Appl Environ Microbiol 43: 517–521
Reid LM, Mather DE, Hamilton RJ, Balton AT (1992) Diallel analysis of resistance in maize to *F. graminearum* infection via the silk. Can J Plant Sci 72(3): 915–923
Schutze R (1988) In vitro selection at the cellular level—a new way of producing disease-resistant plant. Gartenball 33: 166–167
Sharman M, Gilbert J, Chelkowski J (1991) A survey of the occurrence of the mycotoxin moniliformin in cereal samples from sources world wide. Food Addit Contam 8(4): 459–466
Somers DA, Anderson PC (1994) In vitro selection for herbicide tolerance in maize. In: Bajaj YPS (ed) Biotechnology in agriculture and forestry, vol 25. Maize. Springer, Berlin Heidelberg New York, pp 293–313
Tomov N, Ivanova I (1990) Susceptibility to *Fusarium* infection of ear and yield in high-lysine hybrids of maize homozygous for the opaque-2 gene. Rasteniev d nauki 27: 105–110
Ullstrup AJ (1971) Hyper-susceptibility of high-lysine corn to kernel and ear rot. Plant Des Rep 55: 1046
Vesonder RF, Ciegler A, Jensen AH, Rohwedder WK, Weisleder D (1976) Co-identity of the refusal and emetic principle from *Fusarium*-infected corn. Appl Environ Microbiol 31: 28–285
Warmke HE, Schenck NC (1971) Occurrence of *Fusarium moniliforme* and *Helminthosporium maydis* on and in corn seed related to T cytoplasm. Plant Dis Rep 55: 486–489
Wenzel G (1985) Strategies in unconventional breeding for disease resistance. Annu Rev Phytopathol 23: 149–172
Wenzel G, Uhrig H, Burgermeister W (1984) Potato—a first crop improved by application of microbiological techniques? In: Genetic manipulation: impact on man and society. Cambridge University Press, Cambridge, pp 127–137

Wheeler HE, Luke HH (1995) Mass screening for disease-resistant mutants in oats. Science 122: 1229
Wyllie TD, Morehouse LG (1978) Mycotoxic fungi, mycotoxins, mycotoxi-cross: an encyclopedic handbook. Marcel Dekker, New York
Xie YJ, Chen L, Dai RJ (1986) Study of somatic embryogenesis and plant regeneration in maize. Acta Genet Sci 13: 113–119
Zhou HS (1992) Selection of somaclonal variants resistant to toxin of *Fusarium moniliforme* from a maize inbred susceptible to ear rot. Maize Genet Coop Newsl 66: 16
Zhou HS, Shen JY (1993) Selection of specific resistant plants from C-cytoplasmic sterile maize calli. Sci Agric Sin 26(3): 33–38
Zummo N, Scott GE (1992) Interaction of *Fusarium moniliforme* and *Aspergillus flavus* on kernel infection and aflatoxin contamination in maize ears. Plant Dis V76(8): 771–773

# I.5 In Vitro Production of Fall Armyworm (*Spodoptera frugiperda*)-Resistant Maize and Sorghum Plants

B.R. WISEMAN[1], D.J. ISENHOUR[2], and R.R. DUNCAN[3]

## 1 Introduction

Maize (*Zea mays* L.) and grain sorghum, (*Sorghum bicolor* (L.) moench), are two important cereal crops. Maize is grown throughout temperate, subtropical, and tropical latitudes, whereas grain sorghum is usually grown in more arid environments that receive limited rainfall.

The fall armyworm, *Spodoptera frugiperda* (J. E. Smith), is a serious pest of maize and grain sorghum in the Americas (Horovitz 1960; Sparks 1979; Dicke and Guthrie 1988). Sparks (1979) reviewed the biology of the fall armyworm, and Ashley et al. (1989) provided the most current bibliography of this pest. It is one of only a few insect species that disperse and breed throughout the Americas. The fall armyworm is polyphagous and attacks many food crops and grasses. Infestations can limit production of maize and sorghum in various areas of the southeastern United States, Mexico, and Central and South America. Average annual losses by the fall armyworm in the United States exceed $300 million; during particularly severe outbreaks, such as occurred during 1975–1977, losses attributed to this pest exceeded $500 million annually (Mitchell 1979).

The adult of the fall armyworm is nocturnal, and at dusk adults initiate movement to suitable host plants. When fall armyworm populations are high, oviposition is rather indiscriminate; but the female usually oviposits eggs on the lower sides of maize or sorghum leaves. As larvae hatch from the egg, they begin to feed on the host plant. The whorl of young plants is particularly vulnerable. Larvae continue to devour foliage or other host materials until they have completed 6 instars. The 6th instar drops to the ground and pupates ca. 2.5–7.5 cm deep in the soil. The entire life cycle requires ca. 4 weeks in the southeastern United States in the growing season.

Insecticidal control of the fall armyworm often is not justified from an economic standpoint on sorghum or maize, especially in developing nations. A fall armyworm management program relying on cultural practices, biological

[1] USDA-ARS, Insect Biology and Population Management Research Laboratory, Tifton, GA 31793-0748, USA
[2] Pioneer Hi-Bred International, Dept. of Research Specialists, Johnston, IA 50131-0085, USA
[3] Dept. of Crop and Soil Science, University of Georgia, Georgia Station, Griffin, GA 30223-1797, USA

Biotechnology in Agriculture and Forestry, Vol. 36
Somaclonal Variation in Crop Improvement II (ed. by Y.P.S. Bajaj)

control, and crop genotypes with resistance to damage by larvae is an ideal approach. The use of resistant cultivars as the foundation for developing insect management systems has been stressed by various authors (Dahms 1972; Wiseman 1987). Wiseman (1987) stated that plant resistance will be "the foundation upon which most IPM strategies are built" in the future.

Continued progress in a crop breeding program is dependent in large part on the availability of genetic variation. Traditionally, the source for this variation has been plant introductions or exotic germplasm that were incorporated into the breeding programs. Plant tissue culture-induced variation, or somaclonal variation, has been reported for several agronomic crops (Larkin and Scowcroft 1981; Earle and Gracen 1985; Evans and Sharp 1986; Lee et al. 1988; Croughan and Quisenberry 1989; Bajaj 1990). Croughan and Quisenberry (1989) first reported somaclonal variation for resistance to feeding by larvae of the fall armyworm in regenerates of bermudagrass, *Cynodon dactylon* (L.) Pers.

## 2 Establishment of In Vitro Systems

### 2.1 Maize

Maize genotypes with resistance to leaf-feeding by larvae of the fall armyworm were developed by Scott and Davis (1981a, b). Several studies since have shown the effects of resistant genotypes on the growth and development of the fall armyworm (Wiseman et al. 1981; Williams et al. 1983; Isenhour et al. 1985). At the present time, resistant genotypes have a great potential as an alternative to insecticides for fall armyworm control in maize.

The maize genotypes Antigua 2D-118, MpSWCB-4, Mp496, and Pioneer X304C were selected as regeneration candidates because of their ability to express totipotency in cell culture. The $F_1$ progeny of a single cross of A188 x Antigua 2D-118 were also evaluated. A188 was used as a parent because of its high level of regeneration potential from cell culture (Earle and Gracen 1985).

Multiple field plantings of the maize genotypes were made during the summer of 1986–1988 at the Coastal Plain Experiment Station, University of Georgia, Tifton, GA using standard agronomic practices. Plants were selfed by hand pollination to ensure a uniform source of immature embryos.

Callus tissue was initiated from immature embryos using procedures described by Green et al. (1974) and Duncan et al. (1985). These procedures involve the culturing of immature maize embryos on agar-based media supplemented with necessary plant inorganic salts, nutrients, vitamins, and plant growth regulators. Callus tissue can be cultured for several months by transferring it to fresh media. Plant regeneration is achieved by varying the plant growth regulator concentrations of the medium. Regenerated, or $R_0$, plants were transferred from their agar substrate to soil in the greenhouse, allowed to mature, and then selfed. Seeds from these plants, designated as $R_1$ seed, were then planted in the field.

### 2.2 Grain Sorghum

Progress has been limited in the development of sorghum genotypes with leaf-feeding resistance to the fall armyworm. McMillian and Starks (1967) found differences in susceptibility of sorghum to feeding by larvae of fall armyworm under greenhouse and laboratory conditions. Wiseman and Lovell (1988) and Diawara et al. (1990) identified new sources of resistance to larvae of the fall armyworm in sorghums from Africa.

Culture of sorghum tissue for in vitro selection followed the procedures of MacKinnon et al. (1987). Seeds of Hegari were surface-sterilized in a 2% solution of sodium hypochlorite and then soaked for 24 h in sterilized water. Mature embryos from imbibed seeds were dissected and plated on a solid medium containing Linsmaier and Skoog's (1965) media with 2% sucrose, 2 mg/l of 2,4-D, and 0.5 mg/l kinetin. The callus culture from which line TCCP12 originated was initiated in August of 1984 and maintained in dark for nine subcultures of about 4 weeks each. After the fourth passage, each subculture was inoculated with 0.1 g of creamy-white, compact-appearing embryogenic callus. Following the ninth subculture, 0.9% pickling grade NaCl was added to the media before autoclaving. Callus was maintained in in vitro salt stress for five subcultures of 4 weeks each.

Regeneration began after the 14th subculture by moving callus to light and plating onto fresh LS media containing 1 mg/l indoleacetic acid (IAA) and 0.5 mg/l 6-benzyl-amine purine (BAP), but lacking NaCl. Two subcultures on this media were required to produce shoots of 2–3 cm. Shoots were then transferred to a final subculture of LS media containing 3 mg/l indole-3-butyric acid (IBA) to enhance root formation.

After 17 subcultures at 25 °C and a light intensity of ca. 2000–2500 lumen/$m^2$ or lux/$m^2$, plantlets were transferred to the greenhouse. Plantlets were then potted in a soil + less potting mixture and placed in a humidified, shaded plastic tent for 7 days, after which they were sufficiently acclimated and moved to the greenhouse for further development. The $R_0$, or regenerated plants giving rise to the TCCP12 lines were transferred to the greenhouse in December 1985, bagged to ensure selfing in March 1986, and $R_1$ seed were harvested in May of 1986. The number of seeds was increased in the summer of 1986 at the University of Georgia Bledsoe Research Farm at Griffin, GA.

## 3 In Vitro Studies on Fall Armyworm Resistance

### 3.1 Evaluation of Maize for Resistance

Field screening at the whorl stage and laboratory bioassays using meridic diets were the two methods used to measure resistance to the fall armyworm in the progeny of regenerated plants.

Field testing used artificial infestations of laboratory-reared fall armyworm larvae. A fall armyworm colony was initiated with feral larvae collected from maize. The colony was maintained as described by Mihm (1983), and feral larvae were added every summer. Infestation procedures were those described by Wiseman et al. (1980) and Mihm (1983) with an infestation rate of 40 neonates per plant. A randomized complete block design was used with individual plots consisting of a single row 3 m in length. There were ten and six replications for 1988 and 1989, respectively. Plantings were made using standard agronomic practices for sorghum in south Georgia with respect to tillage, herbicides, and irrigation. Leaf-feeding by larvae of the fall armyworm was visually rated using a system with 1 = resistant to 9 = susceptible (Davis et al. 1992).

Meridic diets were made by incorporating oven-dried maize whorl tissue from field-grown plants into a pinto bean diet (Wiseman et al. 1984; Isenhour et al. 1985). Whorl leaves were manually removed from the plants, cut into 4–6-cm sections and oven-dried for 72 h at 42 °C. Leaves were then ground in a Cyclotec 1093 mill (1 mm mesh screen size). Powdered foliage was then held at −20 °C until diets were made.

Powdered maize foliage was incorporated into the pinto bean diet to a final concentration of 45 mg/ml of diet. Diets were dispensed in 10-ml aliquots into 30-ml diet cups. The diet was allowed to solidify and then a single fall armyworm neonate was placed on the diet surface and the cup was capped with a paper lid. Cups were held at 26 °C and a photoperiod of 14L:10D in a growth chamber. A randomized complete block design was used with 25 replications. Weight of larvae was recorded after 8 and 12 days.

### 3.2 Field Evaluation of Sorghum for Resistance

Sorghum lines regenerated via tissue culture were evaluated in the field for resistance to the fall armyworm during the summer of 1986–1989. Sixty-three lines of regenerated sorghums were evaluated for whorl-feeding resistance to larvae of the fall armyworm.

Artificial infestations with laboratory-reared fall armyworm larvae were used to evaluate sorghum lines for resistance to the fall armyworm using procedures developed by Wiseman et al. (1980) and Mihm (1983). Foliage-feeding by fall armyworm larvae was rated at 7 and/or 14 days after infestation using a visual rating system (with 1 = resistant to 9 = susceptible) developed by Davis et al. (1992). Plant height measurements were taken in 1989 on the fifth through ninth plants in each row at physiological maturity.

Sorghum was planted at the Coastal Plain Experiment Station, University of Georgia, Tifton, GA. A randomized complete block design was used for all plantings, except for one of the 1988 studies in which a limited amount of remnant seed was screened in a nonreplicated planting. Individual plots consisted of a single row 3 m in length, spaced 0.76 m apart. Plantings were made using standard agronomic practices for sorghum in south Georgia with respect to tillage, herbicides, and irrigation (Duncan 1984).

### 3.3 Laboratory Evaluation of Sorghum for Resistance to Fall Armyworm

Two studies were conducted to measure the response of fall armyworm to different sorghum lines: a larval developmental study and a free-choice preference test. Entries used in the developmental study and subsequently in the free-choice test were selected at random from the group of lines planted during the 1989 field test. Included were eight regenerated lines, non-regenerated Hegari, and Northrup-King Savanna 5, a high-tannin commercial hybrid. Whorl-stage plants were harvested and removed from the field and the whorl tissues were separated manually, sectioned into 4–5-cm sections, and oven-dried at 42 °C for 72 h. Leaves were then ground (Tecator Cyclotec 1093 sample mill with a 1-mm mesh screen; Fisher Scientific, Atlanta, GA) and held at −20 °C until used.

In the developmental study, oven-dried whorl tissue was incorporated into artificial insect diets at a concentration of 50 mg dried foliage/ml of diet as described by Isenhour et al. (1985). Neonate fall armyworm larvae were placed singly in 30-ml diet cups, each containing ca. 20 ml of the meridic diet. Cups were capped with a paper lid and held in an environmental growth chamber at 26 °C with a photoperiod of 14:10 (L:D). The following developmental parameters were recorded: larval weight at 8 and 12 days of age, days to pupation, pupal weight, and days to adult eclosion. The developmental study was arranged as a randomized complete block design with 25 replications. A replication consisted of one fall armyworm larva per treatment, with 11 treatments, one being a diet-check without dried foliage.

The free-choice test used meridic diets prepared as described above. Sorghum lines were selected for this test based on the performance of the larvae in the developmental study. The two regenerated lines on which fall armyworm development was poorest and two regenerated lines that did not differ significantly from the nonregenerated parent were selected for comparison with Hegari (nonregenerated parent), and a meridic diet-check without dried foliage.

The test arena consisted of a plastic container 20 cm width × 30 cm length × 9 cm diameter with a tight fitting lid. One section (ca. 2 cm diameter × 3 cm height) of each type of the meridic diet was randomly arranged in the bottom of a labeled plastic container. Approximately 200–250 fall armyworm neonates were placed in the center of the container. The capped container was then placed in darkness at 26 °C. After 72 h, the number of larvae on each of the different diets was recorded.

### 3.4 Evaluations of Maize for Resistance to Fall Armyworm

Isenhour and Wiseman (1988) earlier had little success in evaluating callus tissue of resistant and susceptible maize lines in diet mixtures because callus-diet mixtures failed to confer the degree of resistance noted in foliage-diet mixtures. When the resistance mechanism was present in the silk and not the foliage, the callus-diet mixtures failed to exhibit any evidence of resistance. Therefore, it was necessary to regenerate plants for whole-plant analysis.

Successful plant regeneration is defined as the attainment of a viable plant capable of developing to maturity in soil. Plant regeneration was achieved for Antigua 2D-118, MpSWCB-4, Mp496, and A188 × Antigua 2D-118. No plants were regenerated from callus cultures of Pioneer X304C; however, this result may have been due to the technique used rather than to a lack of genotype response.

The number of successful regenerates by genotype for the 1987, 1988, and 1989 seasons are listed in Table 1. Values for "percent efficiency," i.e., the percentage of embryos placed on media that initiated callus formation, are also listed in Table 1. The percent efficiency is perhaps a crude measurement of genotypic capacity to form totipotent callus tissue, but it provides comparison of genotypic regeneration.

Antigua 2D-118 exhibited the highest regeneration efficiency. Even though the greatest number of plants was regenerated from MpSWCB-4, it had the poorest regeneration efficiency. MpSWCB-4 was used as one of the "standards;" therefore, it received more emphasis than other genotypes in the study. The single cross with A188 as the female parent did not express a higher degree of regeneration than its donor male parent, Antigua 2D-118. Similar results were observed in crosses with A188 and Zapalote Chico.

Results of the 1988 field evaluation of the progeny of $R_0$ plants ($R_1$'s) and nonregenerated parental lines are presented in Table 2. Significant differences in resistance to leaf-feeding by larvae of the fall armyworm were observed between the $R_1$ plants and the nonregenerated parental plants. Higher levels of resistance were observed for the $R_1$'s of MpSWCB-4 than in non-regenerated plants of this genotypes. The reverse was observed for A188 x Antigua 2D-118, with the $R_1$ plants showing a greater susceptibility to leaf-feeding by fall armyworm larvae than the $F_1$ of this cross (Table 2).

The $R_2$ plants (Table 3) failed to exhibit an increased level of resistance to leaf-feeding by larvae of the fall armyworm. The greater resistance exhibited by $R_2$ plants of MpSWCB-4 X Mp496-$R_0$ than by $F_2$'s of a comparable single cross was encouraging. Due to a limited amount of seed, this line was not evaluated as an $R_1$ in 1988. This line was derived by crossing a female nonregenerated MpSWCB-4 with pollen from an $R_0$ plant of Mp496. The progeny from this cross were then selfed and designated R since it had a regenerated parent.

**Table 1.** Plants regenerated from callus cultures of maize genotypes resistant to leaf-feeding by larvae of *Spodoptera frugiperda*, 1987–1989, Tifton, GA (Isenhour and Wiseman 1991)

| Genotype | No. of plants | Efficiency[a] (%) |
|---|---|---|
| Antigua 2D-118 | 42 | 53 |
| MpSWCB-4 | 342 | 15 |
| Mp496 | 93 | 25 |
| (A188 × Ant. 2D-118) | 6 | 13 |

[a] Total number of plants obtained divided by the number of immature embryos that were placed on media.

**Table 2.** Damage ratings for leaf-feeding by larvae of *S. frugiperda* at 7 and 14 days after artificial infestation of $R_1$ and nonregenerated parental lines of maize. Tifton, GA. 1988. (Isenhour and Wiseman 1991)

| Genotype | Damage rating[a] | |
|---|---|---|
| | 7 days | 14 days |
| Antigua 2D-118 NR-R[b] | 4a[c] | 6ab |
| MpSWCB-4 R1 | 5abc | 5a |
| MpSWCB-4 NR | 4ab | 6ab |
| (A188 × Ant. 2D-118) R1 | 6c | 8c |
| (A188 × Ant. 2D-118) F1 | 5ab | 7b |
| Pioneer 3369A NR-S[b] | 6c | 7b |

[a] Visual rating scale: 1 = resistant and 9 = susceptible (Davis et al. 1992).
[b] NR: nonregenerated control; R: resistant; S: susceptible.
[c] Means within a column followed by the same letter do not differ significantly at the 0.05 level (Waller and Duncan 1969).

**Table 3.** Damage ratings for leaf-feeding by *S. frugiperda* larvae at 7 and 14 days after artificial infestation of $R_2$ and nonregenerated parental lines of maize Tifton, GA. 1989. (Isenhour and Wiseman 1991)

| Genotype | Damage rating[a] | |
|---|---|---|
| | 7 days | 14 days |
| MpSWCB-4 R2 | 7a[b] | 8a |
| MpSWCB-4 NR-R[3] | 6b | 7b |
| (MpSWCB-4 × Mp496 Ro) R2 | 6b | 7b |
| (MpSWCB-4 × Mp496) F2 | 8a | 8a |
| Pioneer X304C NR-R[c] | 8a | 8a |
| Pioneer 3192 NR-S[c] | 8a | 8a |

[a] Visual rating scale: 1 = resistant and 9 = susceptible (Davis et al. 1992).
[b] Means within a column followed by the same letter do not differ significantly at the 0.05 level (Waller and Duncan 1969).
[c] NR: nonregenerated controls; R: resistant; S: susceptible.

The differences in resistance between $R_2$ and $F_2$ lines of MpSWCB-4 X Mp496 observed in the field were also detected in the diet test (Table 4). A significantly lower larval weight and a longer time to adult eclosion were observed for fall armyworm larvae that were fed the diet with leaf tissue from (MpSWCB-4 X Mp496 $R_0$) $R_2$ than occurred on any other diet.

### 3.5 Field Screening of Sorghum for Fall Armyworm Resistance

None of the 16 $R_1$ lines screened in Tifton in 1986 showed resistance to feeding by larvae of the fall armyworm, and none of the lines was retained for further

**Table 4.** Mean weight of *S. frugiperda* larvae after 8 days feeding on maize leaf tissue from regenerated and nonregenerated maize lines in meridic diets. (Isenhour and Wiseman 1991)

| Genotype | $\bar{x}$ Weight (mg) of larvae at | | Adult (d) |
|---|---|---|---|
| | 8 days | 12 days | Eclosion |
| MpSWCB-4 R2 | 28a[a] | 172b | 31ab |
| MpSWCB-4 NR-R[b] | 28a | 169b | 31ab |
| (MpSWCB-4 × Mp496 Ro) R2 | 19b | 132c | 33a |
| (MpSWCB-4 × Mp496) F2 | 30a | 204ab | 30b |
| Pioneer X304C NR-R[2] | 29a | 184ab | 30b |
| Pioneer 3192 NR-S[2] | 35a | 233a | 30b |

[a] Means within a column followed by the same letter do not differ significantly at the 0.05 level (Waller and Duncan 1969).
[b] NR: Nonregenerated control; R: resistant; S: susceptible.

evaluation. However, nine regenerated lines were selected based on differences observed in naturally occurring fall armyworm damage. $R_2$ plants from these lines were evaluated during 1987 in Tifton using artificial infestations, and one line, TCCP12, showed a marked reduction in defoliation by larvae of the fall armyworm. Plants from this line were selfed, and progeny from individual plants were evaluated.

In 1988, visual ratings 7 days after infesting the non-replicated $R_2$ planting revealed striking differences in fall armyworm damage. Line TCCP12-120 was rated as 1 (having no damage) while other TCCP12 lines were rated either 5 or 6 as compared with 5 for Hegari. The resistant check, 1821 CM, was rated at 2. At 14 days after infestation, TCCP12-120 was rated 7, compared with 9 (susceptible) for the other TCCP12 lines, 8 for Hegari, and 1821CM was rated 3.

One $R_3$ line of TCCP12 line (Table 5), TCCP12–600, had significantly ($F = 2.81$; df = 7, 28; $P < 0.0001$) lower damage due to feeding by larvae of the fall armyworm at 7 days following infestation than Hegari, but its damage did not differ from damage done to 1821CM, the resistant check. However, at the 14-day rating, TCCP12-600 did not differ significantly ($F = 16.5$; dF = 10, 14; $P < 0.0001$) from Hegari and exhibited significantly greater damage than 1821CM.

In 1989, $R_3$ and $R_4$ lines from TCCP12 were compared for their resistance to leaf-feeding by larvae of the fall armyworm (Table 6). Two lines, TCCP12-102 and TCCP12-123, had significantly ($F = 2.81$; d$F$ = 7, 28; $P < 0.0001$) lower damage ratings than Hegari 7 days following infestation. However, as in 1987, no significant differences in injury to foliage were detected after 14 days.

Significant differences ($F = 3.88$; df = 7, 28; $P < 0.0001$) in plant height were observed between the $R_3$ and $R_4$ lines and Hegari (Table 6). Line TCCP12-102, which had the lowest 7-day damage rating, also was among the shortest of the lines evaluated. Shortened plant height is an agronomically desirable trait, especially if a line is to be a parent in hybrid crosses. Decreased plant height aids in mechanical harvesting and lessens the likelihood of plant lodging.

**Table 5.** Damage rating to sorghum leaves by larvae of the fall armyworm at 7 and 14 days following infestation of $R_3$ sorghum lines. Tift Co. GA., 1988 (Isenhour et al. 1991)

| Line | x̄ Leaf damage rating[a] | |
|---|---|---|
| | 7 days | 14 days |
| TCCP12-100 | 7abc[b] | 8a |
| TCCP12-200 | 6bcd | 9a |
| TCCP12-300 | 7abc | 9a |
| TCCP12-400 | 8a | 9a |
| TCCP12-500 | 6bcd | 8a |
| TCCP12-600 | 5e | 8a |
| TCCP12-700 | 6bcd | 9a |
| TCCP12-800 | 7abc | 8a |
| Hegari[c] | 7abc | 8a |
| Huerin[c] | 6bcd | 8a |
| 1821CM[d] | 5e | 5b |
| SEM | 0.6 | 0.5 |

[a] Plants were infested on 22 June with 30 fall armyworm neonates per plant. Visual scale: 1 = resistant, 9 = susceptible.
[b] Means within a column followed by the same letter do not differ significantly at the 0.05 level Waller-Duncan k-ratio $t$ test (Waller and Duncan 1969).
[c] Nonregenerated susceptible checks.
[d] Resistant check.

## 3.6 Laboratory Evaluation of Sorghum for Resistance to Fall Armyworm

The effects of sorghum lines in meridic diets on the growth and development of fall armyworm larvae are presented in Table 7. A significantly lower weight was recorded for larvae feeding on TCCP12–123 and TCCP12–872 at 8 ($F = 3.8$; df = 10, 25; $P < 0.0001$) and 12 ($F = 3.82$; df = 10, 25; $P \leqslant 0.0001$) days of age than occurred for larvae on the nonregenerated parent Hegari. The weight of larvae that fed on a diet containing foliage of Hegari was three and two times greater than the regenerated TCCP12-123 at 7 and 14 days, respectively. Although no significant differences were detected in time to pupation or adult eclosion, larvae feeding on these two lines resulted in a notable 2-day increase in developmental parameters.

Larvae in the free-choice test showed a clear and significant ($F = 6.89$; df = 5, 9; $P < 0.0001$) preference for the diet-check (Table 8), which lacked any sorghum leaf tissue. Significant differences in preference of larvae also were observed among the diets that had sorghum tissue. Line TCCP12-123 had significantly fewer fall armyworm larvae after 72 h than Hegari or the two $R_4$ lines evaluated. The number of larvae on the $R_3$ line, TCCP12-872, was intermediate between the number of larvae preferring to feed on this line. These results are similar to those obtained from the 1989 field study and the developmental study.

**Table 6.** Leaf-damage ratings for sorghum by larvae of the fall armyworm at 7 days following infestation, and mean plant height of $R_3$ and $R_4$ sorghum lines. Tift Co., GA. 1989. (Isenhour et al. 1991)

| Line (generation) | $\bar{x}$ 7-day damage rating[a] | $\bar{x}$ Plant height (cm) |
|---|---|---|
| TCCP12-601 ($R_4$) | 8.0a[b] | 185a–g |
| TCCP12-602 ($R_4$) | 7.9a | 183a–g |
| TCCP12-603 ($R_4$) | 8.0a | 185a–g |
| TCCP12-604 ($R_4$) | 8.0a | 180d–g |
| TCCP12-605 ($R_4$) | 8.0a | 187a–f |
| TCCP12-606 ($R_4$) | 7.9a | 174g |
| TCCP12-607 ($R_4$) | 8.0a | 174g–h |
| TCCP12-608 ($R_4$) | 7.8ab | 161i |
| TCCP12-610 ($R_4$) | 7.0bc | 183a–g |
| TCCP12-611 ($R_4$) | 8.0a | 176e–g |
| TCCP12-612 ($R_4$) | 8.0a | 183a–g |
| TCCP12-613 ($R_4$) | 8.0a | 178e–g |
| TCCP12-102 ($R_3$) | 6.0d | 165hi |
| TCCP12-103 ($R_3$) | 8.0a | 193a |
| TCCP12-121 ($R_3$) | 8.0a | 188a–d |
| TCCP12-122 ($R_3$) | 8.0a | 187a–f |
| TCCP12-123 ($R_3$) | 6.5cd | 191a–b |
| TCCP12-124 ($R_3$) | 8.0a | 176f–g |
| TCCP12-125 ($R_3$) | 8.0a | 187a–f |
| TCCP12-126 ($R_3$) | 7.9ab | 180d–g |
| TCCP12-127 ($R_3$) | 7.9ab | 182b–g |
| TCCP12-871 ($R_3$) | 7.9ab | 188a–d |
| TCCP12-872 ($R_3$) | 7.5ab | 191a–b |
| TCCP12-873 ($R_3$) | 7.0bc | 184a–g |
| TCCP12-874 ($R_3$) | 7.8ab | 186a–f |
| TCCP12-875 ($R_3$) | 8.0a | 189abc |
| TCCP12-876 ($R_3$) | 8.0a | 178d–g |
| TCCP12-877 ($R_3$) | 7.9ab | 182b–g |
| Hegari (nonregenerated)[c] | 8.0a | 193a |
| SEM | 0.29 | 3.71 |

[a] Plants were infested on 15 June with 35 fall armyworm neonates per plant. Visual scale: 1 = resistant and 9 = susceptible.
[b] Means within a column followed by the same letter do not differ significantly at the 0.05 level Waller-Duncan k-ratio *t* test (Waller and Duncan 1969).
[c] Susceptible parent.

Two tissue culture-derived $R_3$ lines, TCCP12-102, TCCP12-123, were the best sources of resistance to leaf-feeding by larvae of fall armyworm. These two tissue culture-derived $R_3$ lines were released in 1990 and subsequently registered in 1991 (Duncan et al. 1991). Based on the laboratory evaluations, resistance to foliage damage by larvae of the fall armyworm observed in TCCP12-123 was due to both antibiosis and non-preference. A significantly lower weight was obtained for larvae in the developmental study, and a significant degree of nonpreference was detected in the free-choice tests. Unfortunately, TCCP12-102 was not one of the lines that had been randomly selected for inclusion in the laboratory study,

**Table 7.** Development of fall armyworm larvae feeding on meridic diets containing dried sorghum leaf tissue from regenerated and nonregenerated sorghum lines. (Isenhour et al. 1991)

| Line (generation) | $\bar{x}$ Larval wt. (mg) 8 days | 12 days |
|---|---|---|
| TCCP12-602 ($R_4$) | 64 ± 10ab[a] | 389 ± 44bcd |
| CCP12-604 ($R_4$) | 67 ± 10ab | 394 ± 42abcd |
| TCCP12-606 ($R_4$) | 83 ± 10a | 400 ± 42abcd |
| TCCP12-608 ($R_4$) | 74 ± 10a | 382 ± 44bcd |
| TCCP12-121 ($R_3$) | 66 ± 10ab | 363 ± 41cd |
| TCCP12-123 ($R_3$) | 26 ± 10c | 204 ± 42e |
| TCCP12-125 ($R_3$) | 65 ± 10ab | 391 ± 41abcd |
| TCCP12-872 ($R_3$) | 52 ± 10bc | 325 ± 41de |
| Hegari[b] | 83 ± 10a | 448 ± 41abc |
| Savannah-5[c] | 90 ± 10a | 512 ± 41a |
| Diet-check | 90 ± 10a | 489 ± 46ab |

[a] Means (± SEM) within a column followed by the same letter do not differ significantly at the 0.05 level; Waller-Duncan k-ratio *t*-test (Waller and Duncan 1969).
[b] Nonregenerated parent.
[c] Commercial susceptible hybrid from Northrup King.

**Table 8.** Feeding preference of fall armyworm larvae exposed for 72 h to selected meridic diets containing dried sorghum whorl tissue. (Isenhour et al. 1991)

| Line (generation) | $\bar{x}$ No. of larvae[a] |
|---|---|
| TCCP12-604 ($R_4$) | 26.9b[b] |
| TCCP12-606 ($R_4$) | 26.1b |
| TCCP12-123 ($R_3$) | 15.0a |
| TCCP12-872 ($R_3$) | 20.9ab |
| Hegari (nonregenerated)[c] | 26.1b |
| Diet-check | 41.6c |
| SEM | 3.8 |

[a] Between 200–250 neonates were released per container.
[b] Means within a column followed by the same letter do not differ significantly at the 0.05 level Waller-Duncan k-ratio *t*-test (Waller and Duncan 1969).
[c] Susceptible parent.

but its 7-day damage rating for 1989 was not significantly different from the damage rating of TCCP12-123.

The lines identified as resistant should be considered as candidates for conventional breeding programs for developing cultivars with improved resistance to larvae of the fall armyworm.

## 4 Release and Registration of In Vitro-Derived Resistant Plants of Maize and Sorghum

Maize germplasm, RGRFAW-15, was developed cooperatively by the Department of Entomology, University of Georgia, Coastal Plain Experiment Station, and the USDA/ARS Insect Biology and Population Management Research Laboratory at Tifton, GA. RGRFAW-15 may be traced to an individual embryo selected 26 June 1987. This germplasm was developed via plant regeneration from callus tissue of MpSWCB-4. RGRFAW-15 was released to the public by the Georgia Experiment Station June 1992. Seed is maintained by sib-pollination by the Georgia Seed Development Commission, Athens, GA and is available to both public and private maize breeders. Seed provided to public breeders is used in closed pedigrees.

Sorghum germplasm lines GATCCP100 (Reg. no. GP-312, PI 537308) and GATCCP101 (Reg. no. GP-313, PI 537309) were released by the University of Georgia Experiment Station and the Tissue Culture for Crops Project at Colorado State University in January 1990 (Duncan et al. 1991). Regeneration began after the 14th subculture, and after 17 subcultures plantlets were transferred to the greenhouse. The regenerated plant was designated TCCP12. Nine regenerated ($R_2$ seed) families from 139 total regenerated were selected in 1986. The $R_2$ plants from 63 selections were evaluated for leaf-feeding damage in 1987. The $R_2$ selection was self-pollinated, and individual panicles were threshed separately for subsequent evaluations. During 1989, $R_3$ and $R_4$ lines from TCCP12 were compared for their resistance to fall armyworm foliage feeding. Both GATCCP100 and GATCCP101 were selected from the $R_3$ lines. Seeds are maintained by R. R. Duncan and may be requested at the University of Georgia. Germplasm amounts (5 to 10 g) are available for distribution. The regenerated germplasm lines should be useful in sorghum forage breeding programs where the fall armyworm problems persist. Hybrid combinations involving some A1 cytoplasmic sterile lines with these two 3-dwarf pollinator lines could provide a good level of resistance to fall armyworm leaf-feeding (Duncan et al. 1991).

## 5 Summary

Plant regeneration of maize and sorghum genotypes known to be resistant to leaf-feeding by the fall armyworm was attempted via somatic embryogenesis. Successful regeneration from callus tissue cultures was achieved for the maize genotypes Antigua 2D-118, MpSWCB-4, Mp496, and some selected single crosses and the Hegari sorghum genotype. Progeny from these regenerates were evaluated under laboratory and field conditions for increased resistance to leaf-feeding by larvae of fall armyworm. Significant differences in resistance to leaf-feeding by the fall armyworm were observed between the regenerated and non-regenerated lines. Two sorghum regenerates (GATCCP100 and 101) with improved resistance to fall armyworm larvae have been released.

## References


Ashley TR, Wiseman BR, Davis FM, Andrews KL (1989) The fall armyworm: a bibliography.Fla Entomol 72: 152–204

Bajaj YPS (ed) (1990) Biotechnology in agriculture and forestry, vol 11. Somaclonal variation in crop improvement I. Springer, Berlin Heidelberg New York

Croughan SS, Quisenberry SS (1989) Enhancement of fall armyworm resistance in bermudagrass through cell culture. J Econ Entomol 82: 236–238

Dahms RG (1972) The role of host plant resistance in integrated insect control. In: Jotwani MG, Young WR (eds) The control of shoot fly. Oxford and IBH, New Delhi, pp 152–167

Davis FM, Ng SS, Williams WP (1992) Visual rating scales for screening whorl-stage corn for resistance to fall armyworm. Miss Agric For St Tech Bull 186: 9

Diawara MM, Wiseman BR, Isenhour DJ, Lovell GR (1990) Resistance in converted sorghums to the fall armyworm. Fla Entomol 73: 112–117

Dicke FF, Guthrie WD (1988) The most important corn insects. In: Corn and corn improvement. Agron Monogr 18, Agron Soc Am, Madison, WI, pp 767–867

Duncan DR, Williams ME, Zehr BE, Widholm JM (1985) The production of callus capable of plant regeneration from immature embryos of numerous *Zea mays* genotypes. Planta 165: 322–332

Duncan RR (1984) Grain sorghum production in the south eastern USA. Soil Conserv Ser Spec Publ, Athens, GA

Duncan RR, Isenhour DJ, Waskom RM, Miller DR, Nabors MW, Hanning GE, Wiseman BR, Petersen KM (1991) Registration of GATCCP100 and GATCCP101 fall armyworm resistant Hegari regenerants. Crop Sci 31: 242–244

Earle ED, Gracen VE (1985) Somaclonal variation in progeny of plants from corn tissue cultures. In: Henke R, Hughes K, Hollaender A (eds) Propagation of higher plants through tissue culture. Plenum Press, New York, pp 139–152

Evans DA, Sharp WR (1986) Applications of somaclonal variation. Bio/Technology 4: 528–532

Green CE, Phillips RL, Kleese RA (1974) Tissue cultures of maize: initiation, maintenance, and organic factors. Crop Sci 14: 54–58

Horovitz S (1960) Trabajosen marcha sobre resistencia a insectos en el maiz. Agron Trop (Venez) 10: 107–114

Isenhour DJ, Wiseman BR (1988) Incorporation of callus tissue into artificial diet as a means of screening corn genotypes for resistance to the fall armyworm and the corn earworm (Lepidoptera: Noctuidae). J Kans Entomol Soc 61: 303–307

Isenhour DJ, Wiseman BR (1991) Fall armyworm resistance in progeny of maize plants regenerated via tissue culture. Fla Entomol 74: 221–228

Isenhour DJ, Wiseman BR, Widstrom NW (1985) Fall armyworm feeding responses on corn foliage and foliage/artificial diet medium mixtures at different temperatures. J Econ Entomol 78: 328–332

Isenhour DJ, Duncan RR, Miller DR, Waskom RM, Hanning GE, Wiseman BR, Nabors MW (1991) Resistance to leaf-feeding by the fall armyworm (Lep: Noct) in tissue derived sorghum. J Econ Entomol 84: 680–684

Larkin PJ, Scowcroft WR (1981) Somaclonal variation—a novel source of variability from cell culture for plant improvement. Theor Appl Genet 60: 197–214

Lee M, Geadelmann JL, Phillips RL (1988) Agronomic evaluation of inbred lines derived from tissue cultures of maize. Theor Appl Genet 75: 841–849

Linsmaier EM, Skoog F (1965) Organic growth factor requirements of tobacco tissue cultures. Physiol Plant 18: 100–127

MacKinnon C, Gunderson G, Peterson KM, Nabors MW (1987) Plant regeneration from salt-stressed sorghum cultures. Plant Physiol 6: 107–109

McMillian WW, Starks KJ (1967) Greenhouse and laboratory screening of sorghum lines for resistance to fall armyworm larvae. J Econ Entomol 60: 1462–1463

Mihm J (1983) Efficient mass-rearing and infestation techniques to fall armyworm, *Spodoptera frugiperda*. CIMMYT, El Batan, Mexico

Mitchell ER (1979) Fall armyworm symposium. Fla Entomol 62: 81

Scott GE, Davis FM (1981a) Registration of MpSWCB-4 population of maize. Crop Sci 21: 148
Scott GE, Davis FM (1981b) Registration of Mp496 inbred of maize. Crop Sci 21: 353
Sparks AN (1979) A review of the biology of the fall armyworm. Fla Entomol 62: 82–87
Waller RA, Duncan DB (1969) A bayes rule for the symmetric multiple comparison problem. J Am Stat Assoc 64: 1484–1499
Williams WP, Davis FM, Wiseman BR (1983) Fall armyworm resistance in corn and its suppression of larval survival and growth. Agron J 75: 831–832
Wiseman BR (1987) Host plant resistance in crop protection in the 21st Century. In: Magallona ED (ed) Proc Int Congr of Plant Protect, Manila, vol 1. Univ Philippines of Los Banos, pp 505–509
Wiseman BR, Lovell GR (1988) Resistance to the fall armyworm in sorghum seedlings from Ethiopia and Yemen. J Agric Entomol 5: 17–20
Wiseman BR, Davis FM, Campbell JE (1980) Mechanical infestation device used in fall armyworm plant resistance programs. Fla Entomol 63: 425–432
Wiseman BR, Williams WP, Davis FM (1981) Fall armyworm: resistance mechanisms in selected corns. J Econ Entomol 74: 622–624
Wiseman BR, Gueldner RC, Lynch RE (1984) Fall armyworm resistance bioassays using a modified pinto bean diet. J Econ Entomol 77: 545–549

# I.6 Somaclonal Variation in Sorghum

T. Cai[1] and L.G. Butler[1]

## 1 Introduction

### 1.1 Botany, Distribution, and Importance of Sorghum

Sorghum, *Sorghum bicolor* (L.) Moench (2n = 20), is the world's fifth most important cereal crop. It is known under a variety of names and has been variously classified. Harlan and de Wet (1972) simply classified cultivated sorghum into five basic races (bicolor, guinea, caudatum, kafir, and durra) and ten intermediate races from combinations between the five races. Modern sorghums arose from the wild *Sorghum bicolor* subsp. *arundinaceum* in the northeast quadrant of Africa at least 5000 years ago and spread to other parts of the world (Doggett 1988). Sorghum is cultivated in temperate and tropical regions, on more than 45 million ha today, mainly in the semiarid areas of the tropics and subtropics. The production of world sorghum has steadily increased (70448 Mt in 1992) by raising the yield (kg/ha) in most countries, or by expanding areas in Africa (FAO 1992).

As a versatile crop, sorghum plays an important role in human life. The grain is used for human food in Africa and India, and for livestock feed elsewhere. Stems and foliage are used for green chop, hay, silage, and pasture, and in some areas for building materials and fuel. There are special types of popping, sweet, waxy, and broom sorghums (House 1984; Doggett 1988). Sorghum can be used for industrial products such as syrup, sugar, brooms, ethanol, paper, plastics, and many new uses (Palmer 1992; York 1993). Among the major cereals sorghum has the widest adaptation to harsh environments and the highest yield under adverse conditions (Doggett 1988). Sorghum has relatively effective chemical defenses against herbivores, pathogens, weeds, and other competitors (Butler 1989). A hardy and dependable crop, it can be cultivated in areas where other cereals either cannot be grown or produce low yields. Its unique survival ability and yield stability makes sorghum a crop of increasing importance with the high pressures of increasing population and decreasing available land in the world today.

### 1.2 Genetic Variability and Breeding Objectives

During its long evolution and development, sorghum has accumulated tremendous genetic variability. Now over 30 000 sorghum accessions are available from

[1] Department of Biochemistry, Purdue University, West Lafayette, IN 47907-1153, USA

Biotechnology in Agriculture and Forestry, Vol. 36
Somaclonal Variation in Crop Improvement II (ed. by Y.P.S. Bajaj)

the National Plant Germplasm System in the United States or the International Crop Research Institute for the Semi-Arid Tropics (ICRISAT) in India (Byth 1993; Dahlberg 1993). A general objective of sorghum breeding programs is to create cultivars, lines, and hybrids with endurable pest resistance, broad adaptation, high and stable yield, and good and acceptable quality for sorghum consumers. Biotic constraints on productivity include various bacterial, fungal and viral diseases; soil, foliage, stem and head insects; nematodes; birds; and the parasitic weed *Striga*. Major abiotic stresses include drought, poor soil fertility, and soil salinity, acidity, and alkalinity (ICRISAT 1980; Teetes et al. 1983; House 1984; Frederiksen 1986; Doggett 1988). Substantial efforts have been made to improve the quality of sorghum products for easy processing and good nutrition (ICRISAT 1982; Axtell 1984). Efforts continue to keep the nutritional value of sorghum equal to or near that of corn, to minimize the replacement of sorghum with corn (Ejeta et al. 1991; York 1993).

Breeding for high yields and good grain quality is more difficult for sorghum than for other cereals because of the relatively coarse nature of sorghum grain and poor environments in which it is generally grown.

### 1.3 Somaclonal Variation for Sorghum Research and Improvement

Although sorghums with superior agronomic traits are known, the germplasm base for sorghum breeding is narrow (House 1984; Doggett 1988). Conventional breeding is slow and costly, involving large-scale field trials. Use of somaclonal variation concurrently with conventional breeding procedures can enhance breeding efficiency by broadening variation and by shortening the breeding period (Evans 1988). The environment in in vitro culture is different from the natural environment where spontaneous mutations and recombinants are produced, and is also different from that employed for induction of mutations by physical and chemical agents. Utilization of these unique conditions offers opportunities to detect variations which otherwise are not discernible. Somaclones which retain their parents' desirable agronomic traits but with diminished limitations can be used in breeding programs directly as new cultivars or indirectly as stock lines. In favorable situations, selective pressure can be utilized to increase the likelihood of obtaining variants with a particular desirable trait.

Sorghum and maize are closely related (Berhan et al. 1993). Like maize, sorghum has many characteristics that make it an attractive object of investigation. It has been suggested that any manipulation of the maize genome can be duplicated in sorghum (York 1993), and that sorghum may even have some advantages due to its smaller genome size (Berhan et al. 1993). Compared to maize, however, relatively little research has been done on sorghum. Somaclonal variants with isogenic backgrounds can be valuable in sorghum research. Moreover, progress in sorghum research can benefit maize and other crops. The unique traits which make sorghum so vigorous durable in adverse conditions may be useful and transferable with developing modern biotechnology.

## 2 In Vitro Culture Studies

### 2.1 Callus Induction and Plant Regeneration

Reports on sorghum callus induction and plant regeneration have been cited in various reviews (Smith and Bhaskaran 1986; Kresovich et al. 1987; Duncan et al. 1991a). Explants for initiating sorghum callus cultures are most conveniently taken from germinated seedlings or soaked mature seeds. The major limitation to using these explants has been the relatively low frequencies of embryogenic callus and subsequent regeneration. Since early reports on the embryogenic capacity of sorghum callus induced from the scutellum of immature embryos (Gamborg et al. 1977; Thomas et al. 1977) and from young inflorescences (Brettell et al. 1980), reports of use of the two explants have been sporadic. Young inflorescences as an explant source seem to have become attractive recently (Duncan et al. 1991a; Kaeppler and Pedersen 1993). In our laboratory over 9000 plants have been regenerated from all 30 genotypes tested. Among the three explant sources (shoot portions of mature embryos, immature embryos, and inflorescences), the inflorescence has been the most readily cultured explant source and bears the highest capability for regeneration. Immature embryos are excellent explants, rapidly developing embryolike structures with high density, with great potenital for regeneration under optimal conditions (Cai et al. 1993a, 1994a).

### 2.2 Cell Suspension, Protoplast, and Other Cultures

Procedures for establishing cell suspensions, suspension-derived protoplasts, and protoclones have been established (Brar et al. 1980; Cai and Butler 1992; Chourey and Sharpe 1985). However, plant regeneration from protoplasts has been reported only by Wei and Xu (1990). Efforts to produce dihaploids using anther culture were not successful (Kresovich et al. 1987; Wen et al. 1991). Recently, Kumaravadivel and Sree Rangaswamy (1994) obtained 12 haploids and 248 dihaploids from anther culture of a sorghum hybrid, CSH5. Few other in vitro techniques, such as protoplast fusion (Murty and Cocking 1988), embryo rescue from wide crosses (Lusardi and Lupotto 1990), or ovary and ovule culture have been utilized for sorghum. Culturing detached sorghum panicles, developed in our laboratory (Cai et al. 1994b), has potential for rescuing late-maturing panicles or for investigation of grain development.

Current advances in sorghum genetic engineering research are gratifying. Stable gene transformations at cellular levels have been reported (Battraw and Hall 1991; Hagio et al. 1991), and transgenic sorghum plants have been obtained after microprojectile bombardment of immature embryos (Casas et al. 1993). Research on sorghum genome mapping and application of molecular markers in breeding is underway (Chen et al. 1990; Berhan et al. 1993; Tao et al. 1993; Vierling et al. 1994).

### 2.3 In Vitro Selection and Studies

Uses of cell and tissue culture research on sorghum in studies of stress physiology and biochemistry and for selection of variation for tolerance or resistance to environmental stress include water stress, salinity tolerance, aluminum toxicity, fall armyworm, and Striga (a root parasitic weed) resistance (see Sec. 3.3). Pathology studies were also reported. Kaveriappa et al. (1980) cultured *Sclerospora sorghi* in sorghum calli to observe the growth of the pathogen. Using sorghum callus cultures, Prabhu et al. (1984) studied the role of phenols in invasion of downy mildew. Bhaskaran et al. (1990) observed development and vegetative growth of cultured sorghum inflorescences directed by $GA_3$ in relation to smut infection. In our laboratory, in vitro studies for Striga resistance have been carried out (Cai et al. 1993b; Butler et al. 1994).

## 3 Somaclonal Variation

*Nomenclature.* We use $R_1$, $R_2$, $R_3$, ... to identify the primary regenerants, the first selfed progeny, and successive selfed generations, respectively, in order to correspond with the $F_1$, $F_2$, $F_3$, ... and $M_1$, $M_2$, $M_3$, ..., accepted genetic usage for selection of variation in cross-breeding and mutation breeding, respectively.

### 3.1 Variability in Regenerated Plants and Cultured Cells

$R_1$ plants regenerated from sorghum cultures are generally fertile, with no phenotypic abnormalities, and have 20 chromosomes with normal 10 meiotic pairing (Gamborg et al. 1977; Brettell et al. 1980; Smith and Bhaskaran 1986). However, significant numbers of chlorophyll-deficient plantlets, including albino, virescent, and striped, appeared among the regenerants (Cai et al. 1990). Albinos have been frequently reported (Smith and Bhaskaran 1986; Ma et al. 1987; Bhaskaran and Smith 1988; George et al. 1989; Wei and Xu 1990). In our $R_1$ plants, lethal variants having dark green, wide and curly or narrow and thick leaves, also often appeared. Other apparent variants were dwarfs and/or had multiple panicles (Figs. 1, 2). Among them a superdwarf had a height of only 30 cm, 15 tillers, small heads, and was completely sterile. Two shrunken endosperm variants (Fig. 3) were found from Tx 623 inflorescence calli which had been cultured for 13 months. These variants had a very low seed set. Wei and Xu (1990) described an unusual phenotype in protoplast-derived plants in which numerous axillary shoots appeared, possibly resulting from use of growth regulators. Ma et al. (1987) noted variations including albino, onion-leaf, satellited panicle branches, chimera, slow growth, poor seed set, mixoploid, as well as tall plant and waxy midrid. Only the last two reappeared in the $R_2$ generation. Although Gamborg et al. (1977) observed varied leaf morphology and growth habits, there was no evidence reported showing the changes to be somaclones rather than $F_2$ segregation of the parental hybrid used to establish the culture.

Cytological studies showed chromosomal variation existing in cultured cells and regenerants. Brar et al. (1979) counted chromosomes in cells from 18 callus cultures of sorghum cv. GPK-168. Among them only two were diploid, one was tetraploid, one mixoploid, and 14 had more than 60 chromosomes. Suspension cells and suspension-derived protoplast cells established from the GPK-168 calli were found to have chromosome counts of 64–66 which remained stable over 80 passages (Brar et al. 1979, 1980). They also found that diploid calli were hard and slow growing as compared to friable and rapidly growing polyploid calli. Meiotic analysis of 16 $R_1$ plants from NK-300 immature embryo calli revealed 11 diploids and 5 tetraploids (Brar et al. 1979). These results indicate that polyploidization seems to be prevalent and be favored over normal diploidy in cultures. However, only a slight variation in chromosome number was found in protoclones maintained for up to 7 years (Chourey et al. 1989). Ma and Liang (1985) observed 19% pollen mother cells showing meiotic abnormalities in a mixoploid regenerant. Minor variations in peroxidase banding patterns were found from assayed $R_1$ plants (Ma et al. 1987). Changed mtDNA as a result of sorghum cell culture was reported (Chourey et al. 1986). Further analysis for characterization of the changed region using five protoclones, along with the original suspension and parental seedling, revealed all protoclones to be different from each other and from the parental suspension and seedling, in their mitochondrial genome. Changes involved loss or rearrangement as well as changes in certain regions. Northern analysis for the *cob* gene showed an altered transcript pattern in one protoclone (Kane et al. 1992). These results clearly demonstrated occurrence of genetic variation in sorghum cultures.

## 3.2 Variations in the Progeny

### 3.2.1 Variations Selected from Somaclones

Ma et al. (1987) evaluated 151 somaclones of sorghum cv C401-1 in the $R_2$ and $R_3$ generations. They found persistence of height and waxy midrib from $R_1$ and $R_2$, and appearance of male- and female-sterile variants and dwarfs in $R_2$ plants. Genetic analysis of the somaclonal variations confirmed the height trait to be controlled by one dominant gene, and short stature and male sterility to be controlled by mutated recessive genes. Bhaskaran et al. (1987) reported somaclonal variations in plant height, tiller number, shoot dry weight, seed size, grain yield, and days to flowering on statistical analysis of the $R_2$ population with 96 $R_2$ plants derived from eight $R_1$ plants (cv. IS3620C). However, in the $R_3$ generation and the $R_2$ generation grown 2 years later, only the changes in enhanced tillering and height reduction were maintained; the others were lost. Duncan et al. (1991a) found somaclonal variations in plant height, maturity, insect resistance, acid soil stress tolerance, drought stress tolerance, and restorer/sterility maintenance in their field stress test. With respect to variation in fertility restoration capacity, Duncan et al. (1991a) reported that among their RTx 430 regenerant selections, many of the "– 44" selections function as maintainer lines while most of the "– 46" selections show restorer line capability.

1

**Figs. 1–3.** Somaclonal variations in primary sorghum regenerants.

**Fig. 1.** $R_1$ variants with short culm derived from a small M90950 180-day inflorescence culture

**Fig. 2.** A variant with multiple panicles from a single stem regenerated from a P898012 120-day inflorescence culture

**Fig. 3.** A rare type of shrunken endosperm (*right*) mutation found in Tx 623 $R_1$ plants derived from inflorescence culture which had been maintained for 13 months

2

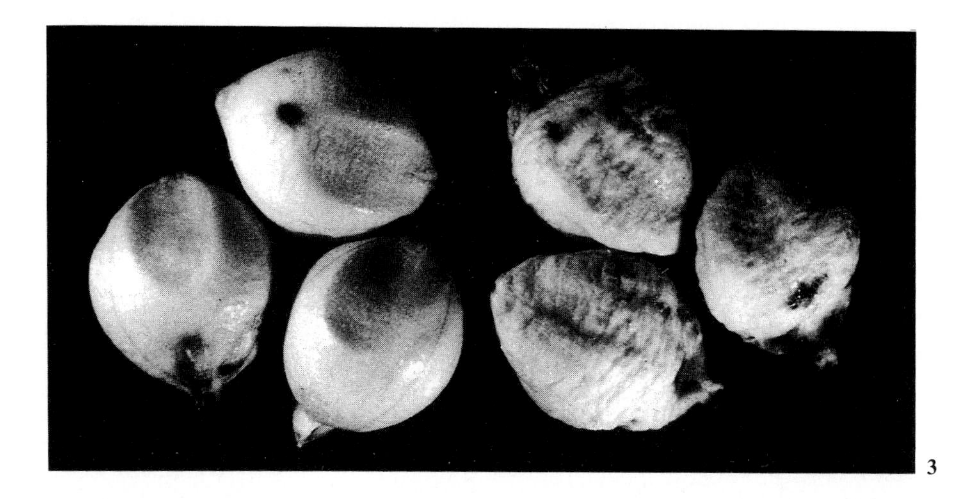
3

The $R_2$ population tested in our field assessment of somaclonal variation consisted of over 60 000 $R_2$ seedlings and 32 800 adult plants, progenies of 1055 high-tannin (Cai et al. 1990) and 186 low-tannin regenerated sorghum plants. Most of the 186 low-tannin sorghum plants were regenerated from selection medium for resistance to the parasitic weed *Striga*. Although the number of somaclones evaluated from the low-tannin group was limited, variations such as albino, yellow and yellow-green seedling, other lethals, short culm, and multiple panicles were found. Because fewer embryogenic cultures were available for maintenance after 120 days in culture, most of these tested somaclones were derived originally from a few callus lines that had been maintained for 7–10 months. The total $R_2$ population was generally uniform and the $R_2$ plants resembled their parental control. A total of 54 variant phenotypes were found, covering a wide spectrum of sorghum variations. The variations which appeared most often were chlorophyll-deficient, short culm or dwarf, narrow leaf, and sterile. Various chlorophyll-deficient variants segregated in individual progeny rows, similar to those described in mutagenesis (Ramulu 1975), including albino (*albina*), lethal or viable yellow-green (*vividis*), yellow (*xantha*), greenish white (*chlorina*), longitudinal striping leaf and/or stem (*striata*), red spotting (*maculata*), green and white transverse striping (*tigrina*), and a rare type of red seedlings. No positive correlation in albino emergence was observed between the $R_1$ and $R_2$ generation (Cai et al. 1990). The red type seedlings showed purple red leaves and very slow growth, but survived for 2 months, much longer than albino or yellow seedlings survived. Short culm mutants were found in both tall and short genotypes, and some of them were very good. Among them, one super dwarf failed to head, indicating a possible photosensitive mutation. Several types of short variation were observed from a single genotype (Cai et al. 1990). Narrow leaf or sterile variations were usually accompanied by shortened plant height. Other variations found included abnormalities such as lethal seedling, shoot or leaf necrosis and late flowering. A tall variation which appeared in the $R_2$ generation was shown in the $R_3$ generation to be controlled by a dominant gene.

Culture-derived dominant height of sorghum has been reported by Ma et al. (1987). Dominant mutations were rare in physical and chemical mutagenesis (Gaul 1964). In addition to shrunken endosperm and red seedling, other rare types of variations observed were ragged-leaf, abnormal panicles, and *Hydra*. The

4

5

6

**Figs. 4–6.** Some rare types of sorghum somaclonal variations

**Fig. 4.** Another multibranched panicle grown in field, derived from a 270-day IS 8260 inflorescence culture

**Fig. 5.** A similar panicle variant with long and helical rachilla and lower fertility grown in the field, derived from a 270-day IS 8260 inflorescence culture

**Fig. 6.** *Hydra* shows disordered growth and development. The variants had abundant tiller and leaves, multibranched shoots and many independent panicles. Its leaves showed various abnormalities: small, short, narrow, wide, creased, and curly. Its panicles were smaller than those of the parent and had normal fertility. The *Hydra* variation was derived from 120-day and 180-day IS 4225 mature embryo cultures

former two derived from 270- and 300-day IS8260 inflorescences cultures and *Hydra* came from 120- and 150-day IS4225 cultures initiated from mature embryos. The ragged leaf variant had sawtooth leaves and shorter culms than parental controls; it produced pollen but set few selfed seeds. One type of panicle variation observed (Fig. 4) was multibranched, and another had long and helical rachiela, partially sterile (Fig. 5). The *Hydra* variation showed disordered growth and development, first appearing as abnormal dwarfs in the $R_2$ generation. In the $R_3$, *Hydra* developed abundant tillers and leaves, multibranched shoots, and numerous independent panicles with normal fertility (Fig. 6). The leaves showed various abnormalities, such as small, short, narrow, wide, creased, and curly. *Hydra's* height was shorter and panicles were smaller than those of the parents. The multiple panicle variation (Fig. 2) observed in P898012 R1 plants showed lower degrees of abnormality in the $R_2$ and $R_3$ generations.

Most of these somaclonal variants were different from their parent controls in more than one character, such as narrow leaf with short culm, dwarf with sterility, etc. Sequential occurrence and accumulation of mutations were observed. Table 1 shows the distribution of 21 variations covering four different chlorophyll variations and seven viable variations in a clonally related population derived from IS 8260 immature embryo culture. Up to six different variation phenotypes could be detected in plants derived from a single callus. Phenotypically similar variations can reappear later in the same callus line population, or segregate in progeny of different callus line populations.

### 3.2.2 *Factors Affecting Production of Variation*

Somaclonal variation occurs randomly at different rates depending upon genetic background, character of the mutation, and culture conditions. Failure to detect variation may be due in part to an insufficient number of $R_1$ plants regenerated and of their progenies evaluated. The size of the $R_2$ population, including the

**Table 1.** Distribution of 21 variations including chlorophyll (CV) and viable variations (VV) in a clonally related population derived from IS 8260 immature embryo culture

| | Culture source | Days in culture 210 | 240 | 270 | 300 | Total[b] |
|---|---|---|---|---|---|---|
| No. of Total $R_1$ families | Callus-1 | | 16 | 16 | 20 | 52 |
| | Callus-3 | | 16 | 75 | 20 | 111 |
| | Callus-12 | 11 | 16 | | | 27 |
| | Multiple | 29 | 12 | 9 | | 50 |
| CV and VV appeared[a] | Callus-1 | | V1 | C4, V1, V2 V5 V7 | C2 | 7(6) |
| | Callus-3 | | C2, V7 | V7 V2 V3 V3 V6 | C6 | 8(5) |
| | Callus-12 | C3, V4 | C1 C1, V3 | | | 5(4) |
| | Multiple | | | V7 | | 1(1) |

[a] C1 ··· C4, and V1, V2, ··· V7 represent four and seven different chlorophyll and viable variation phenotypes, respectively.
[b] Number in parentheses indicates the number of variation phenotypes.

number of $R_1$ progeny rows and the number of $R_2$ plants in each row, determines emergence of variation carried by the $R_1$ plants. Parental genotype plays a decisive role in the rate and range of mutations that occur. As shown in Table 2, the $R_2$ population of IS4225 and the group of low-tannin sorghums had three-to fourfold greater SVR (somaclonal variation rate, see Sect. 3.5) than other populations tested. Most of the variations carried by IS 4225 $R_1$ plants were undesirable, such as shoot or leaf necrosis, and sterility. IS 8260 produced the largest number of variant phenotypes, while SRN-39 had only chlorophyll variations.

Under culture conditions, clonal proliferation of a mutated cell leads to production of phenotypically similar variants. Emphasis should be on increasing the types of the variation rather than on producing larger numbers of variants with similarly changed phenotypes. Variation production appears to increase with increased time in culture. In our conditions, calli cultured 210–300 days produced higher SVR and more variant phenotypes than those cultured 120–180 days (Cai et al. 1990). Although no apparent difference in SVR were found among the three explant types used in our cultures (Cai et al. 1990), more chlorophyll-deficient variations were recovered from immature embryo cultures, and inflorescence cultures produced the highest number of novel variant phenotypes such as red seedling, ragged leaf, shrunken endosperm, and panicle variants. Because some tissues undergo more changes than others during differentiation (Karp 1991), it is likely that young inflorescences consist of a larger number and degree of differentiated tissues and may produce more variations than the scutellum tissue of immature embryos.

Use of chemical or physical mutagens can enhance the production of variation. Wang et al. (1988) reported that corn plants regenerated from irradiated calli had a two- to tenfold increase in mutations over plants regenerated from unirradiated control calli. We found that $R_1$ progeny rows derived from calli treated with a chemical mutagen (sodium azide) showed 5.6% SVR (4 of 72) and four different chlorophyll-deficient variations compared to 1.3% (1 of 75) and only one from untreated control calli. The fact that the viable variations could not

**Table 2.** Variant phenotypes observed in $R_1$ families in $R_2$ generation

| Genotypes[a] | $R_1$ family tested | No. of variant phenotypes[b] CV | VV | Total | SVR (%) |
|---|---|---|---|---|---|
| High tannin (8) | 1055 | 19 | 25 | 44 | 11.3 |
| IS 8260 | 480 | 8 | 12 | 20 | 10.0 |
| IS 8768 | 353 | 5 | 5 | 10 | 9.3 |
| IS 4225 | 112 | 4 | 7 | 11 | 31.3 |
| Others (2) | 110 | 2 | 1 | 3 | 2.7 |
| Low tannin (6) | 186 | 6 | 3 | 9 | 11.3 |
| SRN-39 | 147 | 5 | 0 | 5 | 4.1 |
| Others (3) | 39 | 1 | 3 | 4 | 38.5 |
| Total | 1241 | 25 | 28 | 53 | 11.3 |

[a] Number in parentheses shows the number of genotypes involved.
[b] CV = chlorophyll variations; VV = viable variations.

appear until the third generation from the sodium azide-derived somaclonal population indicated mutation occurrence with small sector and/or low level chimerism.

## 3.3 Somaclonal Variants for Stress Tolerance

### *3.1.1 Drought Tolerance*

Work on physiological and biochemical changes of cultured callus under water stress indicated that proline concentration is an incidental consequence of stress (Bhaskaran et al. 1985). However, Yang et al. (1990) found that proline accumulation in callus may be a response to severe stress to tissues. Newton et al. (1986) reported that soluble carbohydrate concentrations may be correlated with osmotic adjustment. Smith et al. (1985) measured callus growth from ten tolerant to susceptible sorghum cultivars with increasing levels of polyethylene glycol (PEG) as an osmoticum in the medium. They suggested that PEG-induced osmotic stress on callus cultures can be used to screen sorghum cultivars for potential preflowering drought tolerance. Field evaluation of tissue culture-derived sorghum lines has been carried out and five lines were identified as having improved drought tolerance (Waskom et al. 1990).

### *3.3.2 Salinity Tolerance*

Growth of sorghum callus on salinized media was reported as early as 1968 (Strogonov et al.) and plants were regenerated from sodium chloride-selected calli (Bhaskaran et al. 1983, 1986; MacKinnon et al. 1987). However, selection for salt-tolerant plants has been unsuccessful. Cultured cells can be used as a model to elucidate and assess modes of cellular adaptation to salinity stress for a better understanding of stress physiology and biochemistry (Kresovich et al. 1987). Yang et al. (1990) reported that salt tolerances of whole plants and of calli are positively correlated. Calli growth and Na/K ratios could be used as indicators of whole plant salt tolerance in sorghum.

### *3.3.3 Aluminum Toxicity*

Smith et al. (1983) selected calli from four sorghum genotypes on media containing 0, 100, 200, and 400 $\mu$M aluminum for 3 to 12 months. Differences were observed among the cultivars in response to aluminum toxicity. The regeneration capacity was inhibited at 400 $\mu$M aluminum. Plants were obtained from only one cultivar (IS3620C). No field test was reported. Sorghum lines from tissue cultures of Hegari and Tx430 were selected from field evaluation (Waskom et al. 1990) and advanced test (Miller et al. 1992) for increased tolerance to acid soils. One Hegari (Duncan et al. 1991c) and two Tx430 (Duncan et al. 1992) regenerants have been registered as acid-soil tolerant lines. Acid soil- and drought stress-tolerant variants were obtained from both nonstressed and stressed cultures. There was no significant effect of an in vitro stress treatment on the incidence of variants with acid soil or drought tolerance expressed in the field (Waskom et al. 1990).

#### *3.3.4 Fall Armyworm Resistance and Other Environmental Stress*

Two regenerated Hegari lines were identified (Isenhour et al. 1991) as having a significantly higher level of resistance to field armyworm and were registered (Duncan et al. 1991b). The two resistant lines were unexpectedly derived from NaCl-stressed callus fortuitously detected in a natural infestation of fall armyworms in 1986 (Duncan et al. 1991b; Isenhour et al. 1991). Nadar et al. (1975) studied promotion of callus growth by s-Triazine herbicides which may act as a hormone and most likely kinetin-like. Using sorghum embryogenic calli, selection under toxic *Striga* extract and chlorsulfurontech herbicide has been carried out in our laboratory for screening sorghum plants for resistance to *Striga*, a root parasitic weed, and the herbicide, respectively. Plants have been regenerated from the screened calli. Further tests will be conducted in a *Striga*-infected field for the weed resistance and in progenies of regenerants for herbicide resistance.

### 3.4 Somaclonal Variation in Grain Polyphenols

High-tannin sorghums have production advantages such as resistance to molds and birds, but have harmful nutritional effects (Butler 1989). We are attempting to utilize in vitro tissue culture as new approach for overcoming problems associated with utilization of high-tannin sorghums. From the morphological and developmental variations observed in field assessment, we expected to find variations in grain polyphenols in somaclones from high-tannin sorghums. Because many sorghums produce no tannin at all, we hoped to obtain variants blocked at different steps in biosynthesis of the polyphenols, and correspondingly improved in nutritional quality. From 487 $R_1$ families, over 6000 plants separately harvested from the $R_1$, $R_2$, $R_3$ generations and corresponding control plants have been analyzed (Cai et al. 1995). The regenerated plants for the assay were uniform and morphologically resembled their parents. Three grain polyphenol characteristics: total phenols, flavan-4-ols, and proanthocyanidins (tannins) were assayed using our routine procedures. All data were analyzed statistically. The results obtained were unexpected. We could not identify a single somaclone in which the capacity to produce these polyphenols was lost or greatly reduced. However somaclones with statistically increased or decreased levels of total phenols, flavan-4-ols, and proanthocyanidins in the grain were detected. Most of the polyphenol variants produced more, rather than less, polyphenols than their parental control. Variants with decreased levels of polyphenols were relatively little changed from their parental controls and most of them reverted back to normal levels in subsequent generations. In contrast, variants with increased levels of polyphenols had a higher percent change from parental controls and the changes were generally retained in subsequent generations. Table 3 shows selected somaclones which in the $R_3$ generation maintained significantly higher content of these grain polyphenols than their parental control.

Sorghum condensed tannins are oligomers of flavan-3-ols. Anthocyanidins, the major pigments of sorghum grain, are alternative endproducts, with tannin, of flavonoid metabolism in the seed coat. The occurrence of tannin in sorghum grain is controlled by two complementary dominant genes B1 and B2 (Rooney et al. 1980). A possible explanation for our failure to find tannin-free variants from high-tannin parents is that sorghum has multiple metabolic pathways leading to production of polyphenolic materials we assayed as tannins. Loss of one of these pathways or only one step of a pathway may have little effect on overall polyphenol content. Our results seem to indicate that mutations of genes controlling early key steps such as phenylalanine ammonia lyase or chalcone synthase are relatively rare. The conditions we used for regeneration may have favored somaclones which produce increased levels of polyphenols including tannin, rather than decreased levels. By way of contrast, many deleterious variations were found, including various chlorophyll-deficient seedlings and sterile plants. Chlorophyll and fertility, although readily lost by somaclonal variation, are essential for growth, development, and proliferation, whereas the nonessential tannin is not.

Somaclonal variation in some characters seems to be quickly lost in subsequent generations. Variants with decreased levels of polyphenols reverted back to normal levels in the $R_2$ and $R_3$ generations. The multiple panicle trait observed in the P898012 $R_1$ variant (Fig. 2) gradually diminished in the following generations. Bhaskaran et al. (1987) reported sorghum variations in $R_2$ plants which were lost in the late $R_2$ generation after 2 years' storage of the $R_2$ seeds. This loss may be attributed to reverse mutation or to temporary memory of epigenetic variations occurring during tissue culture. Duncan et al. (1991a) pointed out that physiological adaptation from exposure to culture media or in vitro selection pressure can result in nongenetic physiological changes within early generation plants that may mask true genetic effects.

**Table 3.** Polyphenol variation in $R_3$ generation

| Name | Total phenols $A_{720}$ | Flavan-4-ols $A_{550}$ | Proanthocyanidins $A_{550}$ |
|---|---|---|---|
| IS 8260 | 264 | 0.310 | 1.534 |
| SC | 338 | 0.563 | 1.966 |
| IS 4225 | 138 | 0.417 | 0.867 |
| SC | 221 | 0.618 | 1.514 |

### 3.5 Methods for Measuring Somaclonal Variation

De Klerk (1990) described several methods for assessing the extent of somaclonal variation: percentage of aberrant plants; cytological and molecular assays such as chromosome number and structural integrity, and DNA restriction fragments, electrophoresis patterns of proteins and isozymes. All these methods have been used for characterizing cultured sorghum cells and regenerants, e.g., observation of chromosome number and structure (Brar et al. 1979; Ma et al. and Liang 1985; Ma 1987), DNA characterization (Kane et al. 1992), protein and isozyme patterns (Ma et al. 1987; Wozniak and Partridge 1988), and statistically analyzing quantitative variations (Bhaskaran et al. 1987). The following methods were used for calculation of the percentage of variant regenerants. They are similar to those used in plant mutation breeding (Gaul 1964). Somaclonal variation rate (SVR) based per 100 $R_1$ plants and somaclonal variant frequency (SVF) based per 100 $R_2$ plants were calculated to determine the extent of genotypic variability of somaclones. SVR describes the probability of variation of given somaclones during tissue culture. SVF refers to the proportion of detectable variants in a given somaclonal population. The SVR is a function of the rate of occurrence of genetic events in vitro but, unlike mutagenesis, the rate is complicated by similar variations resulting from independent mutation events as well as from proliferation of single mutated cell clones. The formulas are:

$$\text{SVR}\ (\%) = \frac{\text{No. of variations found in } R_1 \text{ families}}{\text{No. of total } R_1 \text{ families}} \times 100\%.$$

$$\text{SVF}\ (\%) = \frac{\text{No. of } R_2 \text{ variants}}{\text{Total no. of } R_2 \text{ plants}} \times 100\%.$$

Here, variations refer to individual deviant phenotypes from the parental phenotype, such as albinism, male-sterility, height, short culm, etc., while variants refer to individual plants that are different from the parental plants in one or more characteristics. Seeds from individual $R_1$ plants were planted in single rows (called $R_1$ progeny rows or $R_1$ family rows). From each row, the number of visually detectable variations and variants along with $R_2$ plants were recorded at various stages. If several plants of the sample phenotypic variant were found in a single row, they were counted individually for SVF calculation but only once for SVR. The SVR and SVF value of each somaclone population group can be obtained by adding data from all $R_1$ family rows of the population group.

Table 4 is a comparison of differences in SVR and SVF between polyphenol variation and chlorophyll and other morphological and developmental variation. The polyphenol characteristics were determined by biochemical and statistic analysis (Cai et al. 1995), and the other variations were assessed in the field (Cai et al. 1990). The table shows that polyphenol variation had six to tenfold greater values for the SVR and SVF than the chlorophyll and other morphological and developmental variations. Mutations have been classified into macro and micro mutations (Gaul 1964). These polyphenol variations appear here as microvariations involving quantitative changes within a relatively small range,

**Table 4.** Somaclonal variation rate (SVR) and somaclonal variant frequency (SVF) for polyphenols and other characteristics

| Characteristic | SVR (%) | SVF (%) |
|---|---|---|
| Total phenols | 37.3[a] | 5.3[b] |
| Flavan-4-ols | 37.3[a] | 4.8[b] |
| Proanthocyanidins | 40.7[a] | 7.8[b] |
| Chlorophyll deficit[c] | 4.6 | 0.72 |
| Morphological and developmental characters[c] | 6.7 | 0.87 |

[a] Determined with IS8260, IS8768, and IS4225. The SVR value is the number of polyphenol variations detected from the assayed $R_1$ families is divided by the total number of $R_1$ families assayed.
[b] Determined with IS3150, IS6881, and IS0724. The SVF value is the number of variants with the variation detected from the assayed $R_2$ plants is divided by the total number of $R_2$ plants assayed.
[c] Data from field assessment of $R_2$ generation.

while the chlorophyll and other characters are macrovariations appearing as qualitative changes. The higher SVR and SVF value for the polyphenol variations than for the morphological variations found in our study is consistent with the estimation that micromutations occur five to ten times more frequently than chlorophyll mutations and other vital macro-mutations (Gaul 1964). It is likely that the macro-mutations involving qualitative changes occur under simple, but strong genetic control resulting in distinct phenotypic changes and occur relatively infrequently, and conversely that micromutations affect characters controlled by many genes with weak function resulting in minor phenotypic changes, and occur relatively frequently.

The size of a mutated sector (S) carried by mutant $R_1$ panicle can be estimated by the formula $S = Q/K$, where Q is the frequency of variants in individual $R_1$ panicle progeny rows. In order to avoid under estimating the size of variations found, K (the expected frequency) was assigned a value of 0.25 in our study, assuming recessive mutations to be preponderant (Cai et al. 1990).

## 3.6 Some Genetic Phenomena in Somaclonal Variation

### *3.6.1 Broad Genetic Mechanisms for Somaclonal Variation*

Although studies on the genetic mechanisms of sorghum somaclonal variation have been limited, it has been demonstrated that somaclonal variation can result from a variety of genetic events (Bajaj 1990). Karp (1991) has separated genetic mechanisms into cytological, molecular, cytoplasmic and epigenetic mechanisms. Considerable differences among individual $R_1$ family rows in segregation behaviour of somaclonal variations observed in our field assessment (Cai et al. 1990), with segregation ratios (normal plants: variants) ranging from less than 1:1 up to 35:1 (Table 5), partly reflect the broad genetic mechanisms contributing to

**Table 5.** Somaclonal variations segregated in $R_1$ families

| Phenotyes of variation[a] | No. of $R_1$ families segregating variations Segregation ratio (normal:variant) | | | | | |
|---|---|---|---|---|---|---|
| | Total | 0 | 1:1-2:1 | 2:1-5:1 | 5:1-10:1 | > 10:1 |
| Albino | 23 | | 1 | 12 | 6 | 4 |
| Yellow, yellow-green | 15 | | 1 | 10 | 3 | 1 |
| Striping | 11 | | | 3 | 2 | 6 |
| Red seedling | 1 | | | 1 | | |
| Other CV | 3 | | | 2 | | 1 |
| Lethal | 5 | | 2 | | 3 | |
| Total CV | 58 | | 4 | 28 | 14 | 12 |
| Tallness | 4 | | | | 3 | 1 |
| Shorter, narrow leaf | 28 | 8 | 2 | 2 | 5 | 11 |
| Sterility | 21 | | 8 | 3 | 3 | 7 |
| Necrosis | 15 | | 1 | 3 | 6 | 5 |
| Ragged leaf | 3 | | | | 2 | 1 |
| Panicle variation | 7 | 1 | 2 | 4 | | |
| *Hydra* | 6 | | 3 | 3 | | |
| Shrunken endosperm | 2 | 2 | | | | |
| Total VV | 86 | 11 | 16 | 15 | 19 | 25 |
| Total variation | 144 | 11 | 20 | 43 | 33 | 37 |

[a] CV = chlorophyll variations; VV = viable variations.

the somaclonal variations. About 30% (43 of 144) of the variant $R_1$ plants segregated variants with a ratio close to 3:1, indicating that recessive mutations were common among the observed variations such as chlorophyll mutations, short culm, narrow leaf, sterility, and necrosis. Changes in chromosomes are also possible for the abnormalities such as dwarf, lethal seedling, and semisterility. Plastid mutations might be involved in the striping variations. Table 5 also shows unusual ratios, nonsegregating homozygotes, and excessive or deficient recessive mutants. About 25% of the variant $R_1$ plants showed a very low frequency of variant $R_2$ plants, with ratios from 10:1 to 35:1. These segregation behaviors are directed by the genetic types of mutation occurring in cultured cells as well as the ontogenetic types of $R_1$ plants carrying the mutated sector(s). In the following sections we will discuss these phenomena with respect to the occurrence, transmission, and appearance of the somaclonal variation. Understanding these phenomena would be helpful to increase our knowledge about somaclonal variation and to provide a scientific basis for the screening and utilization of these variations.

### *3.6.2 Chimeras, Genetic Mosaic $R_1$ Plants*

There is a strong possibility that a plant regenerated from tissue cultures will not be genetically homogeneous because tissue of higher plants is ultimately derived from more than one apical initial cell and apical cell layer (Marcotrigiano 1990). Commonly, a chimeric or mosaic $R_1$ plant consists of parental type of cell and tissue and a mutated type of cell and tissue. However, an $R_1$ chimera may contain two or more types of mutated cells and tissue resulting from independent mutation events. The formation of chimera can occur at the plant, panicle, spike, spikelet, or anther level (Lindgren et al. 1970). The size of the mutated sectors carried by variant $R_1$ plants estimated in our field assessment (Cai et al. 1990) ranged from the whole plant to less than 3% of a single panicle. The average proportion of mutated $R_1$ panicles carrying large (80–100%), medium (40–80%), and small ($<40\%$) mutated sectors that may have originated from one, two, and three or more cells was 38.7, 26.0, and 35.3% respectively. Moreover, some sector mutations did not appear until the $R_3$ generation (Cai et al. 1990). These results demonstrate that chimerism does exist in $R_1$ plants. The frequency of chimeric regenerants estimated in our study ranged from a low of 10–20% to a high of 60–70%. These small-sectored $R_1$ variants and/or low level chimeras may lose their variations when a small number of panicles and seeds are taken from $R_1$ plants to produce their offspring. Even when the mutated sector involves only one of the male or female gametes, the recessive homozygotes fail to appear in the $R_2$ generation, yet the mutated sector may be large. The existence of mosaic $R_1$ plants seems to demonstrate their multicellular origin. It should be noted, however, that occurrence of mutations in cultured materials is a random event. There is no surety that mutations never occur in a single cell-derived somatic embryo after division of the proembryo cell to the formation of spike initial cells. Moreover, a mutation event occurring at the chromatid or subchromatid level in a proembryo cell also leads to mosaic formation. Therefore, the early statement that somatic embryos originating from a single cell exclude the possibility of chimerism (Wang and Vasil 1992) seems to be questionable. Cytoplasmic differences and culture-induced genetic instability may also cause chimerism.

### *3.6.3 Genetic Types of Somaclonal Variation*

As mentioned above, in cultured somatic embryos mutation can occur randomly in the spike initial cells. In most cases only one mutated cell line will be involved in spike formation. The genetic type formed with this pattern would be a simple monolocus mutation, transmitted as a homogeneous, nonchimera, or mosaic along with the parental cell line(s). More than one mutated cell line may be involved in spike formation, and multiple mutations may occur in a single cell or cell lines. The nature and direction of the mutations vary, including dominant, recessive, deletion, translocation, etc., and multiple mutations could result in the same or different phenotypes. The distribution of the multiple mutation events could be linked or nonlinked. Depending upon the type of transmission of the mutated cell line(s), nonmosaic and different levels of chimera may be possible. Therefore, various possible genetic types could be constructed. The complex segregation behaviors observed in our $R_2$ population resulted from the diverse genetic types of culture-induced variations. Here we describe two examples of multiple mutations borne by a single panicle found in our study. A total of 12 short-culm variants (Fig. 2) regenerated from a small callus of cv. M90950 which survived stress by Striga toxin. In the $R_2$ generation, 8 of 12 $R_1$ plant rows were homozygous with no segregation. The other four rows had uniformly short culms but segregated albinos with a ratio close to 3:1. This result indicated occurrence of two mutations: the homozygous short culm which could result from a somatic cross-over or similar event (see Sect. 3.6.4), and a monolocus mutation controlling the albinos. Another example involves multiple unlinked genetic changes. The two panicle variations in Figs. 4 and 5, segregated from six $R_1$ family rows derived from 270- and 300-day IS 8260 inflorescence culture. Three of the six rows simultaneously segregated the two variations and in the other three rows the two variants segregated independently. Although genetic analysis of progenies has not been done to elucidate the genetic type of the multiple mutations, this result may be due to either mosaic mutations independently occurring in two cell lines involving formation of mosaic plants, or multiple but nonlinked mutation events occurring in a single cell.

### *3.6.4 Some Unusual Segregation Behaviors of Somaclonal Variation*

Four unusual phenomena, dominance not expressed in $R_1$ plants, nonsegregating homozygotes, excess recessive, and deficient recessive, were observed in segregation behaviours in our study of somaclonal variation. First, the dominant character does not always appear in the $R_1$ plant carrying the dominant mutation. In the $R_2$ generation we found that a height variation segregated in four $R_1$ family rows, three of them derived from IS 8260 240- day inflorescence culture, and the remaining one from the same culture but after 300 days in culture. The segregation ratio (normal: taller) showed 27:3, 28:3, 30:3, and 13:1, respectively, indicating small mutated sectors contained in the four variant $R_1$ panicles. The tall variants were similar phenotypically to their parents except for height. Three tall variants from a single $R_1$ family were grown in the $R_3$ generation. One of them was uniformly tall, and the other two segregated, with a ratio 3:1 (45 tall: 15 parental height), indicating occurrence of a dominant mutation. Tracing back to the $R_1$ plants, no tall variants were recorded. The phenomenon of the unexpressed dominant is attributed to chimerism. The sector size of the tall mutation in the four $R_1$ panicles (one panicle taken from each plant) was 13.3, 12.9, 12.1, and 9.5%, respectively. Expression of a character is conditioned by the whole plant metabolism. When a mutant sector occupies only a small sector of a plant or is not involved in the tissue-directing expression of the character, the dominant character does not show up in the $R_1$ plant.

Occurrence of homozygosity has been considered to be common in regenerated plants (Karp 1991), but rare in $M_1$ plants from mutagenesis. When the homozygous mutated cell line occupies a whole $R_1$ plant, variant $R_1$ plants appear, e.g., albinos, and their progeny rows do not segregate, e.g., the eight short culm variants. When the homozygous mutation sector exists as a mosaic form, chimerism in the $R_1$ plants would hide nonsegregation phenomenon for which the homozygous sector is responsible. If an $R_1$ plant carrying a homozygous sector of a recessive mutation is a chimera, recessive excess (Table 5) would appear in the progeny row from the $R_1$ plant when the mutated sector is large. The regenerant would be considered to be a nonsectored plant when the sector is small. Not until the $R_3$ generation could these possibilities be distinguished by whether heterozygous segregation occurs. A homozygous mutation sector contains no heterozygote. Homozygous mutation could result from mitotic crossing-over. Larkin et al. (1984) explained this phenomenon as a result of cycles of monosomy/disomy or trisomy/disomy and of nonreciprocal transfer of genetic information between repeated DNA sequences. Cytoplasmic genome mutation could also result in homozygote formation, e.g., albinism. The phenomenon of recessive deficiency is common in induced mutation, and was considered to be partly due to lower competence of mutated gametes than normal ones. The $R_1$ family showing recessive deficiency in our field rows (Table 5) was most likely due to the effect of a small mutated sector in the mosaic $R_1$ plants.

### *3.6.5 Pleiotropic Effect of Somaclonal Variation*

As reported in mutation breeding (Gaul 1964), pleiotropic effects of mutations are also apparent in somaclonal variation. The somaclone variants are often different from their parental controls in more than one characteristic. A variant may have changed from an unfavorable trait to a desirable one, but synchronously, some other trait(s) also changed. Unfortunately, these companion trait(s) are mostly undesirable. The shrunken endosperm characteristic found in our study is useful for nutritional study, but its sterility is deleterious. This pleiotropic effect has been an obstacle to application of induced mutation and somaclonal variation. Three possible explanations for this effect include pleiotropy of a single mutation gene, a minute deficiency involving a small group of genes, and closely linked genes are mutated (Gottschalk 1976). To overcome the pleiotropic effect, backcrossing would be useful if the two changed traits can be separated by recombination. However in most cases these mutants may be improved only by cross-breeding, i.e., as materials introduced into breeding programs.

## 3.7 Procedures for Selection of Somaclonal Variation

### *3.7.1 Choice of Genotypes for Parental Sources*

Choosing parental genotypes as source material is critical for successful regeneration and subsequent recovery of somaclonal variation. There is certainly no positive relationship between plant regeneration and production of variation. Selected lines resistant to fall armyworm and tolerant to acid soil and drought were derived from RTx 430 and Hegari, poor producers of embryogenic calli, instead of SC283 and M35-1, which more readily formed embryogenic calli than did RTx 430 and Hegari (Duncan et al. 1991a). Even a parental genotype bearing both advantages of facile regeneration and production of numerous variations, such as IS8260, which in our project produced 1288 $R_1$ plants and 20 variant phenotypes, produces mostly useless variations. The extreme difficulty is to produce desirable variations. A large percentage of variations resulting from tissue culture are deleterious or unfavorable (Miller et al. 1991). Genotypes introduced into an in vitro culture program should be agronomically important, well-adapted cultivars having a broad genetic background, and improvement should be sought in only one or a few traits. It would facilitate selection if the trait to be modified is a simple qualitative characteristic, e.g., from tall to short. Duncan et al. (1991a) found that in vitro regeneration and subsequent stress tolerance have been most successful with well-adapted, agronomically desirable genotypes from a heterogeneous background rather than those from a more homogenous background.

### *3.7.2 Selection Procedures for Variations with Changes in Qualitative Characters*

Generally, selection for somaclonal variation is not carried out in the $R_1$ generation. Some variations which appear in $R_1$ plants are not heritable (Ma et al. 1987) and represent only physiological adaptation (Duncan et al. 1991a). However, dominant height (Ma et al. 1987) or recessive short culms (Fig. 1) may appear in $R_1$ plants. Through screening for grain polyphenol variation, we found that selection in the $R_1$ generation is unnecessary for recovering the polyphenol variation. However, the selected $R_1$ families showed an increase in the frequency of polyphenol variations (Cai et al. 1995).

Qualitative somaclonal variations are highly heritable, with a relatively simple mode of inheritance directed by one or a few genes. They are mainly selected in the $R_2$ generation. Because of the low SVR and SVF produced by some genotypes, as well as chimeric $R_1$ plants and the small mutation sector of some $R_1$ plants, the population for selecting somaclonal variation should be as large as possible. Generally from each genotype 500–1000 $R_1$ families would be needed, possibly even more for recovering useful agronomic variations. Using a field cultivar as the parent source and considering both variation selection and grain production are practical means of saving land and labor costs. For genetic analysis, the $R_1$ family pedigree method is recommended in $R_2$ field planting. Heterozygous $R_2$ plants in the family rows can be harvested and grown in the $R_3$ generation to identify the inheritance of selected somaclonal variations. Otherwise, backcrossing is required for the identification.

Progenies of each selected $R_2$ variant need to be grown in individual $R_2$ family rows in the $R_3$ generation to increase their homozygous seeds, to evaluate their stability, and to determine the utility of selection. Because of various possible mechanisms for the somaclonal variation (Bajaj 1990), sometimes simple recessive or dominant models are inadequate. The genetic analysis needs to be continued to later generations. After stability identification and inheritance characterization, the variant somaclones, as breeding material, can be incorporated into conventional breeding procedures. Generally, a somaclonal line with a simple changed qualitative trait can be generated in three or four generations, which is shorter than the seven to eight generations of selection required from cross-breeding. After linking to the conventional breeding program there is no time-saving advantage of tissue culture selection (Miller et al. 1991).

Because of chimerism and small mutation sectors of $R_1$ plants, some somaclonal variation does not appear until the $R_3$ generation (Cai et al. 1990). The necessity of selecting somaclonal variation in the $R_3$ generation is determined by whether the spectrum of hidden variations in the $R_2$ generation is similar to that of observed variations in the same $R_2$ population. If the two spectras are same, it would be more efficient to enlarge the $R_2$ population rather than to grow somaclones in the $R_3$ generation for selection of variations similar to those selected in the $R_2$ generation.

### *3.7.3 Selection Procedures for Quantitative Variations*

Selection of quantitative traits is more difficult than selecting qualitative ones. Obtaining sorghum somaclone lines with improved tolerance to acid soils, drought, and insects shows that agronomically important quantitative characters can be successfully selected. The procedures and problems for field stress testing have been described (Duncan et al. 1991a; Miller et al. 1991), including selection, field test, and linkage to conventional breeding programs. The problems they encountered in early generations of field selection were lack of seeds due to less vigorous and low seed set of $R_1$ plants regenerated from stress selection, masking effects resulting from physiological changes during adaptation to culture and to in vitro selection pressure, and unequal synchronization of growth rates between regenerants and parental controls. In order to overcome some of these early generation obstacles, seed increase in early generations and bulk harvest from each $R_1$ family were adopted to maximize genetic diversity and to provide adequate seed for verification of subsequent tolerance in later generations. They found that family selection and bulking in early generations ($R_1$-$R_4$) worked better than single-seed descent or individual selection tracking using the pedigree method (Duncan et al. 1991a). Selection strategies and field test designs are important for obtaining superior lines from tissue culture materials. Several selection methods were suggested by the authors. Selection for specific stress straits should be conducted under proper environments with sufficient selection pressure to screen out the susceptible genotypes but not too high to eliminate all or most regenerated materials. When selecting for multiple characters, several diverse environments would be optimal. An alternative approach for selecting multigenic traits such as salt or drought tolerance would be to first select under optimum conditions in two or three early generations for good agronomic characteristics, then to test the selections under multiple environmental stresses. This approach assumes that after the first selection step these advanced lines will be more homozygous, more genetically stable for expression of the multigenic traits, and more agronomically desirable (Miller et al. 1991). If selection for complex traits in early generations is necessary, one would apply moderate pressure in the $R_2$ and $R_3$ generation, with increasing pressure in later generations leading to retention of the maximum amount of germplasm with potentially useful traits (Miller et al. 1991). Their field test design contains multiple sites in several generations with necessary replicates, along with control plots including nonregenerated parents, susceptible and tolerant controls, as well as suitable statistical techniques. Linkage to conventional breeding programs is the key for transferring in vitro technology to the field. The linkage will insure that the tissue culture material developed has the agronomic performance that is essential for success as a commercial cultivar or as breeding material (Miller et al. 1991).

## 4 Summary

A wide range of somaclonal variation, qualitative and quantitative, morphological and developmental, have been found in sorghum regenerants. Some variations observed in primary regenerants ($R_1$) were epigenetic, although heritable variants also appeared in $R_1$ plants. Cytological and molecular studies showed changes in chromosome and mitochondrial genome existing in cultured cells and regenerants. Apparent genetic variations found in progeny include various chlorophyll-deficient forms, height, dwarf, short culm, sterility, waxy midrib, narrow leaf, maturity, and fertility restoration capacity. Some rare types of variations were also found: red type seedling, ragged-leaf, shrunken endosperm, variations in abnormal panicles, and disordered growth and development. Some quantitative variations selected were readily lost in subsequent generations. Results for screening grain polyphenol variation from high-tannin sorghum somaclones showed that no somaclone had lost or greatly reduced its content of polyphenols, although variants with increased or decreased levels of grain polyphenols were selected. Sorghum lines with increased tolerance to fall armyworm, acid soils and drought stress have been generated from in vitro.

## References

Axtell JD (1984) Improving nutritional quality and food grain quality of sorghum. In: Winn JF (ed) Fighting hunger with research. INTSORMIL a 5-year technical research report of the grain sorghum/pearl millet collaborative research support program. University of Nebraska, Lincoln, pp 157–161

Bajaj YPS (1990) Biotechnology in agriculture and forestry, vol 11. Somaclonal variation in crop improvement I. Springer, Berlin Heidelberg New York, pp 3–48

Battraw MJ, Hall TC (1991) Stable transformation of *Sorghum bicolor* protoplasts with chimeric neomycin phosphotrans-ferase II and $\beta$-glucuronidase genes. Theor Appl Genet 82: 161–168

Berhan AM, Hulbert SH, Butler LG, Bennetzen JL (1993) Structure and evolution of the genomes of *Sorghum bicolor* and *Zea mays*. Theor Appl Genet 86: 598–604

Bhaskaran S, Smith RH (1988) Enhanced somatic embryogenesis in *Sorghum bicolor* from shoot tip culture. In Vitro Cell Dev Biol 24: 65–70

Bhaskaran S, Smith RH, Schertz D (1983) Sodium chloride tolerant callus of *Sorghum bicolor* (L.) Moench. Z Pflanzenphysiol 112: 495–463

Bhaskaran S, Smith RH, Newton RJ (1985) Physiological changes in cultured sorghum cell in response to induced water stress, I. Free proline. Plant Physiol 79: 266–269

Bhaskaran S, Smith RH, Schertz D (1986) Progeny screening of *Sorghum* plants regenerated from sodium chlofide-selected callus for salt tolerance. J Plant Physiol 122: 205–210

Bhaskaran S, Smith RH, Paliwal S, Schertz D (1987) Somaclonal variation from *Sorghum bicolor* (L.) Moench cell culture. Plant Cell Tissue Organ Culture 9: 189–196

Bhaskaran S, Smith RH, Frederiksen RA (1990) Gebberellin $A_3$ reverts floral primordia to vegetative growth in sorghum. Plant Sci 71: 113–118

Brar DS, Rambold S, Gamborg O, Constabel F (1979) Tissue culture of corn and sorghum. Z Pflanzenphysiol 95: 377–388

Brar DS, Rambol S, Constabel F, Gamborg OL (1980) Isolation, fusion and culture of *Sorghum* and corn protoplasts. Z Pflanzenphysiol 96: 269–275

Brettell RIS, Wernicke W, Thomas E (1980) Embryogenesis from cultured immature inflorescences of *Sorghum bicolor*. Protoplasma 104: 141–148

Butler LG (1989) Sorghum polyphenols. In: Cheek PR (ed) Toxicants of plant origin, vol IV, Phenolics. CRC Press, Boca Raton, pp 95–121

Butler LG, Cai T, Ejeta G, Babiker AGT (1994) Possible growth factors for the parasitic weed *Striga* from sorghum host plants. In Agronomy Abstracts, 1994 Annu. Meet, ASA, CSSA, SSSA, Nov 1994, Seattle. Washington, 123 pp

Byth DE (1993) An over-view of the ICRISAT cereals improvement program and genetic resources unit. In: Grain sorghum, Proc 18th Biennial Grain Sorghum Research and Utilization Conference, Lubbock, Texas, Feb 1993, pp 29–32

Cai T, Butler LG (1992) Sorghum suspension cultures. Sorghum Newsl 33: 9

Cai T, Ejeta G, Axtell JD, Butler LG (1990) Somaclonal variation in high tannin sorghum. Theor Appl Genet 79: 737–747

Cai T, Ejeta G, Butler LG (1993a) Facile sorghum plant regeneration from PP290, a high-yielding drought-tolerant variety. In: Agronomy Abstracts, 1993 Annu Meet, ASA, CSSA, SSSA, Nov 1993, Cincinnati, Ohio, 83 pp

Cai T, Babiker AG, Ejeta G, Butler LG (1993b) Morphological response of witchweed (*Striga asiatica*) to in vitro culture. J Exp Bot 44: 1377–1384

Cai T, Ejeta G, Butler LG (1994a) Effect of genotype, explant source and medium composition on callusing, embryogenesis and regeneration in sorghum. In: Agronomy Abstracts, 1994 Annu meet, ASA, CSSA, SSSA, Nov 1994, Seattle, Washingtion, 206 pp

Cai T, Ejeta G, Butler LG (1994b) Development and maturation of sorghum seeds on detached panicles grown in vitro. Plant cell Rep 14: 116–119

Cai T, Ejeta G, Butler LG (1995) Screening for grain polyphenel variants from high-tannin sorghum somaclones. Theor Appl Genet 90: 211–220

Cassas AM, Kononowicz AK, Zehr UB, Tomes DT, Axtell JD, Butler LG, Bressan RA, Hasegawa PM (1993) Transgenic sorghum plants via microprojectile bombardment. Proc Natl Acad Sci USA 90: 11212–1121

Chen Z, Liang GH, Muthukrishnan S, Kofoid KD (1990) Chloroplast DNA polymorphism in fertile and male-sterile cytoplasms of sorghum [*Sorghum bicolor* (L.) Moench]. Theor Appl Genet 80: 727–731

Chourey PS, Sharpe DZ (1985) Callus formation from protoplasts of *Sorghum* cell suspension cultures. Plant Sci 39: 171–175

Chourey PS, Lloyd RE, Sharpe DZ, Isola NR (1986) Molecular analysis of hypervariability in the mitochondrial genome of tissue cultured cells of maize and sorghum. In: Mantell SH, Chapman GP, Street PFS (eds) The chondriome-chloroplast and mitochondrial genomes. Wiley, New York, pp 177–191

Dahlberg J (1993) The U. S. sorghum germplasm collection. In: Grain sorghum. Proc 18th Biennial Grain Sorghum Research and Utilization Conference, Lubbock, Texas, pp 21–28

De Klerk GJ (1990) How to measure somaclonal variation. Acta Bot Neerl 39: 129–144

Doggett H (1988) Sorghum, 2nd edn. Tropical Agriculture Series. Longman, Harlow

Duncan RR, Waskom RM, Miller DR, Voigt RL (1991a) Field stress evaluation of tissue culture regenerated sorghums. In: Proc 17th Biennial Grain Sorghum Research and Utilization Conf, Feb 1991, Lubbock, Texas, pp 15–20

Duncan RR, Isenhour DJ, Waskon RM, Miller DR, Nabors MW, Hanning GE, Petersen KM, Wiseman BR (1991b) Registration of GATCCP 100 and GATCCP 101 fall armyworm-resistant Hegari regenerants. Crop Sci 31: 242–244

Duncan RR, Waskom RM, Miller DR, Voigt RL, Hanning GE, Timm DA, Nabors MW (1991c) Registration of GAC 102 acid soil-tolerant Hegari regenerant. Crop Sci 1396–1397

Duncan RR, Waskon RM, Miller DR, Hanning GE, Timm DA, Nabors MW (1992) Registration of GC 103 and GC 104 acid soil-tolerant Tx 430 sorghum regenerants. Crop Sci 32: 1076–1077

Ejeta G, Mertz ET, Rooney L, Schaffert R, Yohe J (1991) Sorghum nutritional quality, Proc Int Conf, Feb 1990, Purdue University West Lafayette, Indiana
Ejeta G, Butler L, Babiker AG (1993) New approaches to the control of striga. Agriculture Experiment Station. Purdue University, Indiana, USA
Evans DA (1988) Somaclonal and gametoclonal variation. In: Biotechnology in tropical crop improvement. Proc Int Biotechnology Workshop, Jan 1987, ICRISAT Center, India, pp 57–66
FAO (1992) Production yearbook, vol 46, FAO, Rome
Frederiksen RA (1986) Compendium of sorghum diseases. American Phytopathological Society, St Paul
Gamborg OL, Shyluk JP, Brar DS, Constabel F (1977) Morphogenesis and Plant regeneration from callus of immature embryos of sorghum. Plant Sci Lett 10: 67–74
Gaul H (1964) Mutations in plant breeding. Radiat Bot 4: 155–232
George L, Eapen S, Rao PS (1989) High frequency somatic embryogenesis and plant regeneration from immature inflorescence culture of two Indian cultivars of sorghum [*Sorghum bicolor* (L.) Moench]. Proc Indian Acad Sci 99: 405–410
Gottschalk W (1976) Pleiotropy and close linkage of mutated genes: new examples of mutations of closely linked genes. In: Induced mutation in cross-breeding. Proc of an Advisory Group, Vienna, Oct 1975, FAO/IAEA, Vienna, pp 71–78
Hagio T, Blowers AD, Earle ED (1991) Stable transformation of sorghum cell cultures after bombardment with DNA-coated microprojectiles. Plant Cell Rep 10: 260–264
Harlan JR, de Wet JMT (1972) A simplified classification of cultivated sorghum. Crop Sci 12: 172–176
House LR (1984) A guide to sorghum breeding, 2nd edn. ICRISAT, Patancheru, AP, India
ICRISAT (1980) Sorghum diseases, a world review. Proc Int Worksh on Sorghum diseases, Dec 1978. ICRISAT (International Crops Research Institute for the Semi-Arid Tropics), Patancheru 502 324, AP, India
ICRISAT (1982) Proc Int Symp on Sorghum Grain Quality. Oct 1981, ICRISAT, Patancheru, AP, India
Isenhour DJ, Duncan RR, Miller DR, Waskom RM, Hanning GE, Wiseman BR, Nabors MW (1991) Resistance to leaf-feeding by fall armyworm (Lepidoptera: Noctuidae) in tissue culture derived sorghums. J Econ Entomol 84: 680–684
Kaeppler HF, Pedersen JF (1993) Media and genotype effects on sorghum callus growth, morphology, embryogenicity, and regenerability. In: Agronomy Abstracts, 1993 Annu Meet, ASA CSSA SSSA, Nov 1993, Cincinnati, Ohio, 178 pp
Kane EJ, Wilson AJ, Chourey PS (1992) Mitochondrial genome variability in *Sorghum* cell culture protoclones. Theor Appl Genet 83: 799–806
Karp A (1991) On the current understanding of somaclonal variation. Oxford Surv Plant Mol Cell Biol 7: 1–58
Kaveriappa KM, Safeeulla KM, Shaw CG (1980) Culturing *Sclerospora sorghi* in callus tissue of sorghum. Proc Indian Acad Sci 89: 131–138
Kresovich S, McGee RE, Panella L, Peilley AA, Miller FR (1987) Application of cell and tissue culture techniques for the genetic improvement of sorghum, *Sorghum bicolor* (L.) Moench: progress and potential. Adv Agron 41: 147–170
Kumaravadivel N, Sree Rangaswamy SR (1994) Plant regeneration from sorghum anther cultures and field evaluation of progeny. Plant Cell Rep 13: 286–290
Larkin PJ, Pyan SA, Brettel RIS, Scowcroft WR (1984) Heritable somaclonal variation in wheat. Theor Appl Genet 67: 443–455
Lindgren D, Eriksson G, Sulovska K (1970) The size and appearance of the mutated sector in barley spikes. Hereditas 65: 107–132
Lusardi MC, Lupotto E (1990) Somatic embryogenesis and plant regeneration in sorghum species. Maydica 35: 59–66
Ma HT, Liang GH (1985) Studies on culture of immature sorghum embryos in vitro and variation of regenerated plants. Acta Genet Sin 12: 350–357
Ma HT, Gu M, Liang GH (1987) Plant regeneration from cultured immature embryos of *Sorghum bicolor* (L.) Moench. Theor Appl Genet 73: 389–394

MacKinnon CG, Gunderson G, Nabors MW (1987) Plant regeneration from salt-stressed sorghum cultures. Plant Physiol 6: 107–109
Marcotrigiano M (1990) Genetic mosaics and chimeras: implications in biotechnology. In: Bajaj YPS (ed) Biotechnology in agriculture and forestry, vol 11. Somaclonal variation in crop improvement I. Springer, Berlin Heidelberg New York pp 85–111
Miller DR, Waskom RM, Brick MA, Chapman PL (1991) Transferring in vitro technology to the field. Bio/Technology 9: 143–146
Miller DR, Waskom RM, Duncan RR, Chapman PL, Brick MA, Hanning GE, Timm DA, Nabors MW (1992) Acid soil stress tolerance in tissue culture-derived sorghum lines. Crop Sci 32: 324–327
Murty UR, Cocking EC (1988) Somatic hybridization attempts between *Sorghum bicolor* (L.) Moench and *Oryza sativa* L. Curr Sci 57: 668–670
Nadar HM, Clegg MD, Maranville JW (1975) Promotion of sorghum callus growth by the s-Triazine herbicides. Plant Physiol 56: 747–751
Newton RJ, Bhaskaran S, Puryear J, Smith RH (1986) Physiological change in cultured sorghum cells in response to induced water stress. II Soluble carbohydrates and organic acids. Plant Physiol 81: 626–629
Palmer GH (1992) Sorghum—food, beverage and brewing potentials. Process Biochem 27: 145–153
Prabhu MSC, Venkatasubbaiah P, Safeeulla KM (1984) Changes in total phenolic contents of sorghum callus resistant and susceptible to downy mildew. Curr Sci 53: 271–274
Ramulu KS (1975) Mutation breeding in sorghum. Z Pflanzenzuecht 74: 1–17
Rooney LW, Blakely ME, Mill FR, Rosenow DT (1980) Factors affecting the polyphenols of sorghum and their development and location in the sorghum kernel. In: Hulse JH (ed) Polyphenols in cereals and legumes. Proc 36th Annu Symp of the Inst of Food Tech, St Louis, June, 1979, IDRC Publication Ottawa, pp 25–35
Rose JB, Dunwell JM, Sunderland N (1986b) Anther culture of *Sorghum bicolor* (L.) Moench. II Pollen development in vivo and in vitro. Plant Cell Tissue Organ Cult 6: 23–31
Smith RH, Bhaskaran S (1986) Sorghum [*Sorghum bicolor* (L.) Moench]. In: Bajaj YPS (ed) Biotechnology in agriculture and forestry, vol 2. Crops I. Springer, Berlin Heidelberg New York, pp 220–233
Smith RH, Bhaskaran S, Schertz (1983) Sorghum plant regeneration from aluminum selection media. Plant Cell Rep 2: 129–132
Smith RH, Bhaskaran S, Miller FR (1985) Screening for drought tolerance in sorghum using cell culture. In Vitro Cell Dev Biol 21: 541–545
Strogonov BP, Komizerko EI, Butenko RG (1968) Culturing of isolated glasswort, sorghum, sweet clover, and cabbage tissue for comparative study of their salt resistance. Sov Plant Physiol 15: 173–177
Tao Y, Manners JM, Ludlow MM, Henzell (1993) DNA polymorphisms in grain sorghum [*Sorghum bicolor* (L.) Moench]. Theor Appl Genet 86: 679–688
Teetes GL, Seshu Reddy KV, Leuschner K, House LR (1983) Sorghum insect identification handbook. Inf Bull 12, ICRTSAT, Patancheru, AP, India
Thomas E, King PJ, Potrykus I (1977) Shoot and embryo-like structure formation from cultivated tissues of *Sorghum bicolor*. Naturwissenschaften 64: 587
Vierling RA, Xiang Z, Joshi CP, Gilbert ML, Nguyen HT (1994) Genetic diversity among elite *Sorghum* lines revealed by restriction fragment length polymorphisms and random amplified polymorphic DNAs. Theor appl Genet 87: 816–820
Wang AS, Cheng DSK, Milcic JB, Yang TC (1988) Effect of X-ray irradiation on maize inbred line B73 tissue cultures and regenerated plant. Crop Sci 28: 358–362
Wang DY, Vasil IK (1982) Somatic embryogenesis and plant regeneration from inflorescence segments of *Pennisetum purpureum* schum. (Napier or elephant grass). Plant Sci Lett 25: 147–154
Waskom RM, Miller DT, Hanning GE, Duncan RR, Voigt RL, Nabor MW (1990) Field evaluation of tissue culture derived sorghum for increased tolerance to acid soils and drought stress. Can J Plant Sci 70: 997–1004
Wei Z, Zu Z (1990) Regeneration of fertile plants from embryogenetic suspension culture protoplasts of *Sorghum vulgare*. Plant Cell Rep 9: 51–53
Wen FS, Sorensen EL, Bamett FL, Liang GH (1991) Callus induction and plant regeneration from anther and inflorescence culture of sorghum. Euphytica 52: 177–181

Wozniak CA, Partridge JE (1988) Analysis of growth in sorghum callus cultures and association with a 27 KD piptide. Plant Sci 57: 235–246

Yang YW, Newton RJ, Miller FR (1990) Salinity tolerance in *Sorghum*. II. Cell culture response to sodium chloride in *S. bicolor* and *S. halepense* Crop Sci 30: 781–785

York V (1993) Sorghum in the 21st century. In: Grain sorghum. Proc 18th Biennial Grain Sorghum Research and Utilization Conf, Lubbock, Texas, pp 9–10

# I.7 In Vitro Production of Late Blight (*Phytophthora infestans*)-Resistant Potato Plants

A.C. CASSELLS[1] and P.T. SEN[1]

## 1 Introduction

New challenges face the potato breeder, particularly in Europe and the former USSR. There is a need to adapt high-yielding Western processing varieties for potential new growing areas in eastern Europe (Bethell 1990). The threat of new genetic variability in the pathogen, *Phytophthora infestans*, resulting in more rapid changes in virulence and fungicide resistance, has increased with the spread of the A2 mating type outside Mexico (Spielman et al. 1991; Umaerus and Umaerus 1994). Resistance to phenylamide systemic fungicides is widespread (Parry 1990; Cooke 1992). There is also the possibility that the dithiocarbamates, the mainstay of chemical late blight control, may be withdrawn on environmental grounds. It has been estimated that in Ireland approx. 80% of the potato seed growers used dithiocarbamates alone or with systemic fungicides in late blight control. Should the dithiocarbamates be withdrawn, the problem of late blight control would be acute with the lack of adequate varietal resistance in commercially important cultivars and the toxicity to the crop of alternative copper- and fentin-based fungicides. These issues have been discussed by Parry (1990) and Cooke (1992).

In view of the great losses caused there is an urgent need to develop new and unconventional methods for the induction of genetic variability for resistance to potato late blight (van der Zaag 1994).

## 2 *Phytophthora infestans*

The disease cycle and control of *P. infestans* are discussed in Hooker (1981), Parry (1990), and Rowe (1993). Late blight of potato is worldwide in distribution (Hide and Lapwood 1978) and has attracted more plant breeding effort than any other potato disease (Howard 1978). *P. infestans* is an obligate pathogen overwintering (surviving the intercrop period) in infected tubers as groundkeepers or in dumps, etc. Control based on attempts to prevent carryover of inoculum by destroying the canopy of infected crops before lifting and by the destruction of infected tubers, the source of primary inoculum for the next season, has been largely

[1] Department of Plant Science, University College, Cork, Ireland

Biotechnology in Agriculture and Forestry, Vol. 36
Somaclonal Variation in Crop Improvement II (ed. by Y.P.S. Bajaj)

unsuccessful. Primary inoculum from the latter sources and from groundkeepers initiates the new disease cycle and further secondary inoculum produced in the growing crop will lead to epidemics if the environmental conditions are suitable and if large stands of susceptible varieties are cultivated (Fry and Doster 1991). The main control is by strategic use of protectant and systemic chemicals based on accurate disease forecasting (see Hooker 1981; Parry 1990; Rowe 1993). The cost of a prophylactic spraying program in Ireland has been estimated at US$ 150/ha or a yield equivalent of 5t/ha.

### 2.1 Lack of Durable Resistance to *Phytophthora infestans*

The discovery of major genes (the R-genes; see van der Plank 1975) in *Solanum demissum* for vertical resistance to late blight and their incorporation into *S. tuberosum* was considered to be a breakthrough in combating this disease (Hide and Lapwood 1978; Jellis and Richardson 1987). Variability, however, in the pathogen leading to the breakdown of resistance, and inability, due to the complexity of the potato's breeding system, to combine resistance with market acceptability, has led to the abandonment of this major gene (race-specific) resistance breeding strategy in favor of a return to breeding for general (minor gene, nonrace-specific) late blight resistance (Howard 1978; Russell 1978; Simmonds 1979).

### 2.2 Screening for Resistance to Foliar Late Blight

Two problems exist in screening for late blight resistance in the field. Firstly, late maturation is generally related to foliar late blight resistance (Toxopeus 1958; Umaerus et al. 1983; Umaerus and Umaerus 1994); secondly, in most regions, screening for late blight major gene resistance is complicated by host/environmental/pathogen interaction and by variable and, or, low inoculum potential of virulent pathogen isolates (Russell 1978). Some 30% of the variability in potato is controlled by the environment (Russell 1978). Both optimum environmental conditions and diversity in the pathogen are to be found in the Toluca valley in Mexico, where much trialing has taken place (Simmonds 1979).

While a positive relationship between foliar late blight resistance and late maturation in potato has been reported (Toxopeus 1958), it is important in screening to note that there is no strong relationship between foliar and tuber late blight resistance (Umaerus and Umaerus 1994).

### 2.3 In Vitro Screening for Resistance to *Phytophthora infestans*

An understanding of the determinants of pathogenicity assists in the development of in vitro selection procedures (Isaac 1992). Fungi in general produce macerating enzymes, specific and nonspecific toxins, and wilt-inducing carbohydrate polymers (Isaac 1992). Toxin selection has been widely used in developing in vitro selection strategies (Jones 1990; Buiatti and Ingram 1991), albeit few pathogens produce *specific* toxins (Scheffer and Livingston 1984). In the case of *Phytophthora* species, it has been concluded that while toxic metabolites have been isolated from *Phytophthora* culture medium these have not been shown to contribute to fungal pathogenesis, i.e. are not host-specific toxins (Keen and Yoshika 1983; Cerato et al. 1993); however, recent work on *Phytophthora* elicitins may cause this view to be revised (Pernollet et al. 1993).

Juvenility in crops grown from potato microplants and in young tuber-derived plants in the glasshouse will give false positive resistance (see Sect. 2.2). Thus caution should be exercised in extrapolating from results obtained from such screening (Russell 1978).

## 3 Conventional Breeding for Foliar Late Blight Resistance

Potato improvement by conventional breeding is complicated by the polyploid nature of the crop, by poor flowering, by low fertility in some varieties and by the requirement to maintain heterozygosity (Howard 1978). When combined with the potato breeder's concern with ca. 30 characters (see Table 1), including late blight resistance, the task of correction of character defects by conventional breeding methods is seen as a difficult and expensive process. The production and release of a new variety by hybridization between selected parental varieties can take 14 years. Reviews of potato breeding, including attempts to increase the incorporation of exotic germplasm and accelerate improvement by the use of dihaploids, are given in Jellis and Richardson (1987) and Bradshaw and Mackay (1994).

Given the above constraints on conventional breeding in potato and the instability of the R-genes, in recent years farmers in the developed countries have come to rely on chemicals as the main defence against foliar late blight (Cooke 1992).

## 4 In Vitro Induction of Resistance to Late Blight

### 4.1 Genetic Transformation for Late Blight Resistance

Genetic engineering offers the opportunity of introducing major gene resistance to correct character defects such as lack of late blight resistance. Bowles (1990) has defined three classes of defence-related gene products in plants: (1) products which strengthen, modify, or repair the plant cell wall; (2) proteinaceous inhibitors and hydrolytic enzymes, e.g., chitinases and (3) pathogenesis-related proteins.

Gene location and transformation protocols are now well established (Gatehouse et al. 1992). There may be problems however, with the expression of transgenes (Meyer 1993) and with background somaclonal variation (Dale and McPartlan 1992). Even if these difficulties can be overcome in a cost-effective way, there remains the problem of the vulnerability of major gene resistance to late blight (see Sect. 2.1) and the risk of possible adverse pleiotropic consequences of the foreign gene (see Cassells and Jones 1995).

## 4.2 Somaclonal Variation and Late Blight Resistance

Several studies on somaclonal variation in relation to late blight improvement in potato have been reported (Behnke 1979, 1980; Tegera and Meulemans 1985; Meulemans and Fouarge 1986; Meulemans et al. 1987; Cassells et al. 1991; also see Bajaj 1990). In the first study, Behnke (1979) used an uncharacterized toxic culture filtrate to select for late blight "resistant" callus from which she regenerated "late blight resistant" lines (somaclones). In subsequent field trials toxin-selected lines were shown to have been overselected and exhibited no improved resistance and adverse pleiotropic changes which resulted in poor field performance (Wenzel and Foroughi-Wehr 1990; Cerato et al. 1993; Umaerus and Umaerus 1994). In a second program, Meulemans and coworkers (Tegera and Meulemans 1985; Meulemans and Fouarge 1986; Meulemans et al. 1987) selected for somaclonal variants by cocultivation of potato and *P. infestans* in vitro. Cassells et al. (1991) used field selection in a somaclonal program that involved preliminary rejection of gross morphological aberrants and somaclones with low vigor.

### *4.2.1 Screening Against Useless Variation*

*1. Aberrant Morphology.* Since most mutations are deleterious, it follows that the cost effectiveness of exploitation of somaclonal variation will be enhanced by the efficient detection and rejection of useless variants. Gross morphological aberrants can be discarded at microplant (somaclone) establishment (Cassells et al. 1987). Selection for vigor at this stage can also be used to select clones with the potential for vigorous growth and, by extrapolation, high yield; but one should be cautious not to overselect at this stage (Simmonds 1979). Polyploids and aneuploids can also be recognized, in general, by leaf shape abnormality.

*2. Cryptic Useless Variation.* Two further categories of useless unstable variants are encountered in potato somaclones, namely, nonmorphologically aberrant aneuploids and maturation mutants (Cassells et al. 1991). The latter are early- or late-flowering and may occur at high frequency. Late maturing somaclones (Cassells et al. 1991), like the naturally occurring giant-hill field mutant (Yarwood 1946), show a pleiotropic increase in late blight resistance. These may be genotrophs as described in flax (Cullis 1990; Nagl 1990), that is, somaclones in which gene amplification or deletion has occurred. Gene amplification and related genome changes, albeit as yet not related to maturation changes, have been reported in potato (Kemble and Shepard 1984; Landsmann and Urhig 1985).

Maturation mutants cannot yet be detected early and, because of their high frequency, present a significant cost penalty in trialing. While genetic fingerprinting techniques are available which might be used to detect such variants, the economics of their use for this purpose may be unfavorable (Smith and Smith 1992).

### *4.2.2 Strategies for the Exploitation of Somaclonal Variation in the Search for Late Blight Resistance*

In the context of this chapter, the objective of studying somaclonal variation was to determine its potential in the correction of a character defect in potato, namely, late blight susceptibility. The issue of 'optimization', somaclonal variation aside (see reviews by Karp 1990, 1991; Kumar 1994), the cost-effective use of variation derived in this way merits consideration. In practice, this means eliminating all but those variants which meet the criterion of having improved durable late blight resistance and retention of the other important varietal characters (see Table 1).

The first option is the use of in vitro toxin selection (Jones 1990). In the case of *Phytophthora infestans*, this strategy has so far been unsuccessful (see Sect. 2.3 and discussion of this work in Cerato et al. 1993; Umaerus and Umaerus 1994).

A second strategy employed was that of Meulemans and coworkers (Tegera and Meulemans 1985: Meulemans and Fouarge 1986; Meulemans et al. 1987) who used cocultivation of *P. infestans* and potato. The epidemiology of *P. infestans* depends on the ability of the pathogen to infect the host, to colonize the host tissues, and to sporulate (Umaerus et al. 1983). The corresponding host defences are preformed and induced barriers (Isaac 1992). A criticism of in vitro selection is that the preformed barriers may not be fully expressed in juvenile

**Table 1.** Characters of importance in potato breeding in Europe. (After Simmonds 1979)

| | |
|---|---|
| Field | Strong sprouts, early growth and good ground cover; appropriate maturity class; resistance to late blight, to viruses (PVX, PVY, PLRV, etc), to wart disease, to eelworms, and to insects |
| Tubers | Good yield, shape, smoothness, regularity of size, color: resistance to mechanical damage; lack of cracking and second growth; appropriate dormancy; resistance to tuber diseases (late blight, scab, gangrene, skin spot, others); good storage characters (delayed sprouting, disease resistance) |
| Quality | Flesh color; flesh texture; lack of enzymic browning; specific gravity (dry matter content); lack of after-cooking blackening; texture on cooking; flavor; chip/crisp color after frying; reduced sugar content; "reconditioning" capacity after cool storage; low glycoalkoloid content |

tissues. It follows that in vitro selection in cocultivation may be narrowly based on ability to overcome induced responses (see Jones 1990).

A third strategy was that used by Cassells et al. (1991), which is essentially that employed by conventional plant breeders, with modifications which reflect the in vitro origin of the material and that are aimed at reducing the useless variation associated with somaclonal variation (Sanford et al. 1984). In this approach, chimeral breakdown is facilitated by a vegetative cycle prior to selection (Cassells and Periappuram 1993; Cassells et al. 1993). Aberrant and degenerate somaclones are selected *against* on the basis of lack of vigor (Cassells et al. 1987). Further useless variants, e.g. polyploids and aneuploids, are selected *against* based on associated changes in leaf morphology, determined by image analysis (Fig. 1). Further selection is based on the elimination of maturation mutants based on flowering date (Cassells et al. 1991; Fig. 2). This is in deference to the relationship between late maturation and increased late blight resistance (Toxopeus 1958; Umaerus et al. 1983; Umaerus and Umaerus 1994) and the requirement that improved clones have the maturation class of the parental variety (Table 1). When the bulk of the useless variation has been eliminated (Fig. 3), positive can begin, i.e., selection for normal phenotype, normal maturation date and improved late blight resistance. In subsequent years' trials, the selected lines are screened for retention of improved blight resistance and retention of the commercial characters of the variety (as listed in Table 1; Fig. 1). It is essential to prevent virus infection of the lines during trialing by geographical isolation from virus vectors or by prophylactic spraying with contact and systemic insecticides.

While somaclones with improved late blight resistance and yield retention have been identified in this program, several years' more trials will be required to confirm line stability, durability of resistance, and retention of the commercially

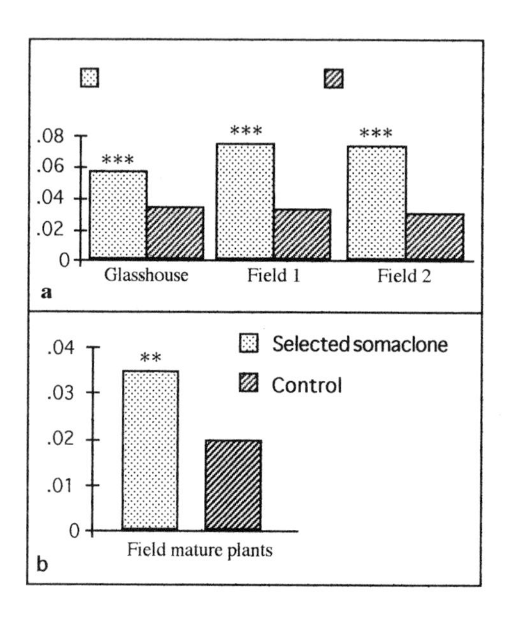


**Fig. 1a,b.** Detection, by leaf image analysis, of cryptic somaclonal variation in potato adventitious regenerants after removal of morphological variants. **a** Detection of significant differences in the standard deviations of terminal leaflet elongation in the microplants after establishment in the glasshouse and in young (*Field 1*) and mature plants (*Field 2*) in the field. **b** Detection of further but reduced cryptic variation in the field after roguing of cryptic aneuploids (methodology as Cassells et al. 1993). The ploidy status of the somaclones was determined by measurement of guard cell length

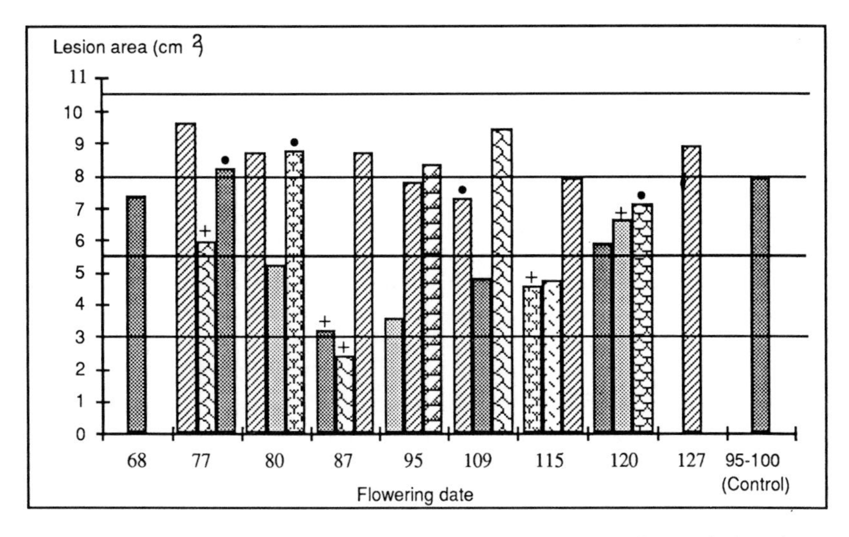


**Fig. 2.** Variation in late blight resistance and maturation date in potato somaclones. The data shown are for the late blight resistance of somaclones with different maturation (flowering) dates in the first year's trials. Tubers from these somaclones were sprouted and grown on in the glasshouse. They were then inoculated with mycelial plugs of *P. infestans* in water agar. The resultant lesion areas were measured by image analysis. The results show the instability of resistance to late blight in the somaclones. Somaclones marked 0 were more susceptible than the controls in year 1, somaclones marked + were more resistant than the control in year 1. Differences of one grid are significant. In the selection scheme outlined in Fig. 3, somaclones with near normal maturation date and improved late blight resistance would be selected for further trialing. Note flowering date in the controls was from Julian day 95 to 100, whereas in the somaclones it was from Julian day 68 to 127

important characteristics of the parent variety. The near-isogenic relationship between such resistant mutants and the parent variety would facilitate the characterization of the resistance by molecular genetic analysis and would create options for genetic transformation of other susceptible varieties (Watanabe 1994).

Work to date on late blight resistance in in vitro culture of potato is summarized in Table 2.

## 5 Summary and Conclusions

Late blight of potato is one of the most destructive diseases of potato, causing heavy losses (van der Zaag 1994). Conventional breeding techniques have not been successful in developing durable resistance (see Sect. 3). In this regard, in vitro induction of genetic variability through somaclonal variation has yielded some interesting data (see Cassells and Jones 1995), and plants showing resistance to late blight have been regenerated in vitro suggesting that spontaneous (i.e.,

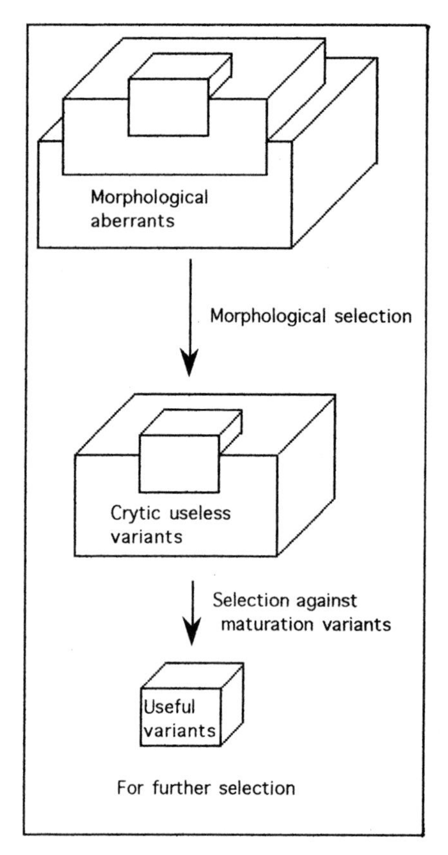


**Fig. 3.** A representation of the problem of selection for the potentially useful random variation in a somaclonal population. Useless variation is removed in stages to leave a pool of random variation which has potential for potato improvement. A prescreening cycle of vegetative propagation facilitates chimeral breakdown (Cassells et al. 1987). Morphological aberrants are removed by visual screening; positive selection for vigor should also be made at this stage (Cassells et al. 1991). Of the remaining cryptic useless variants, some, e.g., cryptic aneuploids, are detectable by image analysis. A further major category of useless variants, the maturation mutants, can, at present, only be identified at flowering time in the first field trials. Removal of the latter facilitates selection for potentially useful, late blight resistance in the somaclonal population

somaclonal variation), and induced mutagenesis, with appropriate selection schemes may have a contribution to make (Cassells and Walsh 1995).

Three strategies have been used by various workers to exploit somaclonal variation in late blight improvement in potato.

1. The use of in vitro toxin selection (Behnke 1979; Wenzel and Foroughi–Wehr 1990; Cerato et al. 1993).
2. Cocultivation of *Phytophthora infestans* and potato in vitro (Meulemans et al. 1987).
3. A strategy which is essentially that employed by conventional plant breeders (Cassells et al. 1991), with modifications reflecting the in vitro origin of the material and aimed at reducing the useless variation associated with somaclonal variation. When the bulk of the useless variation has been eliminated, positive selection begins.

It should be pointed out that juvenility in crops grown from potato microplants and in micro- and mini-tuber derived plants in the greenhouse will give false positive resistance, thus caution should be exercised in extrapolating from results obtained from such screening (Russell 1978).

**Table 2.** Summary of the work done on late blight resistance in in vitro culture of potato

| Model | Observations/Remarks | References |
|---|---|---|
| 1. Callus with toxin selection | Plants from calli resistant to fungal culture filtrates were less susceptible to pathogen in laboratory or glasshouse tests but showed no durable field resistance or lacked vigor in the field | Behnke (1979, 1980)<br>Wenzel and Foroughi-Wehr (1990)<br>Cerato et al. (1993) |
| 2. Co-cultivation of microplants and pathogen in vitro | Microplants derived from protocalli, internode and leaf calli were inoculated in vitro with pathogen; 3–5% showed resistance in vitro independent of origin. Field trials not reported | Tegera and Meulemans (1985)<br>Meulemans and Fouarge (1986)<br>Meulemans et al. (1987) |
| 3. Microplants selected for normal phenotype subjected to field infection | Microplants derived from complex explants were selected for normality; maturation mutants were rejected, resistance found in second year's field trials | Cassells et al. (1991) |

## References


Bajaj YPS (ed) (1990) Biotechnology in agriculture and forestry, vol 11. Somaclonal variation in crop improvement I. Springer, Berlin Heidelberg New York

Behnke M (1979) Selection of potato callus for resistance to culture filtrates of *Phytophthora infestans* and regeneration of resistant plants. Theor Appl Genet 55: 69–71

Behnke M (1980) General resistance to late blight of *Solanum tuberosum* plants regenerated from callus resistant to culture filtrates of *Phytophthora infestans*. Theor Appl Genet 56: 151–152

Bethell J (1990) Trends in potato production. Proc 11th Triennial Conf Eur Assoc Potato Res, Edinburgh, pp 19–27

Bowles A (1990) Defence-related proteins in higher plants. Annu Rev Biochem 59: 873–907

Bradshaw JE, Mackay GR (1994) Potato genetics. CAB International, Wallingford, 576 pp

Buiatti M, Ingram DS (1991) Phytotoxins as tools in breeding and selection of disease-resistant plants. Experientia 47: 811–819

Cassells AC, Periappuram C (1993) Diplontic drift during serial subculture as a positive factor influencing the fitness of mutants derived from the irradiation of in vitro nodes of *Dianthus* 'Mystere'. In: Schiva T, Mercuri A (eds) Creating genetic variation in ornamentals. Proc XVIIth Symp Eur Assoc Res Plant Breed, Istituto Sperimentale per la Floricoltura, San Remo, pp 71–80

Cassells AC, Jones PW (eds) (1995) The methodology of plant genetic manipulation. Kluwer, Dordrecht, 476 pp

Cassells AC, Walsh M (1995) Screening for *Sclerotinia* resistance in *Helianthus tuberosus* L. (Jerusalem artichoke) varieties, lines and somaclones, in the field and in vitro. Plant Pathol 44: 428–437

Cassells AC, Austin S, Goetz EM (1987) Variation in tubers in single cell-derived clones of potato in Ireland. In: Bajaj YPS (ed) Biotechnology in agriculture and forestry vol 3. Potato. Springer, Berlin Heidelberg New York, pp 375–391

Cassells AC, Deadman ML, Brown CA, Griffin E (1991) Field resistance to late blight [*Phytophthora infestans* (Mont.) De Bary] in potato (*Solanum tuberosum* L.) somaclones associated with instability and pleiotropic effects. Euphytica 56: 75–80

Cassells AC, Walsh C, Periappuram C (1993). Diplontic selection as a positive factor in determing the fitness of mutants of *Dianthus* 'Mystere' derived from X-irradiation of nodes in vitro in culture. Euphytica 70: 167–174

Cerato C, Manici LM, Borgatti S, Alicchio R, Ghedini R, Ghinelli A (1993) Resistance to late blight [*Phytophthora infestans* (Mont.) de Bary] of potato plants regenerated from in vitro selected calli. Potato Res 36: 341–352

Cooke LR (1992) The future of potato blight control: a more aggressive pathogen and fewer weapons. In: McCracken AR, Mercer PC (eds) Disease management in relation to changing agricultural practice. Soc Ir Plant Pathol, The Queen's University, Belfast, pp 65–73

Cullis CA (1990) Environmentally induced variation in plant DNA and associated phenotypic consequences. In: Bajaj YPS (ed) Biotechnology in agriculture and forestry, vol 11. Somaclonal variation in crop improvement I. Springer, Berlin Heidelberg New York, pp 224–235

Dale PJ, McPartlan HC (1992) Field performance of transgenic potato plants compared with controls regenerated from tuber discs and shoot cuttings. Theor Appl Genet 84: 585–591

Fry WE, Doster MA (1991) Potato late blight: forecasts and disease suppression. In: Lucas JA et al. (eds) Phytophthora. Cambridge University Press, Cambridge, pp 326–336

Gatehouse AMR, Hilder VA, Boulter D (1992) Plant genetic manipulation for crop protection. CAB international, Wallingford, 270 pp

Hermsen JGTh (1994) Introgression of genes from wild species, including molecular and cellular approaches. In: Bradshaw JE, Mackay GR (eds) Potato genetics. CAB International, Wallingford, pp 515–538

Hide GA, Lapwood DH (1978) Disease aspects of potato production. In: Harris PM (ed) The potato crop. Chapman and Hall, London, pp 407–439

Hooker WJ (1981) Potato diseases. APS Press, St Paul, Minnesota, 141 pp

Howard HW (1978) The production of new varieties. In: Harris PM (ed) The potato crop. Chapman & Hall, London, pp 607–646

Isaac S (1992) Fungal-Plant Interactions. Chapman & Hall, London, 418 pp

Jellis GJ, Richardson DE (1987) The production of new potato varieties. Cambridge University Press, Cambridge, 358 pp
Jones PW (1990) In vitro selection for disease resistance In: Dix PJ (ed) Plant cell line selection: procedures and applications. VCH, Weinheim, pp 113–149
Karp A (1990) Somaclonal variation in potato. In: Bajaj YPS (ed) Biotechnology in agriculture and forestry, vol 11. Somaclonal variation in crop improvement I. Springer, Berlin Heidelberg New York, pp 379–399
Karp A (1991) On the current understanding of somaclonal variation. In: Miflin BJ (ed) Oxford surveys of plant molecular and cell biology, vol 7. Oxford University Press, Oxford, pp 1–58
Keen NT, Yoshika M (1983) Physiology of disease and the nature of resistance to Phytophthora. In: Erwin DC, Bartnicki-Garcia S, Tsao PH (eds) *Phytophthora*: its biology, taxonomy, ecology and pathology. APS Press, St Paul, Minnesota, pp 279–287
Kemble RJ, Shepard JF (1984) Cytoplasmic DNA variation in potato protoclonal population. Theor Appl Genet 69: 211–216
Kumar A (1994) Somaclonal variation. In: Bradshaw JE, Mackay GR (eds) Potato genetics. CAB International, Wallingford, pp 197–212
Landsmann J, Urhig H (1985) Somaclonal variation in potato detected at the molecular level. Theor Appl Genet 71: 500–505
Lawrence WJC (1968) Plant breeding. Edward Arnold, London, 58 pp
Meulemans M, Fouarge G (1986) Regeneration of potato somaclones and in vitro selection for resistance to *Phytophthora infestans* (Mont.) de Bary. Med Fac Landbouww Rijksuniv Gent 51: 533–545
Meulemans M, Duchene D, Fouarge G (1987) Selection of variants by dual culture of potato and *Phytophthora infestans*. In: Bajaj YPS (ed) Biotechnology in agriculture and forestry, vol 3. Potato. Springer, Berlin Heidelberg New York, pp 318–331
Meyer P (1993) Molecular breeding of novel flower colours. In: Schiva T, Mercuri A (eds) Creating genetic variation in ornamentals. Proc XVIIth Symp Eur Assoc Res Plant Breed, Istituto Sperimentale par la Floricoltura, San Remo, pp 1–4
Nagl W (1990) Gene amplification and related events. In: Bajaj YPS (ed) Biotechnology in agriculture and forestry, vol 11. Somaclonal variation in crop improvement I. Springer, Berlin Heidelberg New York, pp 153–201
Parry D (1990) Plant pathology in agriculture. Cambridge University Press, Cambridge, 385 pp
Pernollet J-C, Nespoulous C, Huet J-C (1993) Relationship between the structure, the movement and the toxicity of $\alpha$ and $\beta$ elicitins secreted by *Phytophthora* sp. In: Fritig B, Legrand M (eds) Mechanisms of plant defense responses. Kluwer, Dordrecht, pp 136–139
Peschke VM, Phillips RL (1992) Genetic variations of somaclonal variation in plants. Adv Genet 30: 41–75
Rowe RC (1993) Potato health management. APS Press, St Paul, Minnesota, 168 pp
Russell, GE (1978) Plant breeding for pest and disease resistance. Butterworth, London, 485 pp
Sanford JC, Weeeden NF, Chyi YS (1984) Regarding the novelty and breeding value of protoplast-derived variants of Russet Burbank (*Solanum tuberosum* L.). Euphytica 33: 709–715
Scheffer RP, Livingston RS (1984) Host selective toxins and their role in disease. Science 2333: 17–21
Simmonds NW (1979) Principles of crop improvement. Longmans, London, 408 pp
Smith JCS, Smith OS (1992) Fingerprinting crop varieties. Adv Agron 47: 85–140
Spielman LJ, Drenth A, Davidse LC, Sujkowski LJ, Gu W, Tooley PW, Fry WE (1991) A second worldwide migration and population displacement of *Phytophthora infestans*? Plant Pathol 40: 422–430
Tegera P, Meulemans M (1985) Dual culture of in vitro micropropagated potato plantlets with *Phytophthora infestans* (Mont.) De Bary and assessment of general resistance of potato cultivars to late blight fungus. Med Fac Landbouww Rijksuniv Gent 50: 1069–1080
Toxopeus HJ (1958) Some notes on the relations between field resistance to *Phytophthora infestans* in leaves and tubers and ripening time in *Solanum tuberosum* subsp. *tuberosum*. Euphytica 7: 123–10
Umaerus V, Umaerus M (1994) Inheritance of resistance to late blight. In: Bradshaw JE, Mackay GR (eds) Potato genetics. CAB International, Wallingford, pp 365–401
Umaerus V, Umaerus M, Erjefalt L, Nilsson BA (1983) Control of *Phytophthora* by host resistance: problems and progress. In: Erwin DC, Bartnicki-Garcia S, Tsao PH (eds) *Phytophthora*: its biology, taxonomy, ecology and pathology. APS Press, St Paul, Minnesota, pp 315–326

Van der Plank JE (1975) Principles of plant infection. Academic Press, New York, 216 pp
Van der Zaag DE (1994) The need for international research programmes on some potato pathogens. Potato Res 37: 323–329
Watanabe KN (1994) Molecular genetics. In: Bradshaw JE, Mackay GR (eds) Potato genetics. CAB International, Wallingford, pp 213–235
Wenzel G, Foroughi-Wehr B (1990) Progeny tests of barley, wheat and potato regenerants from cell cultures after in vitro selection for disease resistance. Theor Appl Genet 80: 359–365
Yarwood CE (1946) Increasing yield and disease resistance of giant hill potatoes. Am Potato J 23: 352–369

# I.8 In Vitro Production of *Verticillium dahliae*-Resistant Potato Plants

D. SIHACHAKR[1], R. JADARI[2], A. KUNOTHAI-MUHSIN[3], L. ROSSIGNOL[1], R. HAICOUR[1], and G. DUCREUX[1]

## 1 Introduction

The cultivated potato (*Solanum tuberosum* L.) is a tetraploid ($2n = 4x = 48$) and is thought to have derived from a primitive diploid species, *Solanum stenotomum* ($2n = 2x = 24$). Since tetraploids are more vigorous and more productive, the other degrees of ploidy have been progressively eliminated. The outcome of crosses between tetraploid parental lines is often unpredictable. Since the parents are highly heterozygous, the $F_1$ usually shows a large genetic variation for a number of characteristics, diminishing the chance of finding a new variety. In addition, breeding efforts may be hampered by the fact that sexual incompatibility and male sterility can also occur between and within potato parental lines.

In order to increase genetic variability and to transfer traits of resistance, crosses between potato and its wild relatives have been attempted. In fact, more than 150 wild or primitive species, particularly including *S. tarijense* (Jadari 1993) and *S. torvum* (Yamakawa and Mochizuki 1978; Daunay et al. 1991) with *Verticillium* resistance, have been identified in South America. They constitute important genetic diversity and resources useful for potato breeding. They are more or less close to potato, mostly tuberous and diploid, and carry numerous traits of resistance (Hawkes 1978). Nevertheless, sexual hybridization of European varieties of *S. tuberosum* ($2n = 4x = 48$) with these wild species ($2n = 2x = 24$) is difficult. It is necessary to bring tetraploid potatoes to the diploid level by parthenogenesis resulting from crosses with *S. phureja* (Ivanovskaja 1939; Hougas and Peloquin 1957). Although breeding on the diploid level offers a very interesting means of increasing genetic variability, it still remains difficult to realize. Moreover, at the end of the breeding cycles with diploids, it is necessary to return to the tetraploid level, particularly by using protoplast fusion to recover good vigor and high yields, as well as an optimal expression and regulation of transmitted characters. Potatoes are usually clonally propagated through tubers, a mode of propagation that inevitably results in the extension and introduction of potato pathogens and insects from one area to another, as well as in the overwintering of pathogens and insects in the stored tubers, which may be used for propagative purposes.

[1] Morphogénèse Végétale Expérimentale, Bât. 360, Université Paris Sud, F-91405 Orsay Cedex, France

[2] Institut Agronomique et Vétérinaire Hassan II, BP 6202, Morocco

[3] Southeast Asian Regional Centre for Tropical Biology (Seameo-Biotrop), Jl Raya Km 6, PO Box 116, Bogor, Indonesia

Biotechnology in Agriculture and Forestry, Vol. 36
Somaclonal Variation in Crop Improvement II (ed. by Y.P.S. Bajaj)

In view of the limitations of sexual breeding methods, tissue culture and genetic engineering are being employed in improvement of potato (see Bajaj 1987). In the context of the production of in vitro plants resistant or tolerant to diseases, the exploitation of somaclonal variation offers possibilities for improvement of potato. When plants were regenerated from culture of explants or protoplasts, they frequently exhibited new phenotypes which, if heritable, represented a form of variation that can be incorporated into classical strategies of potato breeding (Rossignol et al. 1984; Evans and Sharp 1986; Fish and Karp 1986; Karp 1989; Zuba and Binding 1989; Wastie et al. 1993). Moreover, stable somaclonal variants with interesting traits may be fixed and accumulated through vegetative multiplication cycles.

In addition to the possibility of inducing important somaclonal variation, protoplast fusion overcomes sexual incompatibility, and also combines genomes from various dihaploid potatoes or different wild species as well. Somatic hybrids resulting from intraspecific cell fusions between different dihaploid clones of potato are brought to the tetraploid and highly heterozygous level which is necessary for recovering hybrid vigor (Debnath and Wenzel 1987; Chaput et al. 1990). Moreover, tolerance to frost from *S. brevidens* (Preiszner et al. 1991) and to salinity from *L. pennellii* (Serraf et al. 1994), as well as resistance to potato viruses (PLRV and PVY) from *S. brevidens* (Austin et al. 1985; Gibson et al. 1988), to insects from *S. berthaultii* (Serraf et al. 1991), and to *Verticillium* wilt from *S. torvum* (Jadari et al. 1992), which is described below, have successfully been transferred into potato through interspecific somatic hybridization.

## 2 Brief Review of *Verticillium* Resistance Studies on Potato

*Verticillium* wilt is one of the most serious diseases of potato and other crops, because it causes heavy yield losses, and affects the quality of potato tubers in particular (Platt and Sanderson 1987). In the past, it has often been confused with *Fusarium* wilt and other wilt diseases. The pathogens responsible for *Verticillium* wilt, *V. albo-atrum* Reinke and Berth, and *V. dahliae* Kleb, have no specific hosts. They affect a wide range of plant species, including tomatoes, eggplants, strawberries, and many species of trees (Rich 1983). The presence of nematodes favors the development of *Verticillium* wilt. Particularly, there is a synergistic action between the nematodes, *Globodera rostochiensis*, and *Verticillium dahliae* (Hide and Corbett 1973).

Both *V. albo-atrum* and *V. dahliae* are members of the Fungi Imperfecti. They survive in soil upon forming persistent structures, microsclerotia and pseudosclerotia for *V. dahliae* and *V. albo-atrum*, respectively. The latter is the dominant species in cool climates, while the former occurs more frequently in slightly warmer climates.

Up to now, there are no totally resistant varieties of potato. Some of them, including Katahdin, Targee, and Désirée, are tolerant to *Verticillium* wilts, and produce satisfactory crops in spite of infection of the fungus; while others, particularly including Kennebec, Russet Burbank, and Nicola, are susceptible.

Although resistance to *Verticillium* and *Fusarium* wilts has been identified in *S. torvum* and *S. tarijense*, wild relatives of eggplant and potato respectively (Ochoa 1990; Daunay et al. 1991), the use of such wild species in potato breeding programs has not been investigated, probably due to partial or total sexual incompatibility. Therefore, attempts at regenerating callus or plants resistant to fungus wilts through tissue culture have been achieved by using spores, filtrate, and toxin as selection pressure (Table 1).

# 3 In Vitro Production of Plants Resistant to *Verticillium*

The production of resistant plants regenerated from callus of various plant explants of potato (Kunothai-Muhsin 1988), and the transfer of resistance to *Verticillium* wilt from *S. torvum* to potato through somatic hybridization (Jadari et al. 1992) are discussed.

## 3.1 Regeneration of Plants from Resistant Callus Derived from Cultured Plant Explants

In order to produce resistant plants, various explants of potato, cv. Nicola, known as very susceptible to *Verticillium dahliae* (Nachmias and Krikun 1984), were induced to form callus, to which a selection pressure using fungus filtrate was applied (Kunothai-Muhsin 1988). Whole leaves, fragments of petioles and stems (1 mm long), and disbudded nodes taken from 4–6-week-old cuttings were induced to form callus on the callus induction medium for 2 weeks, before transfer onto the selection medium containing a lethal dose of sterilized fungus filtrate. The latter was diluted 2.5 times in the culture medium, and prepared from lyophilized mycelium of *Verticillium dahliae* from Morocco. The control was made of calli which were cultured on selection medium but without filtrate. After 2 weeks under selection pressure, the treated and control calli were transferred to the regeneration medium. The basal composition of the culture medium was MS basal medium (Murashige and Skoog 1962) supplemented with vitamins (Morel and Wetmore 1951), 30 g/l sucrose, and solidified with 6 g/l agar. A combination of 2.0 mg/l benzylaminopurine (BAP) and 0.2 mg/l $\alpha$-naphthaleneacetic acid (NAA) was used for callus induction, 2 mg/l BAP and 10 mg/l gibberellic acid ($GA_3$) for selection, and 0.25 mg/l zeatin and 0.25 mg/l kinetin for regeneration.

After a few days on callus induction medium, explants gave rise to green and nodular calli. Their growth was more important when derived from petiole, stem, and particularly from disbudded nodes. Most of the calli, and particularly those smaller in size and derived from whole leaves or petioles, turned brown rapidly and died within a few days on the selection medium. Very little browning was, however, observed on the control calli, but all of them survived. When transferred onto the regeneration medium, some of the treated calli, particularly those derived from disbudded nodes and bigger in size, turned green and regenerated plants within 15 days. This demonstrated that cells from a very susceptible variety

**Table 1.** In vitro production of potato plants resistant to fungus wilts

| Host plant | Causal agent | Selection pressure | Culture response | Reference |
|---|---|---|---|---|
| *S. tuberosum* | *Alternaria solani* | Partially purified toxin | Very good correlation between the fungus susceptibility of potato and the toxin | Matern et al. (1978) |
| *S. tuberosum* | *Phytophthora infestans* | Filtrate | Regenerated plants resistant to filtrate | Behnke (1979) |
| *S. tuberosum* | *Fusarium oxysporum* | Filtrate | Partially resistant plants regenerated from totally resistant callus | Behnke (1980) |
| *S. tuberosum* | *Phytophthora infestans* | Filtrate | Some clones regenerated from protoplast callus resistant to the fungus | Shepard et al. (1980) |
| *S. tuberosum* | *Alternaria solani* | Filtrate | Five out of 500 protoclones are less susceptible than the control, while 4 others are resistant | Shepard et al. (1980) |
| *S. tuberosum* | *Phytophthora infestans* | Spores | Three to 5% of plants regenerated from protoplast and explant callus are resistant | Meulemans and Fouarge (1986) |
| *S. tuberosum* | *Verticillium dahliae* | Toxin | Good correlation between the symptoms observed in the field after fungus inoculation and those obtained in vitro after addition of toxin to the culture | Nachmias and Krikun (1983) |
| *S. tuberosum* | *Verticillium dahliae* | Filtrate | Plants regenerated from explant callus resistant to filtrate | Kunothai-Muhsin (1988) |
| *S. tuberosum* + *S. torvum* | *Verticillium dahliae* | Filtrate | Somatic hybrids resistant to filtrate | Jadari et al. (1992) |

of potato (Nicola) had survived the action of the fungus filtrate. About 20 plants were obtained from callus after application of a selection pressure. The regenerated plants were transferred to the greenhouse for evaluation of *Verticillium* resistance, using both filtrate and conidia.

### 3.2 Transfer of *Verticillium* Resistance from *S. torvum* into Potato by Protoplast Electrofusion

*S. torvum* is a wild relative of eggplant, a nontubering species, and sexually incompatible with potato. It carries numerous interesting traits of resistance, such as that against *Verticillium* (*Verticillium dahliae*), *Fusarium* (*Fusarium oxysporum*), and bacterial (*Pseudomonas solanacearum*) wilts, as well as to nematodes (Daunay et al. 1991). The inheritance of *Verticillium* resistance in *S. torvum* has not been genetically characterized, while in tomato (Schaible et al. 1951) the resistance was controlled by a single dominant gene. It is obvious that the use of this wild species with traits of resistance is of great interest in potato breeding, but limited, however, by strong sexual incompatibility. Therefore, as pointed out above, somatic hybridization provides an effective means of bypassing sexual crosses and of transferring such useful traits into potato.

Mesophyll protoplasts from a dihaploid clone of *S. tuberosum*, cv. Nicola ($2n = 2x = 24$) were electrofused with those from *S. torvum* ($2n = 2x = 24$; Jadari et al. 1992). After one D. C. pulse of 1.2 kV/cm was applied for 40 $\mu$s, the fusion frequency up to 40% was obtained by the addition of 0.5 mM $CaCl_2$ to the electrofusion medium, composed of a 0.5 M mannitol solution. At least 25% of the fused protoplasts were binary fusions. Cell wall formation occurred within 48 h, and the first protoplasts divided on day 5. After 7 days of culture in V-KM medium (Binding et al. 1978) supplemented with 250 mg/l polyethylene glycol (MW 6000), 0.2 mg/l 2,4-dichloropheoxyacetic acid (2,4-D), 0.5 mg/l zeatin, and 1 mg/l NAA, 5% of the cultured protoplasts divided. This was very close to that previously found in intra- (Chaput et al. 1990) and interspecific (Serraf et al. 1991) fusions of potato, using the same electrical apparatus and under the similar cultural conditions. Most dividing cells from fusion experiments gave rise to microcolonies, so that hundreds of calli were recovered after the cultures were highly diluted (eight to ten times), on day 15, with fresh medium containing a combination of 0.2 mg/l 2,4-D and 2 mg/l BAP. Plant regeneration occurred 6–12 weeks after they were transferred onto regeneration medium composed of MS basal medium, vitamins (Morel and Wetmore 1951), 20 g/l sucrose, 2 mg/l zeatin, 0.1 mg/l indol-3-acetic acid (IAA), and solidified with 7 g/l agar.

Early selection of the putative somatic hybrids was based on the apparent difference in the cultural behavior of the parental and hybrid calli (particularly the ability of the latter to regenerate early) in combination with morphological markers such as long single hairs in particular. In fact, protoplast-derived calli of *S. torvum* did not regenerate shoots under the cultural conditions used, while those of potato cv. Nicola did, with a very low frequency (0.5–1%), and 4–6 weeks after regeneration of the first putative hybrids. Finally, using such selection criteria, four putative hybrids were recovered from hundreds of calli, probably resulting from complementation of the two parental genomes. The concept of unilateral regenerative capacity (Adams and Quiros 1985) has successfully been applied to recover somatic hybrids of tomato in which the inherent regeneration capacity was provided by the wild-type side (Kinsara et al. 1986) and also those of potato and eggplant with *S. berthaultii* and *S. torvum*, respectively, provided by the cultivated species (Sihachakr et al. 1989; Serraf et al. 1991).

The chromosome counts made on root tips of the four putative somatic hybrids obtained in this study revealed that they were at the expected tetraploid level ($2n = 4x = 48$ chromosomes). The tetraploid state of the hybrids may be due to the use of differentiated tissues like leaves as protoplast source; somatic hybrids regenerated from cell suspension generally exhibited a higher variation in chromosome numbers (Schieder and Kohn 1986).

Leaves from the four selected putative somatic hybrids and the parental lines were subjected to electrophoresis. Their hybrid nature was confirmed by examining the patterns of four isozyme systems, isocitrate dehydrogenase (Idh; Fig. 1A), malate dehydrogenase (Mdh; Fig. 1B), phosphoglucoisomerase (Pgi; Fig. 1C), and 6-phosphogluconate dehydrogenase (6-Pgd; data not shown), which not only contained the sum of the parental bands, but also an additional band (Idh and 6-Pgd), being specifically relevant to the hybrid nature and not found in the mixed extracts of the parents.

The four somatic hybrid plants were multiplied by clonal propagation. Despite the small number of roots produced and the necrosis of root apices of the hybrid cuttings after 3–4 weeks of subculture, they grew vigorously under in vitro conditions. The hybrid plants exhibited intermediate traits, including leaf form, plant morphology and the presence of anthocyanin. In the greenhouse, the somatic hybrid plants were much less vigorous and more fragile than the parental lines. The development and growth of the four hybrid plants were strongly retarded by difficulties in rooting. Hybrid leaves and stems became dark purple because of an increase in anthocyanin content, and the hybrid plants died within 3–5 weeks. The reason for the difficulty in rooting the hybrids could be the taxonomic distance between the partners, resulting in early events of the limitations to somatic fusion between remote species. Similar problems in rooting were also observed for interspecific somatic hybrids of eggplant with remote species like *S. sisymbriifolium* (Gleddie et al. 1986), *S. khasianum* and *S. torvum* (Sihachakr et al. 1988, 1989), and intertribal somatic hybrids of *Brassica* (Glimelius et al. 1990). In this study, in order to recover normal plant development, the hybrid shoots were grafted on rootstocks of potato or *S. torvum*. Whatever the rootstocks used, the hybrid shoots grew fairly well and were sufficiently developed for morphological observations. Morphological traits intermediate between potato and *S. torvum* provided strong evidence for the hybrid nature of the four selected regenerants (Fig. 2). Under the conditions of our greenhouse, the somatic hybrids set flowers, which aborted precociously.

Tests for *Verticillium* resistance were performed in vitro. Both parental lines and the somatic hybrid plants were multiplied by subculturing one-leaf node cuttings in test tubes containing 20 ml of solid MS medium each. After 3 weeks of culture, 1 ml of a solution composed of either water used as control or 50% fungus extract was added to each culture tube. Two weeks later, foliar symptoms

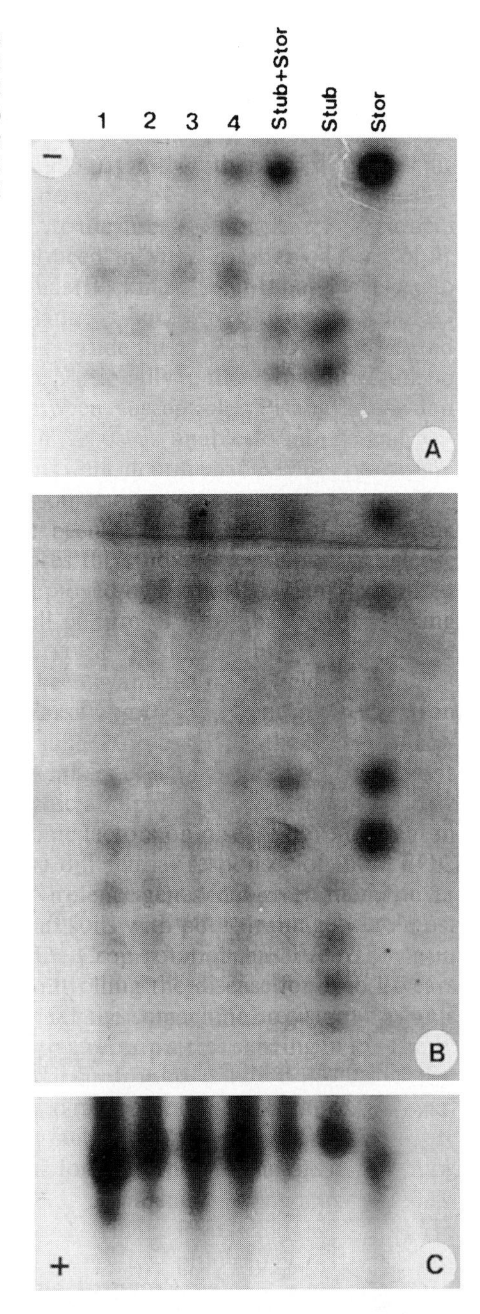


**Fig. 1A–C.** Electrophoresis banding patterns of **A** isocitrate dehydrogenase (Idh) (EC 1.1.1.42), **B** malate dehydrogenase (Mdh) (EC 1.1.1.37), and **C** phosphoglucoisomerase (Pgi) (EC 5.3.1.9.) from four somatic hybrids (*lanes 1–4*), *S. tuberosum* (*Stub*), *S. torvum* (*Stor*), and a mixture of both parents (*Stub* + *Stor*). (Jadari et al. 1992)

were scored by using two indexes according to Beyes and Lafay (1985): (1) propagation symptom index (PSI) defined as [(sum of scores)/(maximum score)] × 100, reflecting the progress of symptoms upward the plant, (2) foliar deterioration index (FDI) defined as [(sum of scores)/(maximum score)] × 100,

**Fig. 2.** Leaves **A** and plants **B** of *S. tuberosum* (*Stub*), *S. torvum* (*Stor*), and their somatic hybrids (*SH*). (Jadari et al. 1992)

expressing wilting degree of leaves. The foliar symptoms were rated from 0 (immune) to 5 (highly susceptible or dead). In fact, the tests showed that all the somatic hybrids were resistant to the fungus filtrate. As shown in Fig. 3, 2 weeks after the fungus filtrate was added to the culture, the symptoms did not progress upward on the hybrid stems (4–7% PSI), and only 2–6% (FDI) of the hybrid leaves had the symptoms (Fig. 3A) like the resistant wild species, *S. torvum*, while potato plants with at least 35% (FDI) of leaves affected developed severe symptoms progressing upward nearly 50% (PSI) of the stems (Fig. 3B). Affected potato plants died within 3–4 weeks, while the somatic hybrids and *S. torvum* continued growing very well. Traits of resistance against *Verticillium* wilt from *S.*

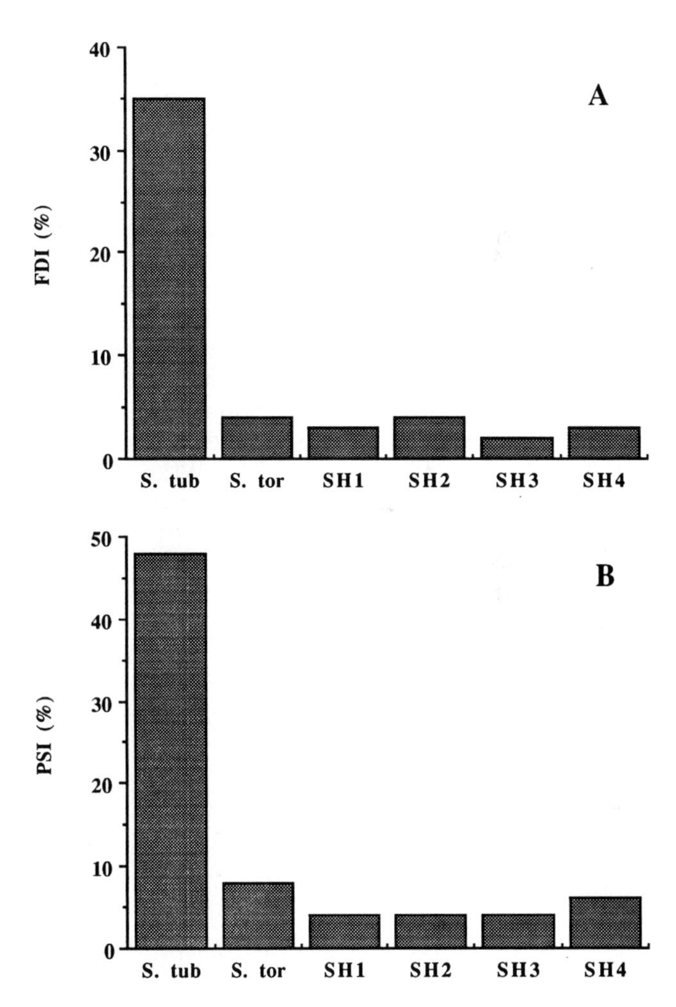


**Fig. 3A, B.** Effects of *Verticillium* filtrate on the parental lines, *S. tuberosum* (S. tub), *S. torvum* (S. tor) and their somatic hybrids (*SH 1–4*). **A** Foliar symptoms expressed as foliar deterioration index (*FDI*). **B** Progress of foliar symptoms upward the plant expressed as propagation symptom index (*PSI*) according to Beye and Lafay (1985). (Jadari et al. 1992)

*torvum* have also successfully been transferred into eggplant by protoplast fusion (Guri and Sink 1988; Sihachakr et al. 1989, 1991, 1994).

## 4 Summary and Conclusions

Whole leaves, fragments of petioles and stems, and nodes, taken from in vitro plants of potato, cv. Nicola, were induced to form callus for 2 weeks, before transfer to selection medium containing *Verticillium* filtrate. Some of the treated calli survived the selection pressure, and regenerated plants when transferred onto the regeneration medium.

Interspecific somatic hybrid plants were regenerated after electrofusion of mesophyll protoplasts with the objective of transferring resistance to *Verticillium dahliae* from *Solanum torvum* into potato. Early selection of the putative hybrids was based on differences in cultural behavior of the parental and hybrid calli (particularly the ability of the latter to regenerate early) in combination with morphological markers. Four putative hybrids were recovered from hundreds of calli, probably resulting from complementation of the two parental genomes. The regenerates were tetraploids ($2n = 4x = 48$ chromosomes) and exhibited intermediate traits including leaf form, plant morphology, and the presence of anthocyanin. The hybrid nature of the four selected plants was confirmed by examining the isoenzyme patterns for isocitrate dehydrogenase (Idh), malate dehydrogenase (Mdh), phosphoglucoisomerase (Pgi), and 6-phosphogluconate dehydrogenase (6-Pgd). While the hybrid plants rooted readily and grew vigorously under in vitro conditions, in the greenhouse their development and growth were retarded by difficulties in rooting. When grafted on potato or *S. torvum* rootstocks, the hybrid plants recovered normal development and growth. Again, they exhibited intermediate morphological traits. Tests for resistance realized in vitro with medium containing 50% *Verticillium dahliae* filtrate showed that all the somatic hybrids were resistant to the fungus filtrate like the wild parent, *S. torvum*, while potato plants died within 3–4 weeks.

The potential usefulness of those somaclones and somatic hybrid plants depends on their stability, and particularly on their fertility and ability to transmit sexually desirable agronomic traits to potato progeny.

## 5 Protocol

### 1. Culture of Plant Explants

Use whole leaves, fragments of petioles and stems (1 mm long) and disbudded nodes taken from 4–6-week-old cuttings.
Incubate plant explants in the callus induction medium (MS + 0.2 mg/l NAA + 2 mg/l BAP).
On day 15, transfer the calli onto the selection medium (MS + 2 mg/l BAP and 10 mg/l $GA_3$ + 1/2.5 sterilized *Verticillium* filtrate).
On day 30, transfer the treated and control calli onto the regeneration medium (MS + 0.25 mg/l zeatin + 0.25 mg/l kinetin).

**2. Procedure for Protoplast Electrofusion**

Adjust the protoplast suspension to the density of $4.5 \times 10^5$/ml with a solution of 0.5 M mannitol and 0.5 mM $CaCl_2$. Equal volumes of each protoplast partner are mixed.
Pipette 500–700 $\mu$l aliquots of the mixture into a $15 \times 50$ mm Petri dish. The protoplasts are allowed to settle on the bottom of the Petri dish for 5 min. The electrode is sterilized in absolute ethanol and well dried, then placed in the Petri dish.
Apply the AC field at 230 V/cm, for 10 s, followed by the application of one DC pulse of 1.2 kV/cm for 40 $\mu$s. Turn off progressively the AC field, and move the electrode to the next Petri dish.
Add progressively 6 ml culture medium (V-KM + 0.2 mg/l 2,4-D, 0.5 mg/l zeatin + 1 mg/l NAA) to the fused protoplast mixture. Keep the cultures in darkness at 27 °C for 7 days, afterwards they are exposed to a 12 h/day illumination at 62 $\mu$M/m$^2$/s.
On day 15, dilute the cultures eight to ten times with the same V-KM medium + 0.2 mg/l 2,4-D + 2 mg/l BAP.
On day 25, pipette the cultures onto the callus growth medium, composed of MS basal medium + vitamins (Morel and Wetmore 1951) + 20 g/l sucrose + 7 g/l agar + 0.5 mg/l NAA + 0.5 mg/l zeatin.
On day 45, transfer individually the calli onto the regeneration medium, whose composition is the same as for the callus growth medium except for growth regulators replaced with 2 mg/l zeatin and 0.1 mg/l IAA. Excise shoots from regenerating calli for clonal propagation by subculture in MS basal medium.

*Acknowledgments.* The authors would like to thank Mrs. A. Servaes for her excellent technical assistance, and Mr Froger for the photography.

# References

Adams TL, Quiros CF (1985) Somatic hybridization between *Lycopersicon peruvianum* and *L. pennellii*; regenerating ability and antibiotic resistance as selection system. Plant Sci 40: 209–219

Austin S, Baer MA, Helgeson JP (1985) Transfer of resistance to potato leaf roll virus from *Solanum brevidens* into *Solanum tuberosum* by somatic fusion. Plant Sci 39: 75–82

Bajaj YPS (ed) (1987) Biotechnology in agriculture and forestry, vol 3. Potato. Springer, Berlin Heidelberg New York

Behnke M (1979) Selection of potato callus for resistance to culture filtrates of *Phytophthora infestans* and regeneration of resistant plants. Theor Appl Genet 55: 69–71

Behnke M (1980) Selection of diploid potato callus for resistant to the culture filtrate of *Fusarium oxysporum*. Z Pflanzenzucht 85: 254–258

Beye I, Lafay JF (1985) Etude de critères pour une réponse générale à la verticilliose chez la tomate. Agronomie 5: 305–311

Binding H, Nehls R, Schieder O, Sopory SK, Wenzel G (1978) Regeneration of mesophyll protoplasts isolated from dihaploid clones of *Solanum tuberosum*. Physiol Plant 43: 52–54

Chaput MH, Sihachakr D, Ducreux G, Marie D, Barghi N (1990) Somatic hybrid plants produced by electrofusion between dihaploid potatoes: BF15 (H1), Aminca (H6) and Cardinal (H3). Plant Cell Rep 9: 411–414

Daunay MC, Lester RN, Laterrot H (1991) The use of wild species for the genetic improvement of Brinjal eggplant (*Solanum melongena*) and tomato (*Lycopersicon esculentum*). In: Hawkes JG, Lester RN, Estrada N (eds) Solanaceae III. Taxonomy, chemistry, evolution. Royal Botanic Garden Kew and Linnean Soc of London, London, pp 389–412

Debnath SC, Wenzel G (1987) Selection of somatic fusion products in potato by hybrid vigour. Potato Res 30: 371–380

Evans DA, Sharp WR (1986) Applications of somaclonal variation. Biotechnology 4: 528–532

Fish N, Karp A (1986) Improvements in regeneration from protoplasts of potato and studies on chromosome stability 1. The effect of initial culture media. Theor Appl Genet 72: 405–412

Frearson EM, Power JB, Cocking EC (1973) The isolation, culture and regeneration of *Petunia* leaf protoplasts. Dev Biol 33: 130–137

Gibson RW, Jones MGK, Fish N (1988) Resistance to potato leaf roll virus and potato virus Y in somatic hybrids between dihaploid *Solanum tuberosum* and *S. brevidens*. Theor Appl Genet 76: 113–117

Gleddie S, Keller KW, Setterfield G (1986) Production and characterization of somatic hybrids between *Solanum melongena* L. and *S. sisymbriifolium* Lam. Theor Appl Genet 71: 613–621

Glimelius K, Fahlesson J, Landgren M, Sjödin C, Sundberg E (1990) Improvements of the *Brassica* crops by transfer of genes from alien species via somatic hybridization. In: Nijkamp HJJ, Van der Plas LHW, Van Aartrijk J (eds) Progress in plant cellular and molecular biology, Kluwer, Dordrecht, pp 299–304

Guri A, Sink KC (1988) Interspecific somatic hybrid plants between eggplant (*Solanum melongena*) and *Solanum torvum*. Theor Appl Genet 76: 490–496

Hawkes JG (1978) History of the potato. In: Harris PM (ed) The potato crop. Chapman and Hall, London, pp 1–14

Hide GA, Corbett DCM (1973) Controlling early death of potato caused by *Heterodera rostochiensis* and *Verticillium dahliae*. Ann Appl Biol 75: 461–462

Hougas RW, Peloquin SJ (1957) A haploid plant of the potato variety Kathadin. Nature 180: 1209–1210

Ivanovskaja EV (1939) A haploid plant of *Solanum tuberosum*. CR Acad Sci URSS 24: 517–520

Jadari R (1993) Contribution à l'étude chez la pomme de terre de la résistance au *Verticillium dahliae* Kleb et à son introgression par hybridation somatique chez un clone sensible. Thèse Doctorat en Sciences, Univ Paris Sud-Orsay, 195 pp

Jadari R, Sihachar D, Rossignol L, Ducreux G (1992) Transfer of resistance to *Verticillium dahliae* Kleb. from *Solanum torvum* SW into potato (*Solanum tuberosum* L) by protoplast electrofusion. Euphytica 64: 39–47

Karp A (1989) Can genetic instability be controlled in plant tissue cultures? IAPTC News Lett 58: 2–10

Kinsara A, Patnaik SN, Cocking EC, Power JB (1986) Somatic hybrid plants of *Lycopersicon esculentum* Mill. and *Lycopersicon peruvianum* Mill. J Plant Physiol 125: 225–234

Kunothai-Muhsin A (1988) Apport de la morphogénèse et de la culture in vitro dans la connaissance de la verticilliose chez la pomme de terre causée par *Verticillium dahliae* Kleb. Essai de mise au point d'un test précoce in vitro de la résistance au filtrat. Thèse Doctorat en Sciences, Univ Paris Sud -Orsay, 134 pp

Matern U, Strobel G, Shepard JF (1978) Reaction to phytotoxins in a potato population derived from mesophyll protoplasts. Proc Natl Acad Sci USA 75: 4935–4939

Meulemans M, Fouarge G (1986) Regeneration of potato somaclones and in vitro selection for resistance to *Phytophthora infestans* Mont. de Bary. Med Fac Landbouww Rijksuniv Gent 51: 533–545

Morel G, Wetmore RH (1951) Fern callus tissue culture. Am J Bot 38: 141–143

Murashige T, Skoog F (1962) A revised medium for rapid growth and bioassays with tobacco tissue cultures. Physiol Plant 15: 473–497

Nachmias A, Krikun J (1983) Screening for *Verticillium* resistance: can bioassay in vitro replace field trial? Potato Res 26: 404

Nachmias A, Krikun J (1984) Transmission of *Verticillium dahliae* in potato tubers. Phytopathology 74: 535–537

Ochoa CM (1990) The potatoes of South America: Bolivia. Cambridge University Press, Cambridge, pp 78–83

Platt HW, Sanderson JB (1987) Comparison of inoculation methods for field studies of varietal response to *Verticillium* wilt potatoes. Am Potato J 64: 87–92

Preiszner J, Fehér A, Veisz O, Sutka J, Dudits D (1991) Characterization of morphological variation and cold resistance in interspecific somatic hybrids between potato (*Solanum tuberosum* L.) and *S. brevidens* Phil. Euphytica 57: 37–49

Rich AE (1983) Potato diseases. Academic Press, New York, pp 71–78

Rossignol-Bancilhon L, Rossignol M, Ducreux G, Nozeran R, Darpas A (1984) Analyse de la variabilité d'individus néoformés in vitro à partir de cals chez la pomme de terre (*Solanum tuberosum* L. var. BF15). Bull Soc Fr Bot 131: 171–190

Schaible L, Cammon OS, Waddoups V (1951) Inheritance of resistance to *Verticillium* wilt in a tomato cross. Phytopathology 41: 986–990

Schieder O, Kohn H (1986) Protoplast fusion and generation of somatic hybrids. In: Vasil IK (ed) Cell culture and somatic cell genetics of plants, vol 3. Academic Press, New York, pp 569–588

Serraf I, Sihachakr D, Ducreux G, Brown SC, Allot M, Barghi N, Rossignol L (1991) Interspecific somatic hybridization in potato by protoplast electrofusion. Plant Sci 76: 115–126

Serraf I, Tizroutine S, Chaput MH, Allot M, Mussio I, Sihachakr D, Rossignol L, Ducreux G (1994) Production and characterization of intergeneric somatic hybrids through protoplast electrofusion between potato (*Solanum tuberosum* L.) and *Lycopersicon pennellii* Corr. Plant Cell Tissue Organ Cult 37: 137–144

Shepard JF, Bidney D, Shahin E (1980) Potato protoplasts in crop improvement. Science 208: 17–24

Sihachakr D, Haicour R, Serraf I, Barrientos E, Herbreteau C, Ducreux G, Rossignol L, Souvannavong V (1988) Electrofusion for the production of somatic hybrid plants of *Solanum melongena* L. and *Solanum khasianum* C. V. Clark. Plant Sci 57: 215–223

Sihachakr D, Haicour R, Chaput MH, Barrientos E, Ducreux G, Rossignol L (1989) Somatic hybrid plants produced by electrofusion between *Solanum melongena* L. and *Solanum torvum* S. W. Theor Appl Genet 77: 1–6

Sihachakr D, Vedel F, Ducreux G (1991) Somatic hybridization by protoplast electrofusion: interest of the Solanaceae. In: de Kouchkovsky Y (ed) Plant sciences today. INRA Publ, Paris, pp 285–286

Sihachakr D, Daunay MC, Serraf I, Chaput MH, Mussio I, Haïcour R, Rossignol L, Ducreux G (1994) Somatic hybridization of eggplant (*Solanum melongena* L.) with its close and wild relatives. In: Bajaj YPS (ed) Biotechnology in agriculture and forestry, vol. 27, Somatic hybridization in crop improvement I. Springer, Berlin Heidelberg New York, pp 255–278

Wastie RL, Rousselle P, Waugh R (1993) A review of techniques for acquiring pest and disease resistance. Proc 12th Triennial Conf EAPR, Paris, France, 18–23 July 1993, pp 57–74

Yamakawa K, Mochizuki H (1978) Studies on the use of disease resistance of wild non-tuberous *Solanum* species in the breeding of eggplants. Abstr no 1674, 20th Congr Hort Meeting, Sydney

Zuba M, Binding H (1989) Isolation and culture of potato protoplasts. In: Bajaj YPS (ed) Biotechnology in agriculture and forestry, vol 8, Plant protoplasts and genetic engineering I. Springer, Berlin Heidelberg, New York, pp 124–146

# I.9 Somaclonal Variation for Salt Tolerance in Tomato and Potato

M. Tal[1]

## 1 Introduction

Currently, genetic improvement of crop tolerance to salinity (and to other prevailing stresses) is considered a major practical alternative for improving agricultural productivity in many arid and semiarid areas in both developed and developing countries. A great effort has been directed toward the development of salt-tolerant crop plants principally through: (1) use of conventional plant breeding (Epstein et al. 1980; Saranga et al. 1992) as well as by more modern molecular techniques (Winicov 1994), both involving the transfer of genes from salt-tolerant plants into the relatively more sensitive ones; (2) use of variability existing or produced in tissue and cell culture.

The unique advantages, the main methodologies and accomplishments, and the potential limitations of the in vitro approach have been reviewed extensively (Chandler and Thorpe 1986; Rains et al. 1986; Collin and Dix 1990; Nabors 1990; Tal 1990, 1993, 1994; Dix 1993). The potential causes suggested as an explanation for the lack of or limited success, which has characterized the in vitro approach for a long time, include (discussed by Tal 1990, 1993, 1994): (1) lack, or loss during selection of regeneration capability; (2) the phenomenon of epigenetic adaptation; (3) lack of correlation between the mechanisms of tolerance operating in cultured cells and those operating in the whole plant; (4) multigenicity of salt tolerance. The recent successful production of healthy, fertile, and genetically stable salt-tolerant regenerants in various species (see Dix 1993; Tal 1993, 1994) suggests that some of the limitations can be overcome and that some of them, possibly, do not exist. Reasons for this success were suggested by Tal (1994): (1) the use of explants with high morphogenetic potential which increases the chance of successful regeneration; (2) the possible operation of similar cellular mechanisms of salt tolerance in cultured cells and in cells of the whole plant; (3) the possibility that the change from salt sensitivity to salt tolerance does not necessarily require mutations in many genes; (4) the use of one-step or short-term selection procedures which may prevent the development of epigenetically adapted cells: such cells may obscure the selection of rare mutants with true, i.e., meiotically inherited, tolerance. Genetically, two different kinds of explants have been used for screening for salt tolerance in these experiments (Ilga Winicov, Department of

[1] Department of Life Sciences, Ben Gurion University of the Negev, P.O. Box 653, Beer Sheva 84105, Israel

Biotechnology in Agriculture and Forestry, Vol. 36
Somaclonal Variation in Crop Improvement II (ed. by Y.P.S. Bajaj)

Biochemistry, University of Nevada, Reno, Nevada, USA, pers. comm.): genetically heterogeneous explants, in which no differentiation was made whether the "mutation" preexisted, or occurred during the culturing process (exemplified by Vajrabhaya et al. 1989), and genetically homogeneous explants from a known salt-sensitive genotype which was maintained as a control, while the "mutants" were induced and selected through cellular amplification of the starting stock by the salt treatment (exemplified by Winicov 1991).

The present discussion concentrates mainly on existing data from in vitro selection of salt tolerance as well as on some related aspects in the two important crop plants, tomato and potato. Various aspects of in vitro manipulation, including applications of different types of cultures, micropropagation, transformation, regeneration, and somaclonal variation, have been reviewed in these two species (Sink and Reynolds 1986; Bajaj 1987; Hille et al. 1989; Buiatti and Morpurgo 1990; Karp 1990; Okamura 1994). The use of variability existing or produced in cultured cells is especially important for potato since, as a tetraploid and vegetatively propagated crop, it poses several problems for conventional breeding (Howard 1978), and its modern cultivars originate from a narrow genetic base (Hawkes 1990). Of these two crop plants, tomato has been suggested for use as a model system for the physiological, genetic, and molecular biologists since (Hille et al. 1989): (1) its genome is relatively small (0.74 pg DNA) and contains a high proportion of unique sequences; (2) it seems to be a true diploid ($n = x = 12$); (3) its conventional and molecular genetic maps are relatively saturated; (4) it can be selfed and outcrossed; (5) its cultivars are usually highly homozygous; (6) it has a relatively short generation; (7) its physiological knowledge is extensive; (8) it is economically important ($50 \times 10^6$ tons/year) and represents an economically important family, the Solanaceae; and (9) it is amenable to in vitro manipulation and therefore can be vegetatively propagated in addition to its usual sexual propagation.

## 2 Common Attributes of Tomato and Potato Plants

Both crops, which belong to two different genera of the Solanaceae, have several attributes in common: (1) the general site of origin—the Andean region (tomato from northern Chile to Equador; potato from southern Bolivia to northern Peru) in South America; (2) systematic origin—the series *Juglandifolia* (in subsection potatoe of the section Petota) is a group most closely related and probably ancestral to *Lycopersicon* (Rick 1979); (3) a general molecular resemblance, e.g., most coding regions and single copy DNA sequences of the two genera are highly conserved (Bonierbak et al. 1988); (4) wide global distribution of their cultivars which results from their economical importance; (5) wild relatives which are characterized by diversified gene pools; these pools contain genes for resistance to various biotic and abiotic stresses, including those of drought and salt stresses. Tomato is one of the genera in which extensive physiological and genetic studies of salt tolerance have been performed. The wild relatives of the cultivated tomato which evolved in dry habitats in

South America have been recommended as a source for genes for improving its tolerance to salt stress (Tal 1971; Rush and Epstein 1976; Dehan and Tal 1978; Saranga et al. 1992). The physiological studies reveal that, relative to the control, the growth of the wild plants is less inhibited by salt as compared with the cultivated species. Plants of the wild species *L. cheesmanii* (Rush and Epstein 1981), *L. peruvianum* and *L. pennellii* (Tal 1971; Dehan and Tal 1978; Tal and Shannon 1983) were found to accumulate more Na and Cl ions when grown together with cultivated species under salinity. The better performance of the wild tomatoes as compared with the cultivated ones under NaCl salinity was suggested to be due, at least in part, to a better osmotic adjustment with less sacrifice of growth. A similar physiological response to high salinity was demonstrated recently by plants of *Solanum kurzianum*, a wild relative of the cultivated potato, which, similarly to the wild relatives of the cultivated tomato, is distributed in dry habitats in South America (Sabbah and Tal 1995).

## 3 Regeneration Capability and Somaclonal Variation

The regeneration of salt-tolerant plants from tolerant cell lines and the demonstration of the sexual inheritance of this tolerance are considered as the critical test and ultimate proof for the isolation of a salt-tolerant genetic mutant (Tal 1990). The lack of large-scale plant regeneration ability from salt-tolerant cultures was considered as one of the major problems in assessing the utility of tissue culture for recovering resistant variants (Chandler et al. 1988).

Tomato plants can be propagated from embryos, meristems, cotyledons, and leaves (Buiatti and Morpurgo 1990). However, regeneration of *L. esculentum* is generally a difficult task in the case of callus, and particularly of cultures kept in an undifferentiated state for a long time. Koornneef et al. (1986) obtained a tomato genotype, superior in regenerating plants from cell cultures, by transferring the regeneration capacity from *L. peruvianum* to *L. esculentum* by classical breeding. According to Buiatti and Morpurgo (1990), the extent and nature of genetic variability in tomato tissue culture and regenerants seem to offer a useful tool for breeding purposes. To their knowledge, up to now attempts to utilize this new source of variability have been mainly directed towards obtaining cultivars resistant to biotic and abiotic stresses. Potato is very amenable to regeneration of whole plants from cultured tissues or cells (Karp 1990). In contrast to regeneration systems that do not involve a stage of disorganized cell growth and therefore are normally associated with genetic stability, regeneration which involves a callus phase can be associated with somaclonal variation which can be utilized for potato improvement.

## 4 The Phenomenon of Epigenetic Adaptation

The term epigenetic adaptation will be used here to define the epigenetic alterations in salt tolerance that are inherited only through mitosis and not through meiosis. It has been suggested that adaptation can be developed in cells exposed gradually to increasing salt stress and therefore can be prevented by exposing the cells directly to sublethal salt concentrations (McHughen and Swartz 1984; Sumaryati et al. 1992). The understanding of adaptation and its control may enable a distinction between mutant and adapted cells and thus may help to improve the efficiency of selection of salt-tolerant mutant cells (Tal 1993). Based on a number of studies of DNA methylation in cultured cells, Ball (1990) suggested that an altered methylation pattern can provide an explanation for epigenetic changes that are sometimes observed in tissue or cell culture. The possibility that adaptation, which is an epigenetic phenomenon, depends on changes in the pattern of DNA methylation (which is also, at least partially, epigenetic) that affects gene expression, is being studied in our laboratory (Sabbah et al. 1995). A higher level of 5-methylcytosine was found in the DNA (including gross DNA as well as that isolated from DNase I-sensitive chromatin) of NaCl-adapted cultured cells of potato as compared with that of control cells. The difference between the two cell types was much more noticeable in the DNA isolated from the DNase I-sensitive chromatin, which presumably represents regions with (relatively to insensitive chromatin) more active genes (Spiker 1985). Additional findings include a much greater decrease in methylation by treating the cells with the methylation inhibitor 5-azacytidine in the adapted than in the control cells, and dry weight, which was similar in control and adapted cells in standard media, was affected (a decrease) by the inhibitor only in the control cells. In order to verify the possibility that adaptation to NaCl and methylation of DNA are causally related, DNA methylation is being studied in sequences of specific salt-responsive genes, using comparative digestion of DNA with methylation-sensitive (isoschizomers) enzymes.

## 5 Do the Mechanisms of Tolerance Operating in Cultured Cells Correlate with Those of the Whole Plant?

Attempts to select salt-tolerant plants in culture are based on the assumption that cultured cells represent physiologically, at least in part, the cells of the whole plant, i.e., that cellular mechanisms which are responsible for salt tolerance in cultured cells operate also in the whole plant and contribute to its tolerance response (Tal 1990). Lack of positive correlation between the mechanisms of tolerance operating in isolated cells and those operating in the intact plant has been suggested as one of the main potential causes for the lack or limited success of the in vitro approach (Dracup 1991). A commonly used test for the existence of a correlation between the mechanisms operating in cultured cells and in the whole plant is based usually on the comparison of salt tolerance on both levels. A positive correlation, where the whole plant and the cultured cells are both tolerant or sensitive to salt, is interpreted as an indication for the operation of similar mechanisms at the two levels. Such a positive correlation between calli

(Tal et al. 1978; Perez-Alfocea et al. 1994a, b) or isolated protoplasts (Rosen and Tal 1981) and whole plants was found in tomato. Taleisnik-Gertel et al. (1983) queried whether the positive correlation in tomato characterizes only relatively undifferentiated dividing cells or whether it occurs also for the more differentiated cells in the whole plant. The results obtained supported the former possibility: detached shoot apices of the two salt-tolerant species *L. peruvianum* and *L. pennellii* grew better than those of the cultivated tomato on saline medium. In contrast, no correlation was found between the response of isolated fully differentiated tissues and the response of the whole plant. Disks or intact leaf segments of fully differentiated leaves of the cultivated tomato survived to the same extent or, in some cases, to an even larger extent than those of the wild species. Contrary to tomato, no correlation was found between the response of cultured cells and that of the whole plant to salt stress in potato: callus derived from plants of the salt-tolerant wild species *S. kurzianum* was not more tolerant than that derived from the relatively sensitive cultivated species (Sabbah and Tal 1995). Should these results be interpreted as an indication for the operation of different mechanisms of salt tolerance on both levels in potato? Doubts have been expressed recently on the validity of the comparison used to test for the existence of a correlation between the mechanisms operating in cultured cells and in the whole plant, since cultured cells are to a large extent in a different environment than the cells of the organized multicellular plant (Dracup 1991). According to Dracup (1991), increased levels of salt tolerance in cultured cells do not tend to be expressed in regenerated plants because of the following reasons: (1) many responses to high NaCl are associated with the integrated functioning of the whole plant, rather than being merely cellular responses to high NaCl; (2) the hormonal and nutritional environments differ between cells in culture and in the intact plant, e.g. while salt exclusion at the cellular level is advantageous in vitro, where carbohydrate supply is unlimited, it can be disadvantageous for leaf cells in vivo where carbohydrate supply (which can be used as an intracellular osmoticum instead) is not unlimited, since ions excluded from them may induce a water deficit and loss of turgor (Oertli 1968). It should be added that such comparisons between cultured cells and intact plants are usually made also under different external (including medium, temperature, light and humidity) conditions. However, even if the comparisons between cultured cells and intact plants are made under similar external conditions (which can be done only under in vitro conditions), the question of whether they can be considered a reliable test would remain open, since the contact and therefore the interaction between the medium (which contains hormones and various nutrients in addition to salt) and cells in culture or with cells in the organized intact plant will always remain different. Contrary to the results of Tal et al. (1978) and Perez-Alfocea et al. (1994a, b), Hanson (1984) found in the same plant material a negative correlation between the response of the whole plant and of cultured cells derived from that plant. In agreement with the above argument, different conditions in the two laboratories could be an explanation for the contradictory results. However, the recent successful production of normal salt-tolerant regenerants and the demonstration of sexual inheritance of this tolerance in various species (summarized by Tal 1993,

1994) suggest that, although the conditions where the tolerance of the cultured cells and of the regenerants have been evaluated cannot be identical, at least some of the cellular mechanisms of salt tolerance operating in cultured cells can and should operate also in the regenerated intact plants and contribute to their tolerance.

# 6 In Vitro Selection and Regeneration of Salt-Tolerant Plants

The main selection procedures that have been used in general for the in vitro production of salt-tolerant cells include (see Tal 1990, 1993, 1994): (1) the one-step or short-term selection procedure; (2) the two-step procedure which includes first the selection of osmotically tolerant cells and only then of salt-tolerant ones: since osmotic effect is part of the effect of salt, osmotically adapted cells can be exposed to conditions of high ionic stress without being osmotically shocked (Harms and Oertli 1985); (3) the long-term stepwise selection procedure; (4) the indirect procedure.

## 6.1 One-Step or Short-Term Selection Procedures

Sterile seeds from three land races (*Salvaja*, *Rusa*, and *Especial*) of tomato from the Canary Islands were germinated in culture media lacking hormones, but supplemented with various concentrations of salt up to about 100 mM NaCl (Garcia-Reina et al. 1988a, b). Organogenic calli from cotyledon, leaf, and shoot apex were generated in media containing hormones and the same NaCl concentrations. After additional manipulations, which included six subcultures and an increase in salt concentration by about fourfold, plantlets tolerant to about 256 mM NaCl were obtained. These plantlets, however, showed slow growth and only some developed roots which were fragile and without root hairs. Consequently, they did not succeed in their growth when transferred to soil. According to Garcia-Reina et al. (1988a) by preselecting seedlings on NaCl, and by direct callus formation on selective media, which avoided the use of previously formed calli in a nonselective medium, they selected for (genetic) NaCl tolerance rather than adapting to progressive salt increments and also maintained the morphogenetic potential. Garcia-Reina et al. (1988b) suggested that the tolerance of the three regenerated genotypes was related to a mechanism that controls the internal osmotic potential (by increasing Cl and Na concentrations while preventing the cells from their toxic effect). It should be pointed out, however, that although regeneration capability was maintained in these experiments, the extended maintenance (7 months) of the calli on the saline media might be the reason for the defective root system of the regenerants, and also that the possibility that the salt tolerance of the plantlets results from true mutation rather than epigenetic adaptation was not demonstrated: as mentioned above, such a demonstration should include the persistence of salt tolerance through addi-

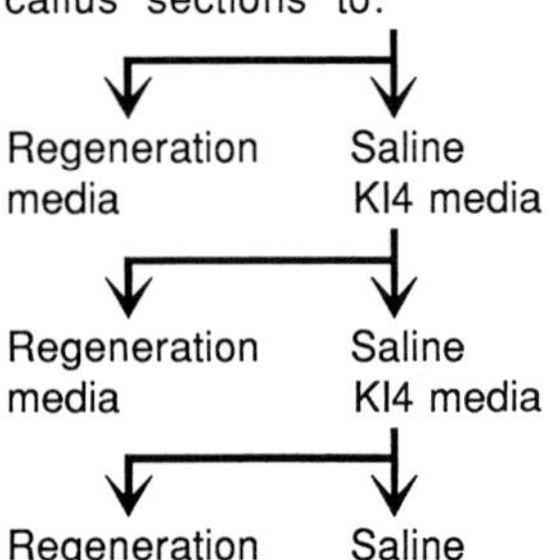


**Fig. 1.** In vitro selection system for NaCl tolerance in the tomato cultivar Improved Pope

tional sexual generations. In another experiment (Zamora 1991), calli induced from leaf segments of aseptically germinated seedlings of the tomato cultivar Improved Pope were subcultured onto media containing sublethal salt (0.30 or 0.35 M NaCl) concentrations (Fig. 1). Pieces of these calli were subsequently subcultured either to fresh saline medium or to a regeneration medium. The best medium found for induction and maintenance of callus was KI4 (MS, Murashige and Skoog 1962, +4 mg/l kinetin +4 mg/l IAA) medium. Regeneration media included either MS medium; MS +1.1 mg/l zeatin; MS +2 mg/l zeatin riboside; MS +2 mg/l BAP + 2 mg/l IAA; or MS + 15 or 20 g/l sucrose. Regenerants ($R_0$) obtained were tested for salt tolerance either under in vitro conditions, in pots irrigated with 0.20 M NaCl solution in the greenhouse, or in a field irrigated with river water in which the salinity level gradually increased during the season. Yield of the latter was found to be either similar to or even higher than that of the control plants. Similar positive results were found in plants of the $R_1$ and $R_2$ selfed sexual generations. Two versions of the one-step selection strategy were used in potato (Sabbah 1994): firstly, cells of cv. Russet Burbank (RB) were spread on filter paper placed on a culture medium containing 400 mM NaCl and covered with feeder cells (cells of cv. Alpha adapted to 200 mM NaCl). The few cells of RB

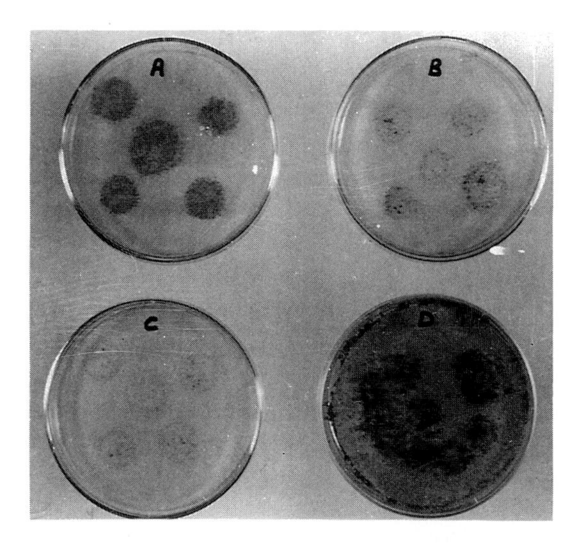


**Fig. 2.** The response of cells of the potato cultivar Russet Burbank to various media. **A** Callus medium: MS mineral salts and myoinositol and (mg/l): thiamine-HCl (10), nicotinic acid (0.5), pyridoxine-HCl (0.5), glycine (2), kinetin (0.5), 2,4-D (5), sucrose 3%, and 1% agar. **B** Callus medium + 300 mM NaCl. **C** Callus medium + 400 mM NaCl. **D** Callus medium + 400 mM NaCl + feeder cells (cells of cv. Alpha adapted to 200 mM NaCl)

remaining alive developed into large calli within 2 to 3 months (Fig. 2). These calli were found to be tolerant to high salinity more than the nonselected control ones. This experiment was not continued because of the unexpected appearance of a bacterial contamination from within the cell. Secondly, calli of cv. RB were exposed directly to a sublethal saline (300 mM NaCl) medium. Calli remaining and developing after this treatment were found to be more tolerant to high salinity (30 mM NaCl) than the control ones. Regenerants of the first ($R_{0-0}$) and the second ($R_{0-1}$) vegetative generations were obtained. The regeneration medium included MS mineral salts, vitamins, and glycine as in the callus medium (Fig. 2), 2 mg/l zeatin riboside, 0.1 mg/l IAA, 1% mannitol, 2% sucrose, and 1% agar. The relative salt tolerance of the regenerants obtained from the salt-selected callus (based on shoot and root dry weight in saline medium, as compared to these weights in control medium) was higher than that of the control regenerants. However, their absolute dry weight (under control as well as under saline conditions) was lower than that of plants produced from standard tubers. At present, attempts are being made to establish an in vitro selection protocol for the production of salt-tolerant potato plants.

## 6.2 Two-Step Selection Procedure

This procedure includes first the adaptation of cells to osmotic stress and only then to salt stress. According to Harms and Oertili (1985), osmotically adapted cells can be exposed to conditions of high ionic stress without being osmotically shocked. Osmotically adapted cells of tomato (Bressan et al. 1981; Handa et al. 1982, 1983) and potato (Sabbah and Tal 1990) showed an increased resistance to salt stress as compared with nonadapted cultures.

### 6.3 Indirect Selection

A nice example for the application of this technique is the development of resistance to an environmental stress in cells selected for proline overproduction. A positive correlation between proline overproduction and salt tolerance in cells selected for resistance to a proline analog or in plants regenerated from these cells was demonstrated in some cases (Tal 1990). One of these demonstrations included potato cells (van Swaaij et al. 1986). Cultured diploid potato cells were selected for resistance to the proline analog hydroxyproline. Sixty-seven hydroxy-proline-resistant ($hyp^r$) cell lines were obtained. Most of the variant lines contained more proline than the wild type. Four of them were found to be also more tolerant to NaCl salinity. One of the latter was regenerated into plants which were found to be both proline overproducers and more tolerant to frost. This phenomenon of cross-resistance, i.e., the resistance to two different stresses (or to another stress), is explained by the existence of common cellular targets or common mechanisms of resistance for different stresses.

### 6.4 Additional Uses of the In Vitro Selection System

Attempts are being made to develop an in vitro technique which will enable the screening of salt-tolerant potato cultivars (Avi Nachmias, Gilat Experimental Station, A.R.O., Israel, pers. comm.). Plantlets are grown in glass tubes containing a mixture of vermiculite: perlite (85:15) together with 75% Hoagland solution, and various concentrations of NaCl (0, 25, 50, and 100 mM). Salt tolerance is estimated by the relative (% of control) decrease in total fresh and dry weights and height. A prerequisite for the application of this technique is that plant response to salinity is positively correlated under in vitro and field conditions. Preliminary in vitro experiments include, therefore, cultivars such as Russet Burbank, Desiree, Nicola, and Alpha, for which salt tolerance under field conditions is already known.

## 7 Summary and Conclusions

In both tomato and potato the improvement of salt tolerance is important because of their wide (tomato) or widening (potato) application in arid or semiarid regions (where salinization in general and the use of brackish water for irrigation tend to increase), as well as in tropical countries characterized by large coastal areas, where the increase in salinity level of the irrigation water is

**Table 1.** Summary of the in vitro work done on salt tolerance in tomato and potato

| Plant species | Explant/ tissue | Salt and conc. (mM) | Results/remarks | Reference |
|---|---|---|---|---|
| Tomato | | | | |
| *L. esculentum*<br>*L. peruvianum*<br>*L. pennellii* | Calli from leaf, stem or root | NaCl (up to 342) | Callus of the two wild salt-tolerant species was more tolerant to NaCl than that of the cultivated species | Tal et al. (1978) |
| *L. esculentum*<br>*L. peruvianum* | Leaf Protoplasts | NaCl (up to 68) | Plating efficiency of protoplasts of the wild salt-tolerant species was higher in saline medium | Rosen and Tal (1981) |
| *L. esculentum*<br>*L. peruvianum*<br>*L. pennellii* | Excised shoot apices and fully developed leaves | NaCl (up to 308) | Unlike fully developed leaves, shoot apices of the two wild salt-tolerant species were more tolerant to NaCl | Taleisnik-Gertal et al. (1983) |
| *L. esculentum*<br>*L. pennellii* | Callus, excised shoot apices and roots | NaCl (up to 205) | Contrary to the results of Tal et al. (1978) and Taleisnik-Gertel et al. (1983), the callus, shoot apex, and root of *L. pennellii* were not more tolerant than those of *L. esculentum* | Hanson (1984) |
| *L. esculentum* | Organogenetic calli from cotyledon, leaf, or shoot apex | NaCl (up to 342) | Plants regenerated from tolerant callus were tolerant to 267 mM NaCl | Garcia-Reina (1988a, b) |
| *L. esculentum* | Callus from leaf | NaCl (up to 350) | $R_0$, $R_1$, and $R_2$ regenerants were found to be salt-tolerant | Zamora (1991) |
| *L. esculentum*<br>*L. pennellii* | Calli from leaf, stem, or root | NaCl (up to 140) | Similar to the results of Tal et al. (1978), callus of the wild species was more tolerant to NaCl | Perez-Alfocea et al. (1994a, b) |

**Table 1.** (*Continued*)

| Plant species | Explant/ tissue | Salt and conc. (mM) | Results/remarks | Reference |
|---|---|---|---|---|
| Potato | | | | |
| *S. tuberosum* | Cell suspension | Hydroxyproline | Few analog-resistant cell lines were found to be NaCl-tolerant | van Swaaij et al. (1986) |
| *S. tuberosum* | Callus from leaf | NaCl (up to 350) | NaCl-adapted callus was obtained by single-step or gradual procedures | Sabbah and Tal (1990) |
| | Cell suspension | NaCl | NaCl-adapted cells were obtained by | |
| | | (up to 100) | – single-step procedure | |
| | | (up to 150) | – gradual procedure | |
| | | (up to 400) | – two-stage procedure—gradual adaptation to 470 mM mannitol, transfer to 150 mM NaCl and gradually to 400 mM NaCl | |
| *S. tuberosum* | Callus from leaves | NaCl (up to 300) | Salt-tolerant callus and $R_0$ plants were obtained | Sabbah (1994) |
| | Suspension cells | NaCl (400) + feeder cells | Salt-tolerant callus was obtained | |
| *S. tuberosum* *S. kurzianum* | Callus from leaves | NaCl (up to 250) | Callus of the salt-tolerant wild species was not more tolerant than that of the cultivated species | Sabbah and Tal (1995) |

a seasonal phenomenon characterizing the dry period of the year. Great effort has been directed in recent years towards the improvement of salt tolerance in crop plants, principally through the use of conventional plant breeding (as well as by more modern molecular techniques) or in vitro techniques. The successful in vitro selection (using one-step or short-term strategies) and regeneration (using highly morphogenetic explants) of healthy, fertile, and genetically stable salt-tolerant plants in several species, together with the fact that both tomato and potato are amenable to various aspects of the in vitro manipulation, both suggest that more can and should be done (starting with highly morphogenetic explants and one-step or short-term selection) in order to apply the variability, existing or produced in tissue or cell culture, for improving salt tolerance (including relative and absolute, vegetative and agronomical yields) in these two important crop plants. The in vitro work already done on salt tolerance in tomato and potato is summarized in Table 1.

## References

Bajaj YPS (1987) Biotechnology in agriculture and forestry, vol 3. Potato. Springer, Berlin Heidelberg New York

Ball SG (1990) Molecular basis of somaclonal variation. In: Bajaj YPS (ed) Biotechnology in agriculture and forestry, vol 11. Somaclonal variation in crop improvement I. Springer Berlin Heidelberg New York, pp 134–152

Bonierbale MW, Plaisted RL, Tanksley SD (1988) RFLP maps based on a common set of clones reveal modes of chromosomal evolution in potato and tomato. Genetics 120: 1095–1103

Bressan RA, Hasegawa PM, Handa AK (1981) Resistance of cultured higher plant cells to polyethylene glycol-induced water stress. Plant Sci Lett 21: 23–30

Buiatti M, Morpurgo R (1990) Somaclonal variation in tomato. In: Bajaj YPS (ed) Biotechnology in agriculture and forestry, vol 11. Somaclonal variation in crop improvement I. Springer Berlin Heidelberg New York, pp 400–415

Chandler SF, Thorpe RA (1986) Variation from plant tissue cultures: biotechnological application to improving salinity tolerance. Biotechnol Adv 4: 117–135

Chandler SF, Pack KY, Pua E-C, Rogolsky E, Mandal BB, Thorpe TA (1988) The effectiveness of selection for salinity tolerance using in vitro shoot cultures. Bot Gaz 149: 166–172

Collin HA, Dix PJ (1990) Culture systems and selection procedures. In: Dix PJ (ed) Plant cell line selection procedures and applications. VCH, Weinheim, pp 3–18

Dehan K, Tal M (1978) Salt tolerance in the wild relatives of the cultivated tomato: response of *Solanum pennellii* to high salinity. Irrig Sci 1: 71–76

Dix PJ (1993) The role of mutant cell lines in studies on environmental stress tolerance: an assessment. Plant J 3: 309–313

Dracup M (1991) Increased salt tolerance of plants through cell culture requires greater understanding of tolerance mechanisms. Aust J Plant Physiol 18: 1–15

Epstein E, Norlyn JD, Rush DW, Kingsbury RW, Kelly DB, Canningham GA, Wrona AF (1980) Saline culture of crops: a genetic approach. Science 210: 399–404

Garcia-Reina G, Moreno V, Luque A (1988a) Selection for NaCl tolerance in cell culture of three canary island tomato and races. I. Recovery of tolerant plantlets from NaCl-tolerant cell strains. J Plant Physiol 133: 1–6

Garcia-Reina G, Moreno V, Luque A (1988b) Selection for NaCl tolerance in cell culture of three canary island tomato and races. II. Inorganic ion content in tolerant calli and somaclones. J Plant Physiol 133: 7–11

Handa AK, Bressan RA, Hasegawa PM (1982) Characterization of cultured cells after prolonged exposure to medium containing polyethylene glycol. Plant Physiol 69: 514–521

Handa AK, Bressan RA, Handa S, Hasegawa PM (1983) Clonal variation for tolerance to polyethylene glycol-induced water stress in cultured tomato cells. Plant Physiol 72: 645–653

Hanson MR (1984) Cell culture and recombinant DNA methods for understanding and improving salt resistance of plants. In: Staples RC, Toenniessen GH (eds) Salinity tolerance in plants. Strategies for crop improvement. Wiley, New York, pp 335–359

Harms CT, Oertli JJ (1985) The use of osmotically adapted cell cultures to study salt tolerance in vitro. J Plant Physiol 120: 29–38

Hawkes JG (1990) The potato. Evolution, biodiversity and genetic resources. Belhaven Press, London

Hille J, Koornneef M, Ramanna MS, Zabel P (1989) Tomato: a crop species amenable to improvement by cellular and molecular methods. Euphytica 42: 1–23

Howard HW (1978) The production of new varieties. In: Harris PM (ed) The potato crop. Chapman and Hall, London, pp 607–646

Karp A (1990) Somaclonal variation in potato. In: Bajaj YPS (ed) Biotechnology in agriculture and forestry, vol 11. Somaclonal variation in crop improvement I. Springer, Berlin Heidelberg New York, pp 379–399

Koornneef M, Hanhart C, Jongsma M, Toma I, Weide R, Zabel P, Hille J (1986) Breeding of a tomato genotype readily accessible to genetic manipulation. Plant Sci 45: 201–208

McHughen A Swartz M (1984) A tissue culture derived salt-tolerant line of flax (*Linum usitatissimum*). J Plant Physiol 117: 109–117

Murashige T, Skoog F (1962) A revised medium for rapid growth and bioassays with tobacco tissue cultures. Physiol Plant 15: 473–497

Nabors MW (1990) Environmental stress resistance. In: Dix PJ (ed) Plant cell line selection procedures and applications. VCH, Weinheim, pp 167–186

Oertli JJ (1968) Extracellular salt accumulation, a possible mechanism of salt injury in plants. Agrochemica 12: 461–469

Okamura M (1994) Pomato: Potato protoplast system and somatic hybridization between potato and wild tomato. In: Bajaj YPS (ed) Biotechnology in agriculture and forestry, vol 27. Somatic hybridization in crop improvement I. Springer, Berlin Heidelberg New York, pp 209–223

Perez-Alfocea F, Guerrier G, Estan MT, Bolarin Maria C (1994a) Comparative salt responses at cell and whole plant levels of cultivated and wild tomato species and their hybrid. J Hortic Sci 69: 639–644

Perez-Alfocea F, Santa-Cruz A, Guerrier G, Bolarin Maria C (1994b) NaCl stress-induced organic solute changes on leaves and calli of *Lycopersicon esculentum*, *L. pennellii* and their interspecific hybrid. J Plant Physiol 143: 106–111

Rains DW, Croughan SS, Croughan TP (1986) Isolation and characterization of mutant cell lines and plants: salt tolerance. In: Vasil IK (ed) Cell culture and somatic cell genetics of plants, vol 3. Academic Press, Orlando, pp 537–547

Rick CM (1979) Biosystematic studies in *Lycopersicon* and closely related species of *Solanum*. In: Hawks JG, Lester RN, Skelding AD (eds) The biology and taxonomy of the Solanaceae. Academic Press, London, pp 667–678

Rosen A, Tal M (1981) Salt tolerance in the wild relatives of the cultivated tomato: responses of protoplasts isolated from leaves of *Lycopersicon esculentum* and *L. peruvianum* plants to NaCl and proline. Z Pflanzenphysiol 102: 91–94

Rush DW, Epstein E (1976) Genotypic response to salinity. Differences between salt sensitive and salt tolerant genotypes of the tomato. Plant Physiol 57: 162–166

Rush DW, Epstein E (1981) Comparative studies on the sodium, potassium and chloride relations of a wild halophytic and a domestic salt-sensitive tomato species. Plant Physiol 68: 1308–1313

Sabbah S (1994) The response of plants and cultured cells of the cultivated potato *Solanum tuberosum* and its wild relative *S. kurzianum* to salt stress. PhD Dissertation (summary in English), Ben Gurion University of the Negev, Beer Sheva, Israel

Sabbah S, Tal M (1990) Development of calli and cells of potato resistant to NaCl and mannitol and their response to stress. Plant Cell Tissue Organ Cult 21: 119–128

Sabbah S, Tal M (1995) Salt tolerance in *Solanum kurzianum* and *S. tuberosum* cvs. Alpha and Russet Burbank. Potato Research (in press)

Sabbah S, Raise M, Tal M (1995) Methylation of DNA in NaCl-adapted cells of potato. Plant Cell Rep 14: 467–470

Saranga Y, Cahaner A, Zamir D, Marani A, Rudich Y (1992) Breeding tomatoes for salt tolerance: inheritance of salt tolerance and related traits in interspecific population. Theor Appl Genet 84: 390–396

Sink KC, Reynolds JF (1986) Tomato (*Lycopersicon esculentum* L.) In: Bajaj YPS (ed) Biotechnology in agriculture and forestry, vol 2. Crops I. Springer, Berlin Heidelberg New York, pp 319–344

Spiker S (1985) Plant chromatin structure. Annu Rev Plant Physiol 36: 235–253

Sumaryati S, Negrutin I, Jacobs M (1992) Characterization and regeneration of salt- and water-stress mutants from protoplast culture of *Nicotiana plumbaginifolia* (Viviani). Theor Appl Genet 83: 613–619

Tal M (1971) Salt tolerance in the wild relatives of the cultivated tomato: responses of *Lycopersicon esculentum, L. peruvianum* and *L. esculentum* minor to NaCl solution. Aust J Agric Res 22: 631–638

Tal M (1990) Somaclonal variation for salt resistance. In: Bajaj YPS (ed) Biotechnology in agriculture and forestry, vol 11. Somaclonal variation in crop improvement I. Springer, Berlin Heidelberg New York, pp 236–257

Tal M (1993) In vitro methodology for increasing salt tolerance in crop plants. Acta Hortic 336: 69–78

Tal M (1994) In vitro selection for salt tolerance in crop plants: theoretical and practical considerations. In Vitro Cell Dev Biol Plant 30P: 175–180

Tal M, Shannon MC (1983) Salt tolerance in the wild relatives of the cultivated tomato: responses of *Lycopersicon esculentum, L. cheesmanii, L. peruvianum, Solanum pennellii* and $F_1$ hybrids to high salinity. Aust J Plant Physiol 10: 109–117

Tal M, Heikin H, Dehan K (1978) Salt tolerance in the wild relatives of the cultivated tomato: responses of callus tissues of *Lycopersicon esculentum, L. peruvianum* and *Solanum pennellii* to high salinity. Z Pflanzenphysiol 86: 231–240

Taleisnik-Gertel E, Tal M, Shannon MC (1983) The response to NaCl of excised fully differentiated and differentiating tissues of the cultivated tomato, *Lycopersicon esculentum*, and its wild relatives, *L. peruvianum* and *Solanum pennellii*. Physiol Plant 59: 659–663

Vajrabhaya M, Thanapaisai T, Vajrabhaya T (1989) Development of salt tolerant lines of KDML and LPT rice cultivars through tissue culture. Plant Cell Rep 8: 411–414

Van Swaaij AC, Jacobson E, Kiel JAK, Feenstra WJ (1986) Selection, characterization and regeneration of hydroxyproline-resistant cell lines of *Solanum tuberosum*: tolerance of NaCl and freezing stress. Physiol Plant 68: 359–366

Winicov I (1991) Characterization of salt tolerant alfalfa (*Medicago sativa* L.) plants regenerated from salt tolerant cell lines. Plant Cell Rep 10: 561–564

Winicov I (1994) Gene expression in relation to salt tolerance. In: Basra AS (ed) Stress-induced gene expression in plants. Harwood Academic Publishers, New York, pp 61–85

Zamora AB (1991) Tissue culture and in vitro selection. In: Final Report to US Agency for International Development (AID), 1 Oct 1986–31 March 1991, Proj No C5-221. Inst Plant Breeding, Univ Philippines at Los Banos, Laguna 4031, Philippines

# I.10 Somaclonal Variation in *Lotus corniculatus* L. (Birdsfoot Trefoil)

M. Niizeki[1]

## 1 Introduction

The genus *Lotus* comprises a heterogeneous assemblage of annual and perennial species numbering close to 200 which are distributed widely throughout the world (Larsen 1958). These species are extremely diverse in morphology and are adapted to a wide range of ecological habitats. The greatest diversity occurs in the Mediterranean basin, which is considered the regional center of origin. A cultivated species, *Lotus corniculatus* L. (birdsfoot trefoil) is a perennial cross-pollinating tetraploid legume ($2n = 4x = 24$; Dawson 1941) and native to Europe and parts of Asia. Nineteenth-century reports show that it grew naturally in many pastures and was good feed for cattle and horses. However, it was not until after 1900 that birdsfoot trefoil was cultivated in Europe (MacDonald 1946).

Birdsfoot trefoil is presently used for pasture, hay, and silage in north central and northeastern United States and eastern Canada. It is fine-stemmed, leafy, slightly decumbent, and does not cause bloat when grazed. It grows on a wide range of soil types and conditions, including moderately alkaline soils, shallow soil, and moderately acid or infertile soil (Rachie and Schmid 1955). The feeding value of birdsfoot trefoil is almost equal to that of *Medicago sativa* L. (alfalfa; Seaney and Henson 1970). In addition to its use as a major pasture legume, either alone or with various grasses, birdsfoot trefoil is used extensively on new highway slopes, for soil improvement, controlling erosion, and beautification (Grant and Marten 1985).

However, birdsfoot trefoil has a number of undesirable characteristics which need to be improved. Among these are lack of seedling vigor, indeterminate flowering, pod shattering (dehiscence), and the presence of hydrocyanic acid (HCN) in the leaves and stems (MacDonald 1946; O'Donoughue and Grant 1988). Modification of undesirable traits, however, has been hampered by the breeding behavior and genetic nature of the species. *Lotus corniculatus* L. is largely an outcrossing species and the characters are mainly inherited tetrasomically (Dawson 1941). In general, there has been increasing interest in the potential use of tissue culture in generating new genetic variability (see Larkin and Scowcroft 1981; Bajaj 1990). The induction of genetic variability for the desirable traits via somaclonal manipulation and its use in breeding programs could be profitable for the improvement of *Lotus corniculatus* L. (Table 1).

[1] Laboratory of Bioscience and Biotechnology, Faculty of Agriculture, Hirosaki University, Hirosaki, Aomori-ken 036, Japan

Biotechnology in Agriculture and Forestry, Vol. 36
Somaclonal Variation in Crop Improvement II (ed. by Y.P.S. Bajaj)

**Table 1.** Summary of the work done on somaclonal variation in *Lotus corniculatus* L.

| Cultivar | Explant or tissue used | Somaclonal variation | Reference |
|---|---|---|---|
| Leo | Internode-derived calli | 2,4-D (2,4-dichlorophenoxyacetic acid)-tolerant calli and regenerated plants | Swanson and Tomes (1980) |
| Leo | Internode-derived calli | 2,4-D-tolerant callus, suspension culture lines, and regenerated plants | Swanson and Tomes (1983) |
| Leo | Internode-derived calli | Chlorophyllous callus | Swanson et al. (1983) |
| Franco | Calli | Agronomical traits such as plant height, dry matter yield, etc. | Damiani et al. (1985) |
| Leo | Hypocotyl-derived calli | Tolerant suspension calli and regenerated plants for 2,4-D and chlorosulfuron (2-chloro-N-[[(4-methoxy-6-methyl-1,3, 5-triazine-2-yl) amino] carbonyl]-benzenesulfonamide) | MacLean and Grant (1987) |
| Leo | Protoplasts of leaves | Morphological characters | Webb et al. (1987) |
| Franco | Leaf-derived calli | Morphological and agronomical traits such as leaflet width and seed yield | Damiani et al. (1990) |
| Viking | Hypocotyl-derived callus | Chromosome structure, agronomical traits and HCN content | Niizeki et al. (1990) |
| Leo | Protoplasts of root hairs | Plant height and stem number | Rasheed et al. (1990) |
| Leo | Hypocotyl-derived calli | Tolerant plants for sulfonylurea herbicide Harmony {DPX-M6316; 3-[[[(4-methoxy-6-methyl-1,3,5-triazine-2-yl) amino] carbonyl] amino] sulfonyl-2-thiophenecarboxylate} | Pofelis et al. (1992) |

## 2 In Vitro Culture Studies

Plant regeneration from birdsfoot trefoil calli was established in an anther culture study by Niizeki and Grant (1971). The plant is easy to regenerate from callus culture through both organogenesis (Swanson and Tomes 1980) and somatic embryogenesis (Mariotti et al. 1984). Plant regeneration from the protoplast-derived calli of birdsfoot trefoil has also been reported (Ahuja et al. 1983). The cultivar Viking produces calli from protoplasts with a high potential for regeneration through adventitious buds (Niizeki and Saito 1986). Therefore, an investigation on somaclonal variation in a large protoplast-derived population could be carried out to determine the stability of such a population (Niizeki et al. 1990). The regeneration capability of *Lotus* cultures has been exploited for the isolation of variants such as those resistant to herbicides, which find application under field conditions (Swanson and Tomes 1980, 1983; MacLean and Grant 1987; Pofelis et al. 1991).

## 3 Somaclonal Variation

### 3.1 Cytogenetical Variation

The calli of *Lotus corniculatus* cv. Viking, were induced from hypocotyls of 10-day-old seedlings on MS medium (Murashige and Skoog 1962) containing 4 mg/l

NAA and 2.5 mg/l kinetin (designated M-1N medium). A solution containing 4% Cellulase Onozuka RS, 1% Macerozyme R-10, 0.2% Pectolyase Y-23, and 0.7M mannitol (pH 5.8) was used to isolate the protoplast from the calli, which were cultured for 2 months. The enzyme was removed by four successive washings with 0.7 M mannitol (pH 5.8) with centrifugation at 80 *g* for 4 min each. The isolated protoplasts ($1 \times 10^4$/ml) were cultured in a thin layer of a KM8p medium (Kao and Michayluk 1975), containing 0.5 mg/l BA instead of zeatin, and no coconut milk. The medium was solidified with 0.6% agar. The colonies derived from single protoplasts could be detected by continuous observation using an inverted microscope. The induced callus lines, all of which were derived from single protoplasts, were again transplanted to the M-1N medium. One of the callus lines that produced numerous shoots was used in this experiment. The shoots were transplanted to the medium of Nitsch and Nitsch (1969) without growth regulators, in order to form complete plantlets. At the time of protoclone acclimatization, a natural population of seed-derived plants was raised by planting seeds in pots.

The regenerated plants which originated from a single protoplast mostly showed 24 chromosomes, the normal tetraploid chromosome number of the species. Among 71 regenerated plants, there was only one octoploid, and one mixoploid plant which had cells with 24 (tetraploid) and 48 (octoploid) chromosomes. No aneuploids were observed and chromosome structural changes were not detected under a light microscope. The ploidy level of the regenerated plants was substantially stable, which suggests that the morphogenetic ability of cells with altered chromosome numbers might be very low. The absence of aneuploids among regenerated plants supports this assumption. The presence of a mixoploid plant could result from endoreduplication during the formation of adventitious buds.

In meiosis of the seed-derived plants, a very small number of PMCs showed abnormalities such as univalents at metaphase I and lagging chromosomes at anaphase I (Table 2). Most of the protoclones, on the other hand, showed a high frequency of meiotic abnormalities, although abnormal somatic chromosomes could not be detected under a light microscope (Fig. 1). It is reasonable to consider that the chromosomal abnormalities resulted from mutation occurring in the course of protoplast or callus culture. The abnormal chromosome set that appeared most frequently at metaphase I generally contained one or two univalents. At diakinesis, asynaptic chromosomes were also occasionally observed. These may have occurred due to deletions or translocations.

In anaphase I and II, bridges and fragments were frequently observed, which may have arisen from a crossover within the inversion. The frequencies of bridges and fragments varied among the protoclones. Besides these chromosome alterations, lagging chromosomes were frequently observed at anaphase I and II. Occurrence of these abnormal chromosome configurations at meiosis seemed to be one of the causes of the decrease of pollen fertility. Indeed, higher frequencies of chromosome abnormalities tended to relate to lower pollen fertility (Table 2). However, no teratogenic phenotypes were observed as a result of these chromosome structural changes. The meiotic analyses in the present study gave evidence for the existence of small structural alterations in the chromosomes of many protoclones. Thus, the presence of somaclonal variants with undetected

**Table 2.** Meiotic configuration and pollen fertility of seed- and protoplast-derived plants. (Niizeki et al. 1990)

| Plant | Metaphase I | | | Anaphase I | | | | | Anaphase II | | | | | Pollen fertility (%) |
|---|---|---|---|---|---|---|---|---|---|---|---|---|---|---|
| | A | B | No. of plants examined | A | C | D | E | No. of plants examined | A | C | D | E | No. of plants examined | |
| Seed-derived | | | | | | | | | | | | | | |
| S10 | 97.5 | 2.1 | 97 | 100 | | | | 50 | 100 | | | | 28 | 98.7 |
| S59 | 100 | | 50 | 96.8 | 3.2 | | | 63 | 100 | | | | 20 | 93.3 |
| Protoplast-derived | | | | | | | | | | | | | | |
| P85 | 100 | | 60 | 93.0 | 4.2 | 2.8 | | 71 | 95.1 | 4.9 | | | 41 | 83.6 |
| P293 | 100 | | 67 | 96.8 | 2.1 | 1.1 | | 94 | 100 | | | | 60 | 78.2 |
| P277 | 96.9 | 3.1 | 32 | 100 | | | | 29 | 100 | | | | 56 | 77.8 |
| P303 | 63.0 | 37.0 | 119 | 79.4 | 20.6 | | | 34 | 76.9 | 23.1 | | | 104 | 72.5 |
| P285 | 66.7 | 33.3 | 24 | 83.8 | 16.2 | | | 31 | 74.1 | 25.1 | | | 27 | 77.3 |
| P32 | 58.3 | 41.7 | 24 | 76.5 | 23.5 | | | 51 | 62.5 | 37.5 | | | 32 | 76.8 |
| P298 | 64.7 | 35.3 | 34 | 82.5 | 17.5 | | | 57 | 75.5 | 24.5 | | | 49 | 76.0 |
| P308 | 59.7 | 40.3 | 67 | 74.0 | 19.0 | 4.0 | 3.0 | 100 | 67.6 | 18.9 | 9.5 | 4.0 | 74 | 72.0 |
| P316 | 57.1 | 42.9 | 49 | 47.4 | 42.1 | 7.9 | 2.6 | 38 | 53.5 | 34.9 | 6.9 | 4.7 | 43 | 64.1 |
| P306 | 38.9 | 61.1 | 36 | 29.8 | 63.8 | 6.4 | | 47 | 7.1 | 71.5 | 14.3 | 7.1 | 28 | 58.2 |
| P296 | 90.7 | 9.3 | 54 | 82.1 | 10.3 | 5.1 | 2.5 | 39 | 84.9 | 15.1 | | | 53 | 70.9 |
| P241 | 77.0 | 23.0 | 61 | 70.1 | 25.4 | 4.5 | | 67 | 71.1 | 28.9 | | | 45 | 53.5 |
| P312 | 55.6 | 44.4 | 36 | 60.9 | 34.8 | | 4.3 | 23 | 11.1 | 88.9 | | | 36 | 68.1 |
| P297 | 67.5 | 32.5 | 83 | 68.6 | 26.9 | 4.5 | | 67 | 69.7 | 30.3 | | | 33 | 69.8 |
| P318 | 41.7 | 58.3 | 60 | 57.9 | 36.8 | | 5.3 | 57 | 44.4 | 44.4 | 3.8 | 7.4 | 27 | 70.6 |
| P299 | 77.2 | 22.8 | 57 | 81.2 | 14.5 | 4.3 | | 69 | 79.5 | 15.9 | 4.6 | | 44 | 69.7 |

A Normal (%). B Univalent (%). C Fragment or lagging chromosome (%). D Bridge (%). E Bridge and fragment or lagging chromosome (%).

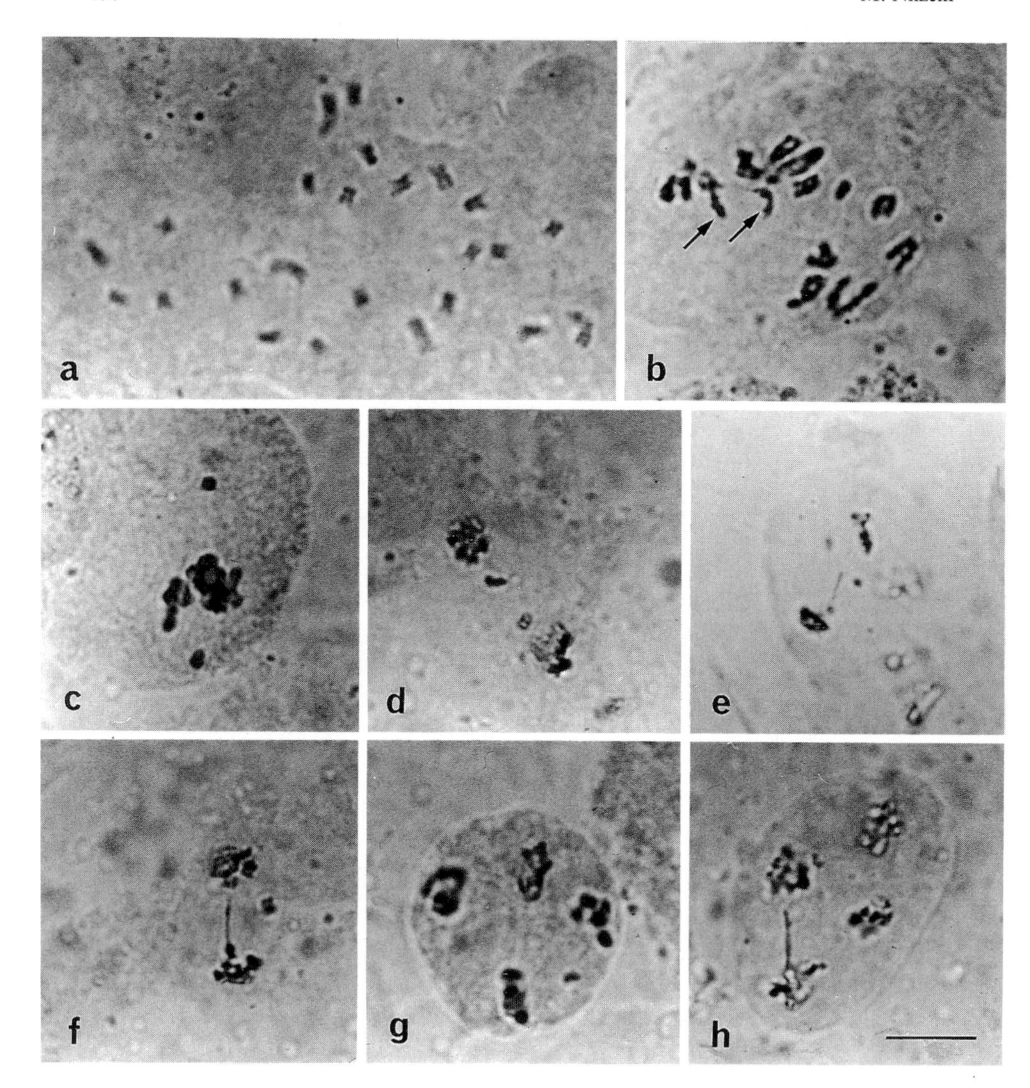

Fig. 1a-h. Mitosis and meiosis in the protoclonal plants. **a** Somatic chromosomes at metaphase of a protoclone in which no chromosome structural changes could be detected, although irregular chromosome configurations were observed in meiosis. **b** Two univalents at diakinesis (indicated by *arrows*). **c** Two univalents at metaphase I. **d** Two lagging chromosomes at anaphase I. **e** A bridge and an undivided lagging bivalent at anaphase I. **g** A lagging chromosome at anaphase II. **h** A bridge at anaphase II. *Bar* 10 µm. (Niizeki et al. 1990)

chromosome structural changes is suspected in the protoclones of *Lotus corniculatus* L.

Generally, there was a large decrease in chromosome abnormalities such as univalents, lagging chromosomes, fragments, and bridges at meiosis in the protoclones of two generations ($P_2$ and $P_3$) after open pollinations of the regenerated protoclones ($P_1$) (Table 3). This may have been caused by the

**Table 3.** Meiotic configurations and pollen fertility of $P_1$, $P_2$, and $P_3$ population (Unpubl. data)

| Plant | Metaphase I | | | Anaphase I | | | | Anaphase II | | | Pollen fertility (%) |
|---|---|---|---|---|---|---|---|---|---|---|---|
| | A | B | No. of cells examined | A | C | D | No. of cells examined | A | C | No. of cells examined | |
| P85 ($P_1$) | 100 | | 60 | 93.0 | 4.2 | 2.8 | 71 | 95.1 | 4.9 | 41 | 83.6 |
| P85 -1($P_2$) | 100 | | 52 | 100 | | | 32 | 100 | | 55 | 96.5 |
| -2($P_2$) | 95.9 | 4.1 | 49 | 94.7 | 4.0 | 1.3 | 75 | 100 | | 56 | 95.9 |
| -1($P_3$) | 96.0 | 4.0 | 49 | 100 | | | 36 | 100 | | 22 | 96.4 |
| -2($P_3$) | 100 | | 21 | 100 | | | 31 | 100 | | 23 | 93.5 |
| P293 ($P_1$) | 100 | | 67 | 96.8 | 2.1 | 1.1 | 94 | 100 | | 60 | 78.0 |
| P293-1($P_2$) | 100 | | 62 | 97.5 | 2.5 | | 82 | 100 | | 76 | 97.1 |
| -2($P_2$) | 95.0 | 5.0 | 40 | 90.0 | 10.0 | | 50 | 100 | | 60 | 94.3 |
| -1($P_3$) | 95.7 | 4.3 | 46 | 95.4 | 4.6 | | 43 | 100 | | 45 | 95.9 |
| -2($P_3$) | 100 | | 47 | 100 | | | 53 | 100 | | 62 | 95.7 |
| P285 ($P_1$) | 66.7 | 33.3 | 24 | 83.8 | 16.2 | | 31 | 74.1 | 25.1 | 27 | 77.3 |
| P285-1($P_2$) | 100 | | 44 | 92.8 | 7.2 | | 42 | 100 | | 60 | 95.3 |
| -2($P_2$) | 80.0 | 20.0 | 55 | 93.3 | 6.7 | | 60 | 90.9 | 9.1 | 44 | 86.6 |
| -1($P_3$) | 100 | | 33 | 100 | | | 24 | 100 | | 28 | 96.0 |
| -2($P_3$) | 94.4 | 5.6 | 71 | 87.8 | 12.2 | 0.5 | 57 | 100 | | 17 | 94.5 |
| P241 ($P_1$) | 77.0 | 23.0 | 61 | 70.1 | 25.4 | 4.5 | 67 | 71.1 | 28.9 | 45 | 53.5 |
| P241-1($P_2$) | 77.4 | 22.6 | 71 | 94.9 | 3.4 | | 59 | 83.3 | 16.7 | 60 | 73.9 |
| -2($P_2$) | 80.0 | 20.0 | 75 | 87.8 | 12.2 | | 41 | 89.4 | 10.6 | 38 | 92.6 |
| -3($P_2$) | 85.3 | 14.7 | 75 | 96.8 | 3.2 | | 62 | 91.2 | 8.8 | 34 | 88.3 |
| -1($P_3$) | 95.5 | 4.5 | 44 | 100 | | | 50 | 100 | | 16 | 88.2 |
| -2($P_3$) | 92.9 | 7.1 | 19 | 94.8 | 5.2 | | 19 | 92.9 | 7.1 | 14 | 86.8 |
| -3($P_3$) | 93.9 | 6.1 | 23 | 100 | | | 37 | 100 | | 40 | 95.0 |

$P_1$ Population of the first protoclones.
$P_2$ Population after open pollination of $P_1$.
$P_3$ Population after open pollination of $P_2$.
A Normal (%). B Univalent (%). C Fragment or lagging chromosome (%). D Bridge (%).

elimination of gametes with abnormal chromosome configurations. Indeed, it was observed that pollen fertility increased drastically in the $P_2$ and $P_3$ generations.

### 3.2 Morphological and Physiological Traits

At the first flowering in the 2nd year, the following characters were evaluated for 60 plants selected randomly from each population consisting of 64 seed- and 71 protoplast-derived plants: (1) plant height, (2) length of the longest internode, (3) diameter of the longest stem, (4) length and width of the leaflet, (5) dry matter yield, (6) pollen fertility, and (7) hydrocyanic acid (HCN) content. HCN content was analyzed according to the procedure of Grant and Sidhu (1967).

Figure 2 shows the frequency and distribution of each trait. The octoploid and mixoploid plants were excluded. The distribution of each character of the protoclonal population is shifted towards values lower than those of the seed-derived population. Mean values of all traits of the protoclonal population were lower than those of the seed-derived population.

The standard deviations of the protoclonal population were also generally smaller than those of the seed-derived population. Protoplast-derived variants with higher values than the highest value of the seed-derived population were not found for any of the traits. However, in all traits except pollen fertility, there were a considerable number of plants with values exceeding those of the initial plant that provided callus for protoplast isolation. With regard to pollen fertility, the protoclonal population was extremely low, and no plant had higher pollen fertility than the plant initially used for protoplast isolation.

The distribution of HCN content in the seed-derived population ranged from the highest to the lowest (Table 4). On the other hand, in the protoclonal population, which was derived from a plant with the highest HCN content, the distribution was limited, although some variants with low HCN content were observed.

In an extensive study of *L. corniculatus*, Damiani et al. (1985) indicated that somaclonal variation took place in callus-derived plants. Likewise, our results showed somaclonal variation among regenerated plants. Most traits of the plant used for protoplast isolation were near the means of the seed-derived population. However, the presence of regenerated plants showing better performance of the traits than the initial plant suggests that tissue culture can produce useful variants, which may be useful for breeding programs for this crop. Moreover, since the induced variants of agronomic traits are genetical (Flashman 1982), it may be possible to produce a superior plant by using other plants which have higher values for the agronomic traits than the mean of the seed-derived population.

In the present material, the observed chromosomal alterations are not accompanied by any drastic phenotypic changes such as teratogenic morphology. A similar case is reported in potato by Secor and Shepard (1981). The quantitative alteration of various characters observed in these studies is probably caused by polygenic changes as well as minor structural alterations in somatic chromosomes. By using a single protoplast-derived population, the present investigation demonstrated that chromosome structural alterations and polygenic changes have occurred during the course of protoplast and callus culture.

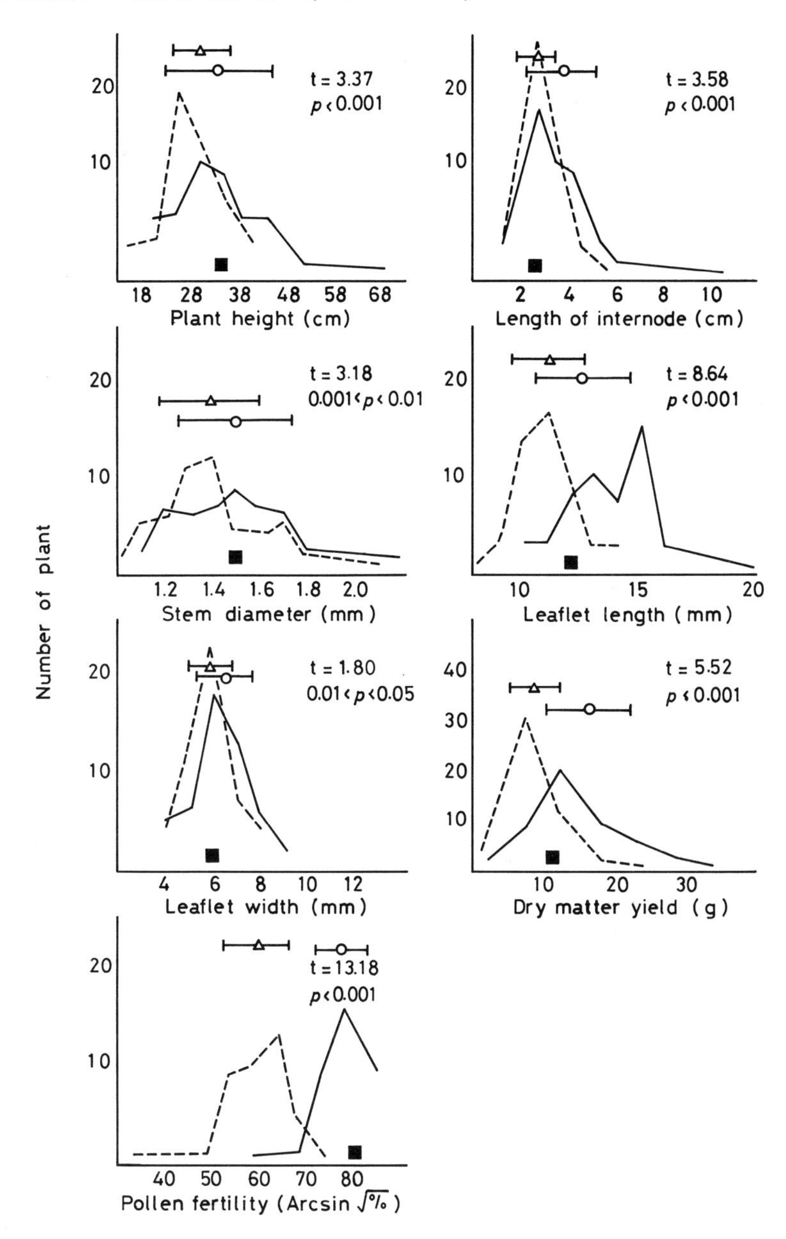


**Fig. 2.** Variation in seven traits of seed-derived (–) and single protoplast-derived (---) populations. ○ and Δ mean value and standard deviation of seed- and protoplast-derived populations, respectively; ■ the value of the plant initially used for protoplast isolation, (Niizeki et al. 1990)

## 3.3 Variation in DNA

*Mitochondrial DNA.* In a preliminary experiment, the mtDNAs of seven callus lines derived from different individual plants were examined. Extracted mtDNAs were digested by a restriction endonuclease, *Eco*RI, and hybridized with three

**Table 4.** Number of plants having HCN graded into five concentrations from – (minimum) to ++++ (maximum). (Niizeki et al. 1990)

| Grade of HCN concentration | Seed-derived population | Single protoplast-derived population |
|---|---|---|
| – | 8 (13.3%) | 0 (0%) |
| + | 11 (18.3%) | 1 (1.7%) |
| ++ | 13 (21.7%) | 5 (8.3%) |
| +++ | 18 (30.0%) | 19 (31.7%) |
| ++++ | 10 (16.6%) | 35 (58.3%) |
| Total | 60 | 60 |

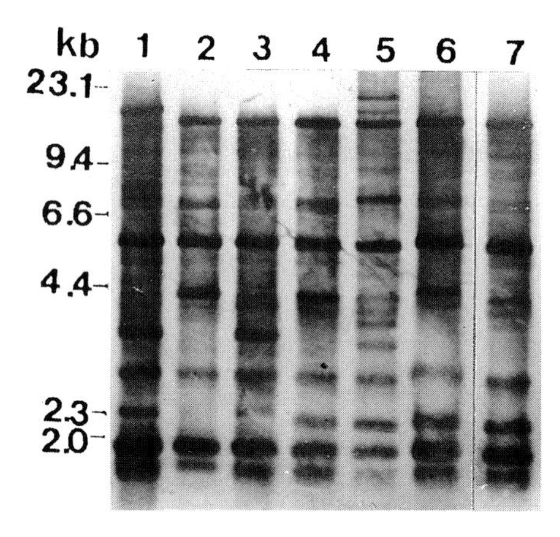


**Fig. 3.** mtDNAs of seven callus lines derived from different individual plants were digested by *Eco*RI and hybridized with mixed probes of *atpA*, *rrn26*, and *coxI*. (unpubl. data)

mixed probes, *atpA*, *rrn26*, and *coxI*. As a result, many different fragment patterns were observed among the callus lines (Fig. 3).

Specific fragments of certain calli may be explained by the polymorphism of mtDNA, because birdsfoot trefoil is a cross-pollinating species and cv. Viking is a synthetic cultivar. However, the possibility that some of the specific fragments may have originated from mtDNA mutation which occurred during callus culture cannot be ruled out.

The mtDNAs of two individual plants derived from seeds and five protoclonal plants in a single protoplast-derived population were analyzed. One gram of leaves was sampled and extracted total DNA was analyzed by the CsCI-EtBr method. Six kinds of restriction endonuclease and 8 mitochondrial genes were used as probes (Table 5). Seed-derived plants, S-3 and 0–61, showed different hybridized fragment patterns in 13% of all the combinations of restriction endonucleases and mtDNA probes (Fig. 4). This result clearly suggests that there is a polymorphism in the mtDNA of birdsfoot trefoil populations derived from seeds. On the other hand, most of the fragment patterns of protoclonal plants analyzed by Southern blotting were the same as those of plant S-3 derived from a seed. Two (4%) of all the combinations of restriction endonucleases and probes showed fragment variations. However, all of the varied fragments were identical to the fragments of plant 0–61.

**Table 5.** Southern blotting of mtDNAs of plants in cv. Viking and protoclonal population. (Unpubl. data)

| RE \ Probe | *rrn26* | *rrn18* | *coxI* | *coxII* | *coxIII* | *Nad1* | *atpA* | *atp9* |
|---|---|---|---|---|---|---|---|---|
| *Eco* RI | ND | ND | S.P | ND | ND | ND | S | ND |
| *Bam*HI | ND | ND | ND | ND | ND | ND | ND | ND |
| *Hind* III | ND | ND | ND | ND | ND | ND | S | ND |
| *Pst* I | ND | ND | ND | ND | ND | ND | S | ND |
| *Sma* I | ND | ND | ND | ND | ND | ND | ND | ND |
| *Sal* I | ND | ND | S.P | ND | S | ND | ND | ND |

S Southern blotting patterns of mtDNAs of two seed-derived plants in cv. Viking are different. P; Protoclonal plants are different from the pattern of seed-derived plants. ND No difference between two seed-derived plants and protoclonal plants. RE Restriction endonuclease.

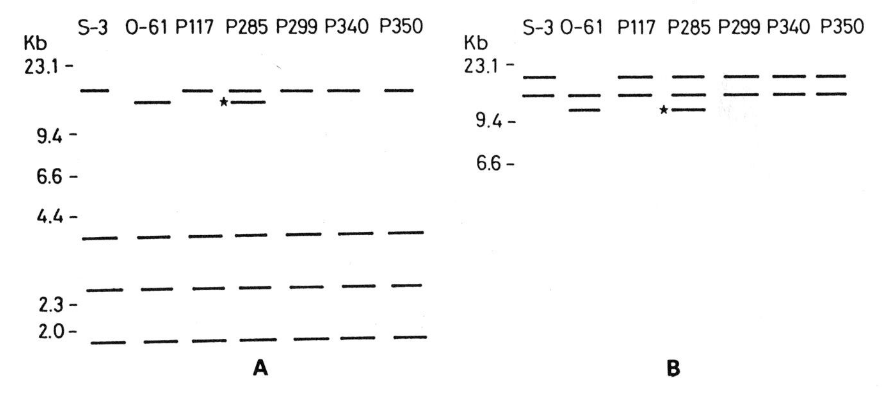


**Fig. 4A, B.** Southern blots of mtDNAs of seed-derived and a single protoplast-derived plants. *S-3* and *0–61* Plants derived from seeds; *P117*, *P285*, *P299*, *P340*, and *P350* protoclones; * variant fragment. **A** and **B** Restriction enzymes are *Eco*RI and *Sal*I, respectively, and probe is *coxI*. (unpubl. data)

These observations indicate that variations may occur frequently at the same location of the mitochondrial genome or else the variations may occur in constant directions. The low occurrence of variation of mtDNA may be caused by a loss of totipotency in callus cells in which mtDNA mutation occurred.

*Chloroplast DNA.* The Southern blots of cpDNA in nine protoclonal plants and ten seed-derived plants were carried out in combinations of three restriction enzymes and five probes of genomic DNA. Specific fragments were found for only one protoclonal plant and one seed-derived plant (Fig. 5). This result indicates that the polymorphism of cpDNA and the occurrence of somaclonal variation in the cpDNA are quite low.

*Nuclear DNA.* The nuclear DNAs were also analyzed by using two restriction enzymes and the probes of major genes, a small subunit of RuBisCO,

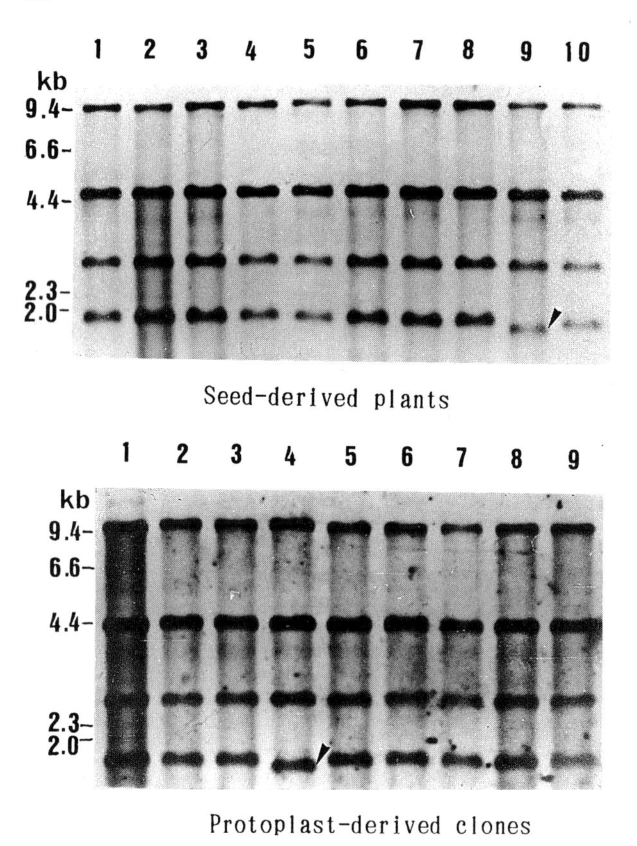


**Fig. 5.** Southern blots of cpDNA in nine protoclones (*below*) and ten seed-derived plants (*above*). Restriction enzyme, *Eco*RI. Probe, tobacco cpDNA *Bam*HI clone, B1. *Arrowhead* indicates variant fragment. (Unpubl. data)

phenylalanine ammonia-lyase, and ribosomal DNA. From this experiment, it was found that these genes were very stable without any variations (Fig. 6). However, there was a considerable amount of variation in six quantitative characters such as plant height and stem diameter, which may be regulated by polygenes.

## 4 Summary and Conclusions

Seventy-one plants, most of which were tetraploid, were regenerated from calli derived from a single protoplast of *Lotus corniculatus* L., cv. Viking. The results concerning variation in the protoclonal population are as follows.

1. The Southern blot of the mitochondrial genome using mitochondrial genes showed that some novel fragments were found among the protoclones. However,

**Fig. 6.** Southern blots of nuclear DNA in nine protoclones and six seed-derived plants (above). Restriction enzyme, *Bam*HI. Probe, ribosomal DNA of pRR217. (unpubl. data)

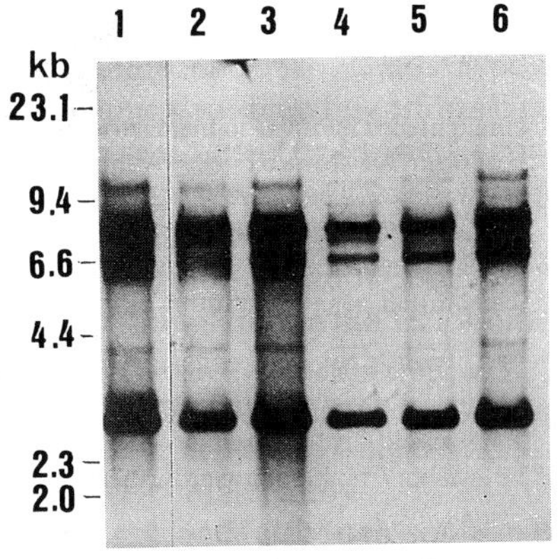


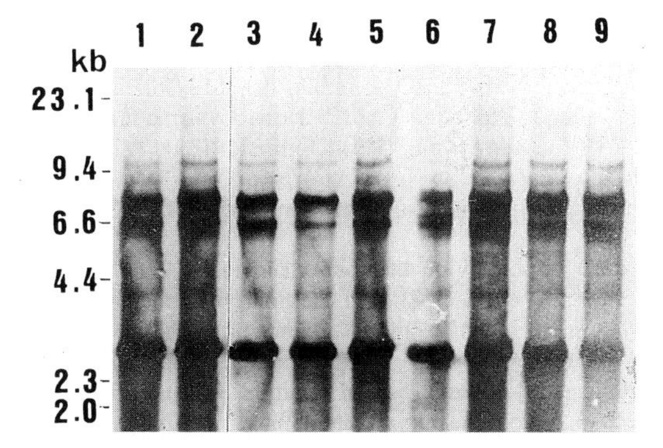


the novel fragments were the same as those fragments found in the polymorphism of the seed-derived population.

2. An analysis of the Southern blot of the chloroplast genome by genomic DNAs, showed that there were very few variations among protoclones.

3. In regard to the nuclear genes, there were no variations in Southern blots when using a small subunit of RuBisCO, phenylalanine ammonia-lyase, and ribosomal DNA as probes. However, there was a considerable variation in traits such as plant height and stem diameter, which may be regulated by polygenes.

4. Abnormal meiotic configurations such as univalents, lagging chromosomes, fragments, and bridges were frequently observed in the PMCs and these tended to be related to low pollen fertility. Because very few physiologically and morphologically abnormal plants were found in the protoclonal population, the mutation of major genes is probably very rare and chromosomal aberrations may occur in that part of the heterochromatin which lacks genetic activity. However, the possibility still exists that plants containing mutations in major genes or with chromosomal aberrations in part of the euchromatin are eliminated during acclimatization. Variation in quantitative characters was inherited by the

progeny, and the elimination of abnormal chromosome configurations resulted in the recovery of plants with high pollen fertility. Therefore, the practicality of a breeding program for quantitative characters including seedling vigor and low HCN by using protoclones in *Lotus corniculatus* L. is affirmed and recommended.

## 5 Protocol

*Lotus corniculatus* seeds from a natural population of cv. Viking were germinated on NN medium (Nitsch and Nitsch 1969) without growth regulators. The calli were induced from hypocotyls of 10-day-old seedlings using MS medium (Murashige and Skoog 1962) containing 4 mg/l NAA and 2.5 mg/l kinetin. The same seedlings from which hypocotyls were partly cut were placed again on NN medium without growth regulators, to induce roots, and then transferred to pots. A solution containing 4% Cellulase Onozuka RS, 1% Macerozyme R-10, 0.2% Pectolyase Y-23, and 0.7 M mannitol (pH 5.8) was used for the protoplast isolation. Calli cultured for 2 months were used in this experiment. The enzyme-callus mixture was incubated at 28 °C for 3–4 h in a shaker bath (60 shakes/min). The protoplasts were separated from undigested cell clumps by filtering through eight layers of cheesecloth. The enzyme was removed by four successive washings with 0.7 M mannitol (pH 5.8) with centrifugation at 80 *g* for 4 min each. The isolated protoplasts ($1 \times 10^4$/ml) were cultured in a thin layer of a modified KM8p (Kao and Michayluk 1975), containing 0.5 mg/l BA instead of zeatin, and without coconut milk. The medium was solidified with 0.6% agar. The culture dishes were kept at 25 °C under light. Divided, single protoplasts grew to form cell clusters consisting of 20–30 cells. After 1 month they developed into globular colonies. Colonies derived from single protoplasts could be detected by continuous observation using an inverted microscope. The colonies were transplanted into the medium that was used for the initial callus induction. One of the callus lines that produced numerous shoots was used in this experiment. Their shoots were transplanted to NN medium, without growth regulators, for the formation of complete plantlets. The plantlets were transferred to 15 × 13 cm pots. Thus, a protoclonal population was obtained from a single protoplast. These seed- and protoplast-derived plants were examined for somatic chromosome number and the structure of the root-tip cells. Meiotic analyses were also carried out on PMCs. At the first flowering in the 2nd year, six quantitative characters such as plant height and stem diameter, HCN content, and pollen fertility were analyzed. Total DNAs were isolated from 1–2 *g* of leaves taken from protoclonal and seed-derived plants and digested with six, three, and two restriction enzymes for the analysis of mtDNA, cpDNA, and nuclear DNA, respectively. DNA fragments were electrophoretically separated in 0.8% agarose gel in a TPE buffer (90 mM Tris-phosphate pH 8.0, 2 mM EDTA) and transferred to a nylon membrane (Hybond – N+, Amersham, UK). After alkali fixation, the membrane was washed in 5x ssc (0.75M NaCl, 75 mM sodium citrate) and 2x ssc for 1 min each. For Southern analysis, eight mitochondrial genes (*rrn26*, *rrn18*, *coxI*, *coxII*, *coxIII*, *nad1*, *atpA*, and *atp9*), three genomic DNAs (*P2*, *P10*, and *B1*) and three nuclear genes (a small subunit of RuBisCO, phenylalanine ammonia-lyase, and ribosomal DNA) were used as probes. Probe labeling, Southern hybridization, and signal detection were performed by using the ECL (enhanced chemiluminescence) gene detection system (Amersham, UK).

*Acknowledgment.* The author expresses his thanks to Dr. W. F. Grant (Department of Plant Science, MacDonald Campus of McGill University) for reviewing this manuscript.

## References

Ahuja PS, Hadiuzzaman S, Davey MR, Cocking EC (1983) Prolific plant regeneration from protoplast-derived tissues of *Lotus corniculatus* L. (birdsfoot trefoil). Plant Cell Rep 2: 101–104

Bajaj YPS (1990) Biotechnology in agriculture and forestry, vol 11. Somaclonal variation in crop improvement I. Springer, Berlin Heidelberg New York

Damiani F, Mariotti D, Pezzotti M, Arcioni S (1985) Variation among plants regenerated from tissue culture of *Lotus corniculatus* L. Z Pflanzenzuecht 94: 332–339

Damiani F, Pezzotti M, Arcioni S (1990) Somaclonal variation in *Lotus corniculatus* L. in relation to plant breeding purposes. Euphytica 46: 35–41

Dawson CDR (1941) Tetrasomic inheritance in *Lotus corniculatus* L. J Genet 42: 49–72

Flashman SM (1982) A study of genetic instability in tobacco callus culture. In Fujiwara A (ed) Plant tissue culture 1982. Maruzen Tokyo, pp 411–412

Grant WF, Sidhu BS (1967) Basic chromosome number, cyanogenetic glucoside variation, and geographic distribution of *Lotus* species. Can J Bot 45: 639–647

Grant WF, Marten GC (1985) Birdsfoot trefoil. In: Health ME, Barnes RF, Metcalfs DS (eds) Forages. Iowa State University Press, Ames, pp 98–108

Kao KN, Michayluk MR (1975) Nutritional requirements for growth of *Vicia hajastana* cells and protoplasts at a very low population density in liquid media. Planta 126: 105–110

Larsen K (1958) Cytotaxonomical studies in *Lotus*. IV. Some case of polyploidy. Bot Tidsskr 54: 44–56

Larkin PJ, Scowcroft WR (1981) Somaclonal variation—a novel source of variability from cell cultures for plant improvement. Theor Appl Genet 60: 197–214

MacDonald HA (1946) Birdsfoot trefoil (*Lotus corniculatus* L.)—its characteristics and potentialities as a forage legume. Cornell Agric Exp Sta Mem 261: 1–182

MacLean NL, Grant WF (1987) Evaluation of birdsfoot-trefoil (*Lotus corniculatus*) regenerated plants following in vitro selection for herbicide tolerance. Can J Bot 65: 1275–1280

Marioti D, Pezzotti M, Falistocco E, Arcioni S (1984) Plant regeneration from leaf-derived callus of *Lotus corniculatus* L. cv. Franco. Genet Agrar 38: 219–223

Murashige T, Skoog F (1962) A revised medium for rapid growth and bioassays with tobacco tissue culture. Physiol Plant 15: 473–497

Niizeki M, Grant WF (1971) Callus, plantlet formation and polyploidy from cultured anthers of *Lotus* and *Nicotiana*. Can J Bot 49: 2041–2051

Niizeki M, Saito K (1986) Plant regeneration from protoplasts of birdsfoot trefoil, *Lotus corniculatus* L. Jpn J Breed 36: 177–180

Niizeki M, Ishikawa R, Saito K (1990) Variation in a single protoplast- and seed-derived population of *Lotus corniculatus* L. Theor Appl Genet 80: 732–736

Nitsch JP, Nitsch C (1969) Haploid plants from pollen grains. Science 163: 85–87

O'Donoughue LS, Grant WF (1988) New sources of indehiscence for birdsfoot trefoil (*Lotus corniculatus*, Fabaceae) produced by interspecific hybridization. Genome 30: 459–468

Pofelis S, Le H, Grant WF (1992) The development of sulfonylurea herbicide resistant birdsfoot trefoil (*Lotus corniculatus*) plants from in vitro selection. Theor Appl Genet 83: 480–488

Rachie KO, Schmid AR (1955) Winter-hardiness of birdsfoot strains and varieties. Agron J 47: 155–157

Rasheed JH, Al-Mallah MK, Cocking EC, Davey MR (1990) Root hair protoplasts of *Lotus corniculatus* L. (birdsfoot trefoil) express their totipotency. Plant Cell Rep 8: 565–569

Seaney RR, Henson PR (1970) Birdsfoot trefoil. Adv Agron 22: 119–157

Secor GA, Shepard JF (1981) Variability of protoplast-derived potato clones. Crop Sci 21: 102–105

Swanson EB, Tomes DT (1980) Plant regeneration from cell culture of *Lotus corniculatus* and the selection and characterization of 2,4-D-tolerant cell-lines. Can J Bot 58: 1205–1209

Swanson EB, Tomes DT (1983) Evaluation of birdsfoot trefoil regenerated plants and their progeny after in vitro selection for 2,4-dichlorophenoxyacetic acid. Plant Sci Lett 29: 19–24

Swanson EB, Tomes DT, Hopkins WG (1983) Modifications to callus culture characteristics and plastid differentiation by the formation of an albino callus of *Lotus corniculatus*. Can J Bot 61: 2500–2505

Webb KJ, Woodcock S, Chamberlain DA (1987) Plant regeneration from protoplasts of *Trifolium repens* and *Lotus corniculatus*. Plant Breed 98: 111–118

# I.11 Somaclonal Variation in *Stylosanthes* Species

I.D. GODWIN[1]

## 1 Introduction

### 1.1 The Genus *Stylosanthes*

The genus *Stylosanthes* (x = 10) belongs to the subfamily Papilionoideae, which includes other domesticated genera such as *Arachis* and *Aeschonymone*. Although a few species are found naturally in Africa, most are endemic to Latin America, where a number of species are important forage legumes. Due to their adaptation to a wide range of tropical and subtropical environments, members of the genus have been introduced to areas including Africa, the Caribbean, Australia and South-East Asia, where they are important legumes for extensive beef production (Burt et al. 1983).

The most important species worldwide include *S. guianensis* (Stylo), *S. humilis* (Townsville stylo), *S. hamata* (Caribbean stylo), and *S. scabra* (shrubby stylo). The use of *S. humilis*, introduced into Australia in the 1920s, was widespread until anthracnose, the fungal disease caused by *Colletotrichum gloeosporoides*, virtually wiped out all introductions. Extensive studies of collections were unable to identify any accessions with natural resistance. *Stylosanthes guianensis* (2n = 2x = 20; Cameron 1967), is a highly variable species, adapted to acid, infertile, P-deficient soils in high rainfall coastal areas (Burt et al. 1983). The species is actually divided into seven distinct taxonomic varieties (Mannetje 1984). The diverse nature of *S. guianensis* has led other taxonomists to treat these varieties as separate species (Costa and Ferreira 1977, 1984). Cultivars have been drawn almost entirely from var. *guianensis*, for use on acid, infertile soils in the high rainfall tropics, with some cultivation of var. *intermedia* (fine-stem stylo), which is better adapted to cooler, subtropical areas (Burt et al. 1983).

Of more importance in Australia and other tropical semiarid areas, however, are the two species most suited to moderately acid soils in the dry to semiarid tropics, *S. hamata* and *S. scabra* (Burt et al., 1974). *Stylosanthes scabra* is an allotetraploid species (2n = 4x = 40), whereas *S. hamata* consists of diploids (2n = 2x = 20) and allotetraploids (2n = 2x = 40). The diploid *S. hamata* types are regarded as one of the diploid progenitors of the tetraploid form, and there is some evidence that the tetraploids should be considered a separate taxon (Stace

[1] Department of Agriculture, The University of Queensland, Brisbane QLD 4072, Australia

Biotechnology in Agriculture and Forestry, Vol. 36
Somaclonal Variation in Crop Improvement II (ed. by Y.P.S. Bajaj)

and Cameron 1988). The value of these short-term perennial species is not just in the provision of nitrogen, but also the fact that they provide quality to the available forage towards the end of the dry season.

## 1.2 Genetic Variation and Breeding Objectives

There are large collections of the genus, most notably at CIAT (International Centre for Tropical Agriculture) in Cali, Colombia and CSIRO (Commonwealth Scientific and Industrial Research Organization) in Brisbane, Australia. These represent a great source of natural variation, and indeed, to date, all existing cultivars are the result of direct introductions from these collections. While important considerations for the release of cultivars have included many agronomic characteristics such as dry matter yield, persistence, seed yield, and drought tolerance, the major objective since the early 1970s has been resistance to the foliar disease anthracnose, caused by the fungus *Colletotrichum gloeosporoides* (*Glomerella cingulata*; Cameron et al. 1984).

There are two quite distinct *Stylosanthes* diseases caused by *C. gloeosporoides*. Type A disease affects *S. scabra*, *S. humilis*, and to a lesser extent, *S. hamata* and *S. guianensis*, while Type B disease affects only *S. guianensis* among the economic species (Irwin and Cameron 1978). Anthracnose can be extremely damaging to *Stylosanthes* pastures and seed crops, particularly after a 2-4-day period of overcast, humid conditions. Regular foliar application of a systemic fungicide will control anthracnose, but this is not economic for broad-scale forage systems, hence genetic resistance represents the only viable option for disease management.

Breeding for resistance has been hampered by the lack of longevity of resistance genes among released cultivars. There are a number of cases in which the resistance has broken down when new cultivars were in prerelease seed increase. The most recent cultivar release in Australia is a *S. scabra* mixture-line, cv. Siran, in an attempt to improve the durability of anthracnose resistance (CSIRO 1990). Other obstacles to conventional breeding are the slow hand pollination techniques and barriers to gene exchange within and between species. In early 1983, a joint UQ-CSIRO project was initiated to induce somaclonal variation (Larkin and Scowcroft 1981) in the *Stylosanthes* genus, in the hope that novel genotypic variation for resistance to anthracnose would be developed. Similar work also commenced at CIAT in Colombia (as reported by Szabados et al. 1986; Miles et al. 1989).

Successful plant regeneration from callus culture had been reported for *S. hamata* (Scowcroft and Adamson 1976). *S. guianensis* (Meijer and Broughton 1981; Mroginski and Kartha 1981) and *S. humilis* (Meijer 1982b), hence it seemed that the generation of somaclonal variants would be achievable. Other agronomic characters, such as yield, phenology, fertility, and growth habit were also monitored. The three species which are of economic importance in Australia, *S. guianensis*, *S. hamata*, and *S. scabra*, were included in the study. The first step in the generation of variants was to test published tissue culture protocols on Australian cultivars of *S. guianensis* and *S. hamata*, or in the case of *S. scabra*, to develop a method of regeneration.

## 2 Tissue Culture and Regeneration Techniques

Cell and tissue culture studies on *Stylosanthes* spp. were earlier reviewed in this Series (Meijer and Szabados 1990). *Stylosanthes hamata* was the first member of the genus to be regenerated from tissue culture (Scowcroft and Adamson 1976). Callus was initiated from seedling explants, encouraged to undergo rhizogenesis, then transferred to a high cytokinin medium for shoot organogenesis. High-frequency regeneration of whole plants was not achieved, however, as there were difficulties in inducing shoots and roots on the same callus.

To gain an accurate assessment of the frequency and scope of somaclonal variation, it is desirable to use a single genotype from each cultivar for explant material. Otherwise, the genetic variation observed may be merely a reflection of the heterogeneity of supposedly inbred material. This precludes the use of seedling explants for a study of any meaningful scope. It is therefore desirable to be able to regenerate explants such as leaves or stems from glasshouse-grown plants. Whole plant regeneration from leaf-derived callus cultures has been reported for *S. guianensis* (Meijer and Broughton 1981; Mroginski and Kartha 1981; Meijer 1982a; Godwin et al. 1987a), *S. humilis* (Meijer 1982b), *S. scabra* (Godwin et al. 1987b) and *S. hamata* (Godwin et al. 1990). In all cases, the frequency of regeneration is genotype-dependent. There are cultivars of *S. guianensis*, *S. scabra*, and *S. hamata* for which 95–100% of leaf-derived calli formed shoots via organogenesis on MS (Murashige and Skoog 1962) supplemented with BAP (Godwin 1987).

The general protocol for regeneration of whole plants from leaf-derived callus cultures of all *Stylosanthes* species employs three steps, viz. callus induction and proliferation, shoot organogenesis, and rhizogenesis on shoots. Callus

**Fig. 1.** Shoot organogenesis on *S. guianensis* cv. Graham calli on MS with 2 mg/l BAP. (Godwin 1987)

induction and proliferation is usually performed on a medium with auxin and cytokinin. For the more amenable genotypes, shoot organogenesis will often occur on the callus proliferation medium, although frequencies are always enhanced by transfer to a medium containing cytokinin only (Godwin 1987; Fig. 1). Morphogenetic potential generally declines with time in vitro (Meijer 1984; Godwin 1987), although some genotypes such as *S. guianensis* cv. Graham have been cultured for periods of 6–12 months without significant decline in regenerative capacity (Godwin 1987). Excised shoots will then readily form roots in the absence of growth regulators, and the addition of any form of auxin is usually deleterious to this procedure (Godwin 1987). Greater detail on medium constituents and recommended growth regulator concentrations are included in Section 6.1.

Dornelas and coworkers (1992) demonstrated regeneration of *S. scabra* via somatic embryogenesis, the only such example within the genus. Using immature cotyledon or zygotic embryo explants, organogenesis was observed on media containing BAP, whereas on media containing 2,4-D only, somatic embryogenesis was obtained. These somatic embryos did not develop into whole plants.

Whole plants of *S. guianensis* were first regenerated by Meijer and Steinbiss (1983) from protoplasts isolated from fine suspension culture cells, but they regenerated at low frequency. This methodology was improved by isolating protoplasts from leaf mesophyll of in vitro-germinated seedlings (Szabados and Roca 1986), and higher regeneration frequencies were obtained by isolating protoplasts from axenic seedling cotyledons (Vieira et al. 1990). Using seedling cotyledons for protoplast isolation also enabled plants of *S. macrocephala* and *S. scabra* to be regenerated (Vieira et al. 1990). In all cases, regeneration from protoplasts was highly genotype-specific, as has been found with all tissue culture methodologies for the *Stylosanthes* genus. Genetic transformation studies have also been successfully conducted on *S. humilis* (see Elliot and Manners 1993).

## 3 Somaclonal Variation

### 3.1 Primary Regenerants from Callus

Primary regenerants ($R_1$ individuals) reveal a number of morphological mutants. Abnormal growth habit, leaf morphology, and plant height have been observed in all species (Godwin et al. 1987a). One of the most common forms of variation is low self-fertility, male sterility, or total sterility (Godwin et al. 1987a; Miles et al. 1989). Only 28.1% of 1024 *S. guianensis* somaclones survived to set seed in one such study (Godwin et al. 1987a). Reasons for this low seed set include the inability of plants to survive the hardening procedure, as well as variation for reduced fertility due to both genetic and physiological factors. This may be associated with gross chromosomal aberrations, changes in chromosome number, or may be epigenetic in nature, possibly in response to the application of plant growth regulators throughout the tissue culture cycle. However, as with other species, it is advisable to assess the extent of somaclonal variation in the selfed progeny. In this way, provided large enough samples are grown, recessive mutations will become apparent. As plants are grown from seed, any physiological effects of growth regulators will be avoided, and it is possible to plant synchronous trials of somaclonal progeny for direct comparison with the parental controls.

### 3.2 Selfed Progenies – Morphological and Agronomic Traits

The most frequent forms of variation seen in *Stylosanthes* somaclonal progenies are reduced fertility, including male or total sterility, dwarfism, aberrant leaf morphologies, and phenology (Godwin et al. 1987a; Miles et al. 1989; Rao et al. 1992). A summary of the range of variation observed may be seen in Table 1. A number of field trials to assess variation among somaclonal progenies of *Stylosanthes* spp. have been reported. The most extensive were grown in 1986 in Australia, and Colombia. In Australia these included 138 *S. guianensis*, 40 *S. scabra, and 38 S. hamata* $R_2$ families (consisting of 18 individuals per family), plus parental controls, a total of 4104 plants (Godwin 1987; Fig. 2). In Colombia, 76 *S. guianensis* $R_2$ families (consisting of 15 individuals per family) were grown, which totalled 840 plants including parental controls.

Among somaclonal families of *S. guianensis* which did set seed, a proportion set seed at a significantly lower rate than the normal parental type. A number of these families had a uniform morphology, and could be characterized by more erect habit, larger leaves (Fig. 3), inflorescences and seeds, reduced fertility, reduced biomass, and a tendency to flower later. Cytological analysis revealed that these were autotetraploids, and these accounted for 7% (Godwin et al. 1987a) or approximately 20% (Miles et al. 1989) of *S. guianensis* families analyzed. The higher incidence of autotetraploidy observed by Miles et al. could be as a result of different in vitro culture conditions or different genotypes used. Autotetraploidy was observed only among somaclones of cv. Graham and CPI 18750, whereas albinism, which segregated as a recessive trait, was observed only among CPI 34911 somaclones, and occurred in 4/37 families derived from this accession (Godwin 1987). It is noteworthy that no autoployploidy was observed among somaclones regenerated from the allotetraploid species, *S. scabra* and *S. hamata*. Autotetraploid *S. guianensis* families did not differ from the parent in total protein or phosphorus contents.

Some of the variation seen in somaclonal families segregated in a Mendelian manner, with the mutant form being recessive, and hence not evident in the primary regenerant. These included traits such as albinism, delayed flowering, reduced cotyledons, wrinkled cotyledons, dwarfism, and tetrafoliate leaves (Fig. 3) in *S. guianensis* (Godwin et al. 1987a), and delayed flowering in *S. scabra* and *S. hamata* (Godwin et al. 1990). Overall, the frequency of morphological/agronomic variants observed accounted for 39.9% of *S. guianensis* families, which was considerably higher than those observed for the allotetraploid *S. hamata* (28.9%) and *S. scabra* (22.5%).

Rao et al. (1992) tested 14 selected *S. guianensis* somaclonal families ($R_4$) for their response to acid soil conditions. Variation was found in carbon partitioning, root biomass production, and nitrogen and phosphorus uptake. Some of the somaclones are being used in continuing experiments to gain an insight into the

**Table 1.** Summary of somaclonal variation studies in the *Stylosanthes* genus

| Species | Explant | Generation | Variants observed | Reference |
|---|---|---|---|---|
| *S. guianensis* | Leaf<br>hypocotyl | Primary | Albinism<br>Loss of apical dominance<br>Leaf and stem morphology | Meijer (1984) |
| *S. guianensis* | Leaf | Primary and $R_2$ | Sterility, male sterility, reduced fertility<br>Leaf and stem morphology<br>Growth habit, dwarfism<br>Tetraploidy (reduced yield, reduced fertility, increased seed size)<br>Albinism Phenology (earlier and later flowering)<br>Dry matter yield reduction | Godwin et al. (1987a) |
| *S. guianensis* | Leaf<br>hypocotyl | Primary and $R_2$ | Tetraploidy<br>Increased anthracnose susceptibility<br>Growth habit<br>Dry matter yield reduction<br>Leaf and flower morphology | Miles et al. (1989) |
| *S. guianensis*<br>*S. hamata*<br>*S. scabra* | Leaf | Primary, $R_2$ and $R_3$ | Increased anthracnose resistance (*scabra* and *guianensis*)<br>Increased anthracnose susceptibility (*guianensis* and *hamata*)<br>Phenology (earlier and later flowering)<br>Sterility<br>Dry matter yield increase (*hamata* only)<br>Dry matter yield decrease | Godwin et al. (1990) |
| *S. guianensis* | Leaf | $R_4$ | Response to acid soil conditions as measured by:<br>Carbon partitioning, root biomass, N and P uptake | Rao et al. (1992) |

**Fig. 2.** Field evaluation of *S. guianensis* somaclonal progenies at Samford in South-East Queensland, planted in January, 1986. The trial consisted of 24 × 60-m rows, a total of 2484 $R_2$ progenies, and 144 parental control individuals. (Godwin 1987)

genetic basis of root production and nutrient uptake. To date, however, there is little evidence that lines with improved agronomic performance on tropical acid soils will be produced using somaclonal variation techniques.

### 3.3 Selfed Progenies – Anthracnose Resistance

*Stylosanthes guianensis* somaclonal progenies have been tested for reaction to Type B anthracnose both in Queensland (Godwin 1987; Godwin et al. 1990) and CIAT (Miles et al. 1989). Miles et al. (1989) reported that 16/70 diploid $R_2$ families were less resistant to the pathogen, and none was significantly more resistant than the parental control. Godwin et al. (1990) found that none of the 264 diploid families differed from the parental controls; however, 8/12 tetraploid families were significantly more resistant. This may be a gene dosage effect, as was reported in hexaploid and octoploid *Medicago sativa* plants with enhanced *Verticillium* resistance (Latunde-Dada and Lucas 1983). However, there is some evidence that the resistance was really a slowing of infection due to greater cuticle thickness. Standard disease testing for anthracnose in *Stylosanthes* relies on visual assessment of damage 10 days after inoculation, at which time the tetraploid families appeared more resistant than the diploid parent. After 14 days, these lines appeared as susceptible as the diploid controls, hence it may be that thicker leaf cuticles and possibly cell walls impeded fungal growth into leaf and stem tissues.

*Stylosanthes scabra* somaclonal progeny from 58 families were tested for resistance to Type A anthracnose. Segregants were isolated from two cv. Fitzroy $R_2$ families with partial resistance, but not total immunity as can be seen with cv.

**Fig. 3.** Leaf morphological variants of **A** *S. scabra* cv. Fitzroy, with tetrafoliate variant and trifoliate parental type; **B** *S. guianensis* cv. Graham *from left to right* normal parental type; recessive variant with lanceolate leaflets; tetraploid variant with wider, slightly wrinkled leaflets. (Godwin 1987)

Seca. These families segregated 17:3 and 14:6 for susceptibility:partial resistance, indicating that the partial resistance may be controlled by a recessive major gene. Selfed $R_3$ progeny from partially resistant individuals of one family were all partially resistant, indicating the trait was highly heritable (Godwin et al. 1990), and it is assumed that the trait was fixed in a homozygous recessive manner. The level of resistance in each family was different, suggesting that they were the result of two different genetic events in vitro. By crossing these families, it may be possible to combine these tolerances to give improved, and perhaps more durable, field resistance. Future work, testing larger numbers of cv. Fitzroy $R_2$ families may identify further sources of resistance.

Sixty-seven *S. hamata* cv. Verano $R_2$ families were tested for resistance to Type A anthracnose, to which the parental control was totally resistant. Five families were significantly more susceptible than the parent, based on the mean. Four of these families were segregating in a 3:1 resistant:susceptible manner, indicating that these had undergone a single gene mutation (Godwin et al. 1990). One $R_2$ family was segregating in a manner which indicated polygenic control of the trait. Single-gene changes in disease resistance have been observed among somaclonal variants and progeny in rice in reaction to *Helminthosporium oryzae* (Ling et al. 1985) and oat in reaction to *Helminthosporium victoriae* and *Puccinia coronata* (Rines and Luke 1985).

## 4 Attempts to Improve Disease Resistance Via In Vitro Selection

Experiments were performed with *S. scabra* calli and Type A *Colletotrichum gloeosporoides*, and *S. guianensis* calli with Type B *C. gloeosporoides*, to test the efficacy of the live pathogen or culture filtrates in eliciting a race-specific response in vitro. This would be valuable information as to whether either selective agent may be useful in an in vitro screening scheme.

Undifferentiated callus cultures (see Sect. 6 for culture conditions) were inoculated with a small amount of hyphae. The extent of hyphal growth and callus fresh weight growth were used to assess differential reaction to the disease. For example, three *S. scabra* genotypes were exposed to races 1 or 4 (type A). On the whole-plant basis, cv. Fitzroy is susceptible to both races, Q10042 is resistant to race 1, and CP193116 is resistant to both races. Results did not clearly demonstrate race-specific responses with either the *S. scabra* x Type A or the *S. guianensis* x Type B interactions. The hyphae quickly overgrew the callus in many cases and spread onto the medium. Benomyl (40 mg/l) was suitable for fungal eradication, and did not adversely affect subsequent regeneration. Nevertheless, it was difficult to envisage how such a system could be usefully modified to operate as an in vitro selection system (Godwin 1987).

Culture filtrates (CF) showed greater promise for in vitro selection purposes, particularly as the selection pressure could be applied more uniformly to the culture, and the culture filtrate would not engulf the medium as a live culture could. Fungal cultures were grown for 5 days in a modified Czapek's (CZ) medium. Filter sterilized CF or CZ was then added to a cooling callus growth medium at 0, 1, 10, 25, or 50% v/v concentration. Actively growing calli were then cultured on these media for 21 days, after which callus fresh weights were measured. For *S. scabra*, culture filtrates of Type A race 1 and race 3 were used with cv. Fitzroy (susceptible to both races) and cv. Seca (resistant to race 1). At 10% CF, there was no significant difference between cv. Fitzroy growth on races 1 and 3. Cultivar Seca had little growth on race 3 CF, but fresh weight increased sixfold on race 1 CF, which correlated well with the whole-plant response (Fig. 4). It was not valid to compare growth of cv. Seca with that of cv. Fitzroy, as the two cultivars behave very differently in culture. On medium containing CZ only, there is a higher level of toxicity than either of the CF treatments, and this may be attributed to the high salt (especially sodium) concentration in CZ, which was depleted in the CF due to the fungal requirement during growth.

A similar experiment was performed with *S. guianensis* x Type B races 1 and 3, using cv. Endeavour (susceptible to both races) and cv. Graham (resistant to race 1 only). Growth of cv. Endeavour callus was inhibited by both CFs at concentrations as low as 1%, whereas cv. Graham grew significantly better on race 1 CF than on race 3 CF (Fig. 5).

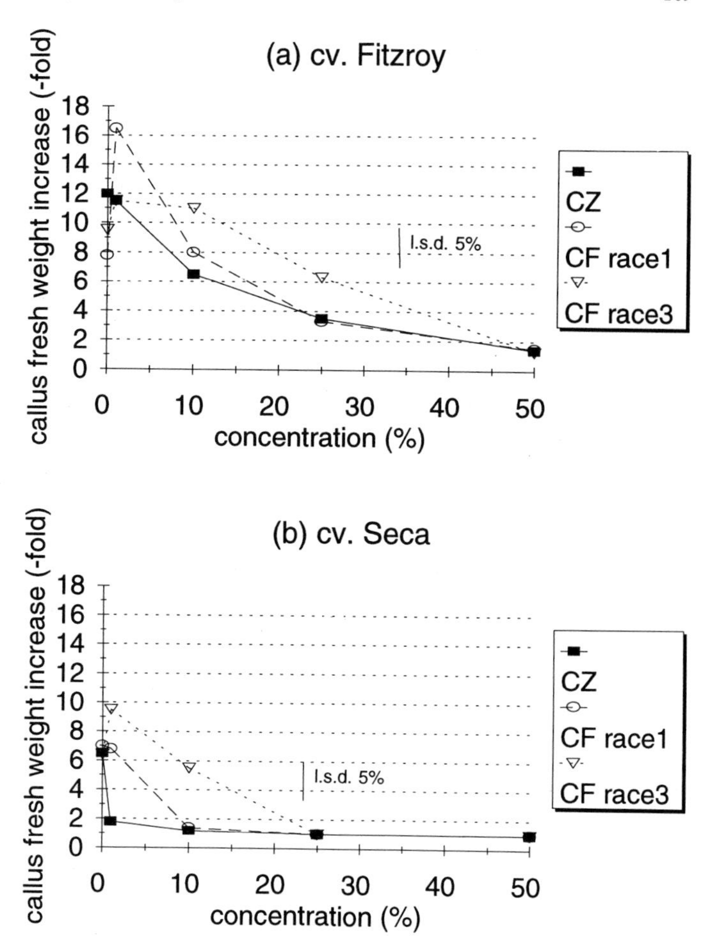


**Fig. 4a, b.** Effect of concentration of Type A anthracnose culture filtrates (*CZ* Control; *CF* race 1; *CF* race 3) on callus fresh weight increase (-fold) over 21 days for *Stylosanthes scabra*: **(a)** cv. Fitzroy and **(b)** cv. Seca. (Godwin 1987)

While there is no published evidence of the involvement of a phytotoxin in the anthracnose x stylo interaction, it can be seen that CFs may have some constituent which could potentially be used as an in vitro selective agent (Godwin 1987).

## 5 Summary and Conclusions

Somaclonal variation can be generated in *Stylosanthes* spp. through a callus culture cycle, or even, it appears, by induction of adventitious shoots directly on leaf explants. In many cases, the genetic basis of this variation has been demonstrated. Useful genetic mutations or alterations, either occurring in vitro, or perhaps preexisting in leaf explants, were observed among somaclonal progenies

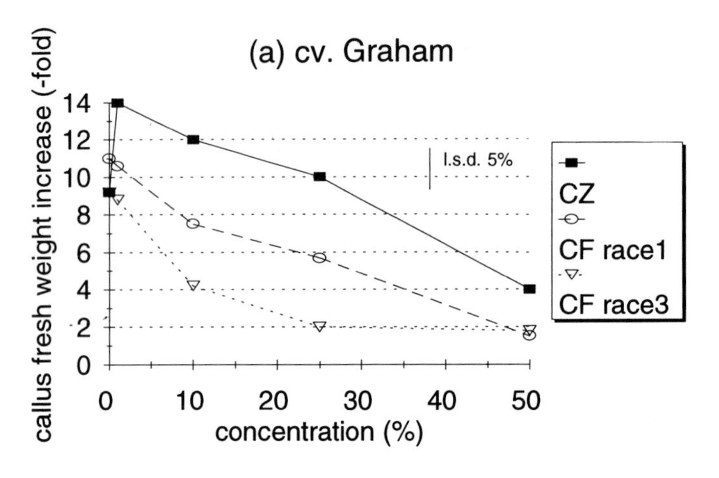


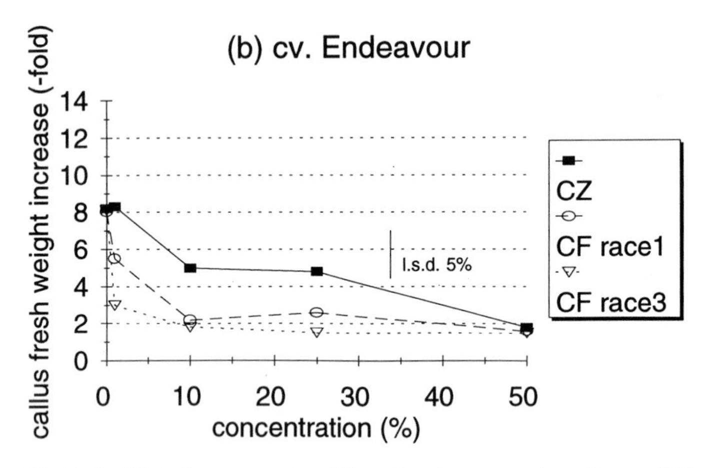


**Fig. 5a, b.** Effect of concentration of Type B anthracnose culture filtrates (*CZ* control; *CF* race 1; *CF* race 3) on callus fresh weight increase (-fold) over 21 days for *Stylosanthes guianensis:* **(a)** cv. Graham and **(b)** cv. Endeavour. (Godwin 1987)

($R_2$ and $R_3$ generations). These included variants with improved dry matter yields, altered growth habit, earlier and later flowering, and increased tolerance to anthracnose. These variants may provide a useful source of genetic variation for existing breeding programs, particularly as, in most instances, variant types retained desirable parental traits and varied in one or two characters only. Analysis of somaclonal variation in the $R_1$ generation does not reveal the full extent of variant types, gives an underestimate of frequency variation, and restricts genetic analysis of variation. By performing experiments on *S. guianensis* $R_2$s (Godwin et al. 1987a), somaclonal variation was observed in 40% of families, twice the frequency reported for the $R_1$ generation (Meijer 1984).

Improved partial resistance to Type A anthracnose was identified in two *S. scabra* families. These have some potential in improving the disease resistance within the species; however, it must be taken into account that other cultivars are already available with equal or greater levels of resistance. Within *S. hamata* somaclones and somaclonal progenies, variants were observed with male sterility, and others with loss of anthracnose resistance. While these characters are not agronomically advantageous, they could be utilized in the *S. hamata* breeding program. Male sterility could be used to implement recurrent breeding scheme as outlined for *S. guianensis* by Miles (1985) and Cameron and Irwin (1987). The anthracnose-susceptible types could be used as susceptible testers in genetic studies, as none of the accessions of *S. hamata* available in Australia is as badly damaged as these somaclonal segregants. These lines could be useful as near-isogenic lines for molecular mapping anthracnose resistance loci, and possibly in eventually cloning such genes.

Within all three species, much of the variation, including dwarfism, albinism, reduced yield, and reduced anthracnose tolerance, diminished the agronomic usefulness of a number of lines. The high levels of somaclonal variation apparent in all three species may create problems for future genetic engineering projects. Transformation protocols have been developed for *S. humilis* (Manners 1988 and Way 1989) but as the most successful system relies on using the hairy root inducing *Agrobacterium rhizogenes*, transformants are morphologically abnormal, and there are no data on how much of this is due to Ri plasmid gene expression and how much is a result of somaclonal variation.

## 6 Protocol for Regeneration of *Stylosanthes* Callus Cultures

Within many species of the *Stylosanthes* genus there are cultivars or accessions which are very easy to regenerate from leaf-derived callus cultures at high frequency. These include *S. guianensis* cv. Graham, *S. scabra* cv. Fitzroy, and *S. hamata* cv. Verano (Godwin 1987). For these genotypes, plant regeneration should be possible on close to 100% of explants. It should be noted that within these species there are also some extremely recalcitrant genotypes.

Donor plants should be grown in controlled environment conditions, with 30 °C day/25 °C night, 14-h photoperiod with regular liquid fertilization. Leaf explants give the best results, and surface sterilization with 1% NaOCl, followed by three sterile water rinses usually yields axenic cultures. MS basal medium (Murashige and Skoog 1962) is suitable for all stages of the regeneration precedure, with 3% sucrose and 0.8% agar. The plant growth regulators will vary depending on the plant species and the stage of culture. Generally, the callus initiation phase requires the use of an auxin and cytokinin, shoot regeneration requires cytokinin only, and roots are formed on shoots in the absence of plant growth regulators. All culturing should be performed at 28 °C, with best results using a 16 h day/8 h night, with 14 $\mu$ mol $m^{-2} s^{-1}$ as minimum light intensity.

Rapidly proliferating callus can be induced and maintained on MS + 1 mg/l 2,4-D + 1 mg/l BAP. This medium is particularly suitable for subsequent shoot regeneration, indeed shoot organogenesis will occur on this medium with the most amenable genotypes. To maintain actively growing callus cultures for a longer term (6 months or more), it is necessary to increase the 2,4-D concentration to 2 mg/l. For high frequency shoot organogenesis, calli should be transferred to MS + 2 mg/l BAP. Shoots can then be induced to form roots on 1/2 MS (1% sucrose) without plant growth regulators.

*Acknowledgments.* The author acknowledges the financial support of the UQ-CSIRO Collaborative Research Fund, and the helpful comments of Drs. S.W. Adkins and C. Rathus regarding this manuscript.

## References

Burt RL, Edye LA, Williams WT, Gillard P, Grof B, Page M, Shaw NH, Williams RJ, Wilson GPM (1974) Small sward testing of *Stylosanthes* in northern Australia: preliminary considerations. Aust J Agric Res 25: 559–575

Burt RL, Cameron DG, Cameron DF, Mannetje L't, Lenne, J (1983) *Stylosanthes*. In: Burt RL, Rotar PP, Walker JL, Silvey MW (eds) The role of *Centrosema*, *Desmodium* and *Stylosanthes* in improving tropical pastures. Westview Press, Boulder pp 141–180

Cameron DF, (1967) Chromosome number and morphology of some introduced *Stylosanthes* species. Aust J Agric Res 18: 375–379

Cameron DF, Irwin JAG (1987) Use of natural outcrossing to improve the anthracnose resistance of *Stylosanthes guianensis*. Special Publ Agron Soc NZ No 5, Proc NZ Plant Breeding Conf pp 224–227

Cameron DF, Hutton EM, Miles JW, Brolmann JB (1984) Plant breeding in *Stylosanthes*. In: Stace HH, Edye LA (eds) The biology and agronomy of *Stylosanthes*. Academic Press, Sydney, pp 589–606

Costa NMS, Ferreira MB (1977) O Genero *Stylosanthes* no Estado de Minas Gerais. EPIMAG, Belo Horizonte 42 pp

Costa NMS, Ferreira MB (1984) Some Brazilian species of *Stylosanthes*. In: Stace HH, Edye LA (eds) The biology and agronomy of *Stylosanthes*. Academic Press, Sydney, pp 23–48

CSIRO (1990) Shrubby stylo (*Stylosanthes scabra*) cv. Siran. Plant Variet J 3(4): 33–35

Dornelas MC, Vieira MLC, Appezzato-da-Gloria, B (1992) Histological analysis of organogenesis and somatic embryogenesis induced in immature tissues of *Stylosanthes scabra*. Ann Bot 70: 477–482

Elliot AR, Manners JM (1993) Transformation of *Stylosanthes* species. In: Bajaj YPS (ed) Biotechnology in agriculture and forestry, vol 23. Plant protoplasts and genetic engineering IV. Springer, Berlin Heidelberg New York, pp 361–374

Godwin ID (1987) Somaclonal variation in the plant improvement of tropical pasture legumes of *Stylosanthes*. PhD Thesis, The University of Queensland, Brisbane, Austrilia, 180 pp

Godwin ID, Gordon GH, Cameron DF (1987a) Callus culture-derived somaclonal variation in the tropical pasture legume *Stylosanthes guianensis* (Alubl.) Sw. Plant Breed 98: 220–227

Godwin ID, Gordon GH, Cameron DF (1987b) Plant regeneration from leaf-derived callus cultures of the tropical pasture legume *Stylosanthes scabra* Vog. Plant Cell Tissue Organ Cult. 9: 3–8

Godwin ID, Gordon GH, Cameron DF (1990) Variation among somaclonal progenies from three species of *Stylosanthes*. Aust J Agric Res 41: 654–656

Irwin JAG, Cameron DF (1978) Two diseases of *Stylosanthes* spp. caused by *Colletotrichum gloeosporoides* in Australia, and pathogenic specialisation within the causal organisms. Aust J Agric Res 29: 305–317

Larkin PJ, Scowcroft WR (1981) Somaclonal variation – a novel source of variability from cell cultures for plant improvement. Theor Appl Genet 60: 197–214

Latunde-Dada AO, Lucas JA (1983) Somaclonal variation and reaction to *Verticillium* wilt in *Medicago sativa* plants regenerated from protoplasts. Plant Sci Lett 32: 205–211

Ling DH, Vidhyasekeran P, Borromeo ES, Zapata FJ, Mew TW (1985) In vitro screening of rice germplasm for resistance to brown spot disease using phytotoxin. Theor Appl Genet 71: 133–135

Manners JM (1988) Transgenic plants of the tropical pasture legume *Stylosanthes humilis*. Plant Sci 55: 61–68

Manners JM, Way H (1989) Efficient transformation with regeneration of tropical pasture legume *Stylosanthes humilis* using *Agrobacterium rhizogenes* and a Ti plasmid binary vector system. Plant Cell Rep 8: 341–345

Mannetje, L't (1984) Considerations on the taxonomy of the genus *Stylosanthes*. Some Brazilian species of *Stylosanthes*. In: Stace HH, Edye LA (eds) The biology and agronomy of *Stylosanthes*. Academic Press, Sydney, pp 1–21

Meijer EGM (1982a) Shoot formation in tissue cultures of three cultivars of the tropical pasture legume *Stylosanthes guianensis* (Aubl.) Sw. Z Pflanzenzuecht 89: 169–172

Meijer EGM (1982b) High frequency plant regeneration from hypocotyl- and leaf-derived tissue cultures of the tropical pasture legume *Stylosanthes humilis*. Physiol Plant 56: 381–385

Meijer EGM (1984) Some aspects of long-term culture of *Stylosanthes guianensis* (Aubl.) Sw. (Leguminosae). J Plant Physiol 117: 131–135

Meijer EGM, Broughton WJ (1981) Regeneration of whole plants from hypocotyl-, root-, and leaf-derived tissue cultures of the pasture legume *Stylosanthes guianensis* (Aubl.) Sw. Physiol Plant 52: 280–284

Meijer EGM, Steinbiss HH (1983) Plantlet regeneration from suspension and protoplast cultures of the tropical pasture legume *Stylosanthes guianensis.* Ann Bot 52: 305–310

Meijer EGM, Szabados HN (1990) Cell and tissue culture of *Stylosanthes* spp. In: Bajaj YPS (ed) Biotechnology in agriculture and forestry, vol 10. Legumes and oilseed crops I. Springer, Berlin Heidelberg New York, pp 312–322

Miles JW (1985) Evaluation of potential genetic marker traits and estimation of outcrossing rate in *Stylosanthes guianensis.* Aust J Agric Res 36: 259–265

Miles JW, Roca WM, Tabares E (1989) Assessment of somaclonal variation in *Stylosanthes guianensis*, a tropical forage legume. In: Mujeeb-Kazi, A Sitch LA (eds) Review in advances in plant biotechnology, 1985–88. 2nd Int Symp Genetic manipulation in crops CIMMYT and IRRI. Mexico City, Mexico, pp 249–257

Mroginski LA, Kartha KK (1981) Regeneration of plants from callus tissue of the forage legume *Stylosanthes guianensis.* Plant Sci Lett 23: 245–251

Murashige T, Skoog F (1962) A revised medium for rapid growth and bioassays with tobacco tissue cultures. Physiol Plant 15: 473–497

Rao IM, Roca WM, Ayarza MA, Tabares E, Garcia R (1992) Somaclonal variation in plant adaptation to acid soil in the tropical forage legume *Stylosanthes guianensis.* Plant Soil 146: 21–30

Rines HW, Luke HH (1985) Selection and regeneration of toxin-insensitive plants from tissue cultures of oats (*Avena sativa*) susceptible to *Helminthosporium victoriae.* Theor Appl Genet 71: 16–21

Scowcroft WR, Adamson JA (1976) Organogenesis from callus cultures of the legume *Stylosanthes hamata.* Plant Sci Lett 7: 39–42

Stace HM, Cameron DF (1988) Cytogenetic review of the taxa in *Stylosanthes hamata* sensu lato. Trop Grassl 21: 182–188

Szabados L, Roca WM (1986) Regeneration of isolated mesophyll and cell suspension protoplasts to plants in *Stylosanthes guianensis.* A tropical forage legume. Plant Cell Rep 3: 174–177

Szabados L, Tabares E, Lopez J, Miles J, Lenne J, Roca W (1986) Variability in Stylosanthes cell and tissue cultures. In: Somers DA, Gengenbach BG, Biesboer DD, Hackett WP, Green CE (eds) VI Int Congr Plant tissue and cell culture abstracts, IAPTC, Minneapolis, 224 pp

Vieira MLC, Jones B, Cocking EC, Davey MR (1990) Plant regeneration from protoplasts isolated from seedling cotyledons of *Stylosanthes guianensis, S. macrocephala* and *S. scabra.* Plant Cell Rep 9: 289–292

# I.12 Somaclonal Variation in Banana and Plantain (*Musa* Species)

O. REUVENI[1], Y. ISRAELI[2], and E. LAHAV[1]

## 1 General Account

Banana and plantain (*Musa* sp.) are an important food crop, especially in the tropics. The world produces more than 70 million tons of banana, plantain, and cooking bananas, but most of these are used for domestic consumption. Bananas are produced in the tropical and subtropical regions and the supply is available to the consumer throughout the year. About 80% of the world's export bananas comes from the Americas (Central, South, and the Caribbeans). About 10 million tons of bananas enter the world market and serve as an important factor in the economics of many developing countries.

There are several traditional methods of field propagation of banana plants, which are aimed at producing a maximum number of rhizomes per plant. Various methods were used at one time to stimulate adventitious bud production on the rhizomes (Hamilton 1965; Mendez and Loor 1979), but they have been replaced by mass multiplication in vitro. The earliest application of in vitro culture for *Musa* was reported by Cox et al. (1960), who successfully cultured zygotic embryos. The first in vitro clonal propagation was reported by Ma and Shii (1972, 1974) in Taiwan. Since then, the in vitro method has been developed as a basic technique for *Musa* propagation (Hwang et al. 1984; Cronauer and Krikorian 1986; Krikorian 1989; Vuylsteke 1989) and also for conservation and movement of germplasm (De Langhe 1984; Vuylsteke 1989; De Smet and Van den Houwen 1991).

Improvement programs initiated in the history of banana research have met with limited success in achieving commercially acceptable cultivars with a wide genetic background, which might overcome the monoculture of closely related clones. This situation might be explained by the inability to develop a good long-term functional research system (Buddenhagen 1993), and the complicated genetic system of bananas. Sterility, interspecific hybrid constitution, heterozygosity, and polyploidy are common in many clones. The complexity of banana genetics justifies the use of more sophisticated biotechnological techniques to support conventional breeding programs (Novak 1992). New research initiatives

Contribution from the Agricultural Research Organization, The Volcani Center, Bet Dagan, Israel, No. 1405-E, 1994 series

[1] Institute of Horticulture, Agricultural Research Organization, The Volcani Center, P.O. Box 6, Bet Dagan 50250, Israel
[2] Jordan Valley Banana Experiment Station, Zemach 15132, Israel

Biotechnology in Agriculture and Forestry, Vol. 36
Somaclonal Variation in Crop Improvement II (ed. by Y.P.S. Bajaj)

during the past several years have entered a new period of remarkable progress by using conventional and biotechnological techniques (De Langhe 1993).

An unexpected source of variability arose concomitantly with the use of in vitro regeneration of bananas (Côte et al. 1993b). On the other hand, it is obvious that for the production of novel plants by transformation and protoplast fusion or micropropagation, instability is undesirable. Conversely, it was found in bananas that changes can occur in beneficial agronomic traits such as disease tolerance and fruit size.

With the increased use of shoot-tip culture for *Musa* micropropagation, somaclonal variation was detected as a most important limiting factor. Despite the many advantages of shoot-tip propagation – availability of plants, precocity, high production, uniformity, accelerated growth, pest and disease-free plants – somaclonal variation is the major obstacle preventing a rapid expansion of the in vitro method.

## 2 In Vitro Regeneration and Somaclonal Variation

There are a few fields where in vitro culture techniques can be involved in banana improvement (Novak 1992). These include in vitro culture of embryos to improve the recovery of hybrid plants (Cox et al. 1960; Tapia and Pons 1989). Shoot-tip culture of *Musa* has been developed and applied extensively for micropropagation, transfer of germplasm, and medium-term conservation under minimal growth conditions (Vuylsteke 1989). Disease and pest elimination can be obtained by using healthy suckers selected for micropropagation (Jones and Tezenas du Montcel 1993). Elimination of cucumber mosaic virus (CMV) after thermotherapy followed by in vitro culture of shoot tips (Berg and Bustamante 1974) and in vitro shoot-tip culture has been used for mutation induction (Omar et al. 1989; Novak et al. 1990).

The formation of embryogenic cultures derived from different primary explants, such as zygotic embryos, meristematic layers of proliferating shoot tips, basal leaf sheats of rhizome tissue, and young male flowers, has been reported. Somatic embryos obtained served to establish embryogenic suspensions from which plants were regenerated (Escalant et al. 1993; Novak et al. 1989). Promising results were obtained also for somatic hybridization and protoplast culture (Haicour et al. 1993; Panis et al. 1994). Preliminary results show that androgenesis and induced parthenogenesis are applicable to banana, which will make it possible to obtain homozygotic plants more rapidly (Bakry et al. 1993). Transformation of cells and protoplasts and the regeneration of transgenic bananas have been reported (Haicour et al. 1993; Sagi et al. 1994). Cryopreservation studies have also been conducted successfully (Panis and Swennen 1995).

### 2.1 Somaclonal Variation in Micropropagation

Somaclonal variation in bananas has been reported by many researchers (Table 1). In the Cavendish subgroup, the frequency of off-types varies from as

**Table 1.** Rate of common somaclonal variants among AAA banana vitroplants

| Type of variation | Cultivar | Variants (%) | Reference |
|---|---|---|---|
| Dwarfism | Williams | 2.2–17.3 | Reuveni et al. (1986) |
| " | " | 7–90 | Walduck et al. (1988) |
| " | " | 50 | Daniells (1988) |
| " | " | 4.1–31.7 | Smith (1988) |
| " | " | 10.2–17.3 | Reuveni and Israeli (1990) |
| " | " | 4–20 | Daniells and Smith (1993) |
| " | " | 2–6 | Daniells and Bryde (1993) |
| " | " | 3.4 | Arias (1993) |
| " | " | 6–10.3 | Johns (1994) |
| " | " | <0.1 | Israeli et al. (1995) |
| " | Grand Nain | 3.4 | Reuveni et al. (1986) |
| " | " | 12.5 | Stover (1987) |
| " | " | 0.1 | Arias and Valverde (1987) |
| " | " | 3.4–20 | Reuveni and Israeli (1990) |
| " | " | 3–63.3 | Carrieres (1991) |
| " | " | 0 | Daniells and Smith (1993) |
| " | " | 0–1 | Daniells and Bryde (1993) |
| " | " | 2.6–3.2 | Arias (1993) |
| " | Mons Mari | 17–22 | Walduck et al. (1988) |
| " | " | 0.2–28 | Johns (1994) |
| " | Chinese Cavendish | 19 | Walduck et al. (1988) |
| " | " | 8.6 | Johns (1994) |
| " | New Guinea Cav. | 3 | Drew and Smith (1990) |
| Extra Dwarfism | Nathan | 0.6–11.7 | Reuveni et al. (1986) |
| " | Grand Nain | 2.5 | Stover (1987) |
| Gigantism | " | 1 | Arias and Valverde (1987) |
| " | " | 2.5 | Stover (1987) |
| " | " | 0.9 | Arias (1993) |
| Mosaic-like | Williams | 0.2–2.8 | Reuveni et al. (1986) |
| " | " | 0.1–1.9 | Reuveni and Israeli (1990) |
| " | " | 1.3–2.6 | Johns (1994) |
| " | " | 0.8 | Israeli et al. (1995) |
| " | Grand Nain | 3.8 | Reuveni et al. (1986) |
| " | " | 1–9.2 | Reuveni and Israeli (1990) |
| " | Mons Mari | 0.4–35 | Johns (1994) |
| " | Chinese Cavendish | 2.6 | Johns (1994) |
| " | Nathan | 0.6–0.8 | Reuveni et al. (1986) |

low as 1–2% to as high as 50% or even more among plants of the same cultivar propagated by the same laboratory (see Table 1 and references cited therein). According to Stover (1987), less than 5% off-types is the commercially acceptable rate. In many places where bananas are micropropagated, this goal is actually achieved at present (Hwang et al. 1984; Arias and Valverde 1987; Arias 1993), but in a few cases a high frequency which might limit the commercial use of micropropagated plants is still observed (Carrieres 1991; Daniells and Smith 1993).

In plantain, the rate of variation is highly clonal-dependent (Table 2), and may range from 0 to 100% (Vuylsteke et al. 1991; Krikorian et al. 1993). Exceptionally high frequency of variation was reported for Lady's Finger (AAB;

**Table 2.** Rates of inflorescence reversion to French type and abnormal inflorescence among AAB plantain vitroplants

| Type of plantain | Cultivar | Variants (%) | Reference |
|---|---|---|---|
| French Horn | Bise Egome 2 | 69.1 | Vuylsteke et al. (1991) |
| False Horn | Maricongo | 21.0 | Ramcharan et al. (1987) |
| ,, ,, | ,, | 0–100 | Krikorian et al. (1993) |
| ,, ,, | Dwarf | 38.0 | Ramcharan et al. (1987) |
| ,, ,, | Agbagba | 3.2 | Vuylsteke et al. (1991) |
| ,, ,, | Big Ebanga | 35.0 | Vuylsteke et al. (1991) |
| ,, ,, | Falso Cuerno | 1.4 | Sandoval et al. (1991) |
| Horn | Ubok Iba | 5.6 | Vuylsteke et al. (1991) |

Daniells and Smith 1993) and Figue Sucree (AA; Carrieres 1991); on the other hand, high stability was recorded for Saba (ABB; Stover 1987).

Most of the variations in banana and plantain may be grouped into changes in plant stature, leaf structure, pseudostem pigmentation, or inflorescence and fruit morphology.

### *2.1.1 Plant Stature*

Variations include dwarfism, gigantism, and a variety of intermediate heights. Actually, all height classes of the Cavendish subgroup may be found among in vitro-propagated Cavendish plants (Stover 1988). The variation is not only in plant height but in a complex of structural and agronomical characteristics, including leaf morphology, rate of leaf emission, shape and size of the bunch and single fruit, deciduous or persistent flowers, etc. (Israeli et al. 1991). All these variations occur also in conventional field propagation of the Cavendish bananas (Stover and Simmonds 1987), but at a much lower rate. Dwarfism is the most common variation, accounting for 75% (Stover 1987) to 90% (Israeli et al. 1991) of the total variants. It is characterized by both high incidence and high rate (Israeli et al. 1991). Gigantism is less common in banana than dwarfism; however, in some cases, this was reported to be the main variation (Arias and Valverde 1987). Occasionally an extra-dwarf variant (Fig. 1) may occur. It is characterized by very short stature and uncommercial fruit. It was recorded in a few Cavendish clones, and resembles very closely the Extra-Dwarf Cavendish clone (Israeli et al. 1991).

In plantain, variation in stature is less common, but does occur (Sandoval et al. 1991; Vuylsteke et al. 1991).

### *2.1.2 Leaf Structure*

The most common change in leaf structure is the mosaic-like variation. These plants are characterized by thick, narrow leaves, with a variable degree of pale green mottling (Fig. 2) which is easily detected when the lamina is backlit. The leaf

**Fig. 1.** Extra-dwarf vitroplant variant with a very short stature and a choked bunch at flowering time

surface is covered with depressions and protuberances, and the margins are wavy (Israeli et al. 1991). The fruits of these plants are small and have no commercial value. The plants are aneuploids, with a probably variable number of extra chromosomes which affect the intensity of symptoms (Reuveni et al. 1986). This type of variation has a high incidence, occurring in almost every population of shoot-tip-propagated Cavendish plants. However, the percentage rate in the population is not very high. It was found in the Red banana (Israeli et al. 1991) and was also observed in plantain (Israeli, unpubl.). At least some of the "virus-like" symptoms noted by Krikorian et al. (1993) in the Maricongo plantain were most probably this type of variation.

Less common foliage variations include variegation; changes in coloration (which are usually connected with changes in pseudostem coloration; see below), increased waxiness, drooping leaves, and deformed lamina (Smith 1988; Daniells and Smith 1991; Israeli et al. 1991; Daniells 1993). Some of these variations (e.g., the deformed lamina plants) are of epigenetic origin, since they revert very quickly (in the first or second vegetative generation) to a normal plant. Others are presumably of chimeric nature, since normal and mutated suckers occur concurrently (e.g., the variegated off-type; Israeli et al. 1991). Of the latter, the incidence of the variegated lamina off-type is intermediate (it shows up in most of the batches), but its rate is very low. The other foliage variations are very minor in occurrence.

**Fig. 2.** Mosaic-like vitroplant variants at the nursery stage with pale green mottling

### *2.1.3 Pseudostem Pigmentation*

Greenish, brown-black, or reddish pseudostems, with variable patterns, have been reported (Daniells and Smith 1991, 1993; Israeli et al. 1991), usually associated with changes in petioles and midrib coloration. None of these is of high incidence or rate except for the green-red variation, which occurs at a high percentage when Red bananas are micropropagated. This is explained by the periclinal chimera structure of the mother plants (Stover and Simmons 1987), which split into green and red pseudostems.

### *2.1.4 Inflorescence and Fruit*

Changes in plant stature are usually associated with changes in the inflorescence and fruits, as expected. However, changes specific to the reproductive organs have been recorded also. These are nipple-like fruit tips and bunches with only male flowers (Stover 1987), hairy fruit, exeptionally short or long fingers, long-peduncle (Smith 1988), persistent flowers, and split fingers (Israeli et al. 1991), and variation in the morphology and/or coloration of the bracts and male bud (Stover 1987; Daniells and Smith 1991). The variations in the reproductive organs of the Cavendish bananas are usually of low incidence and rate. In Figue Sucree (AA), a very high rate of degeneration of the male bud was reported (Carrieres 1991).

In plantain, variations in type of inflorescence are the most common. The Horn or False Horn plantains change into the French type (Ramcharan et al. 1987; Vuylsteke et al. 1991; Krikorian et al. 1993). The occurrence of variations varies considerably among clones: some are completely stable while others are most unstable (Table 2 and references therein).

## 2.2 Somaclonal Variation and Disease Response

The occurrence of somaclonal variation during culture has led to interest in the possibility of selecting disease-resistant plants through in vitro culture. Of the many banana diseases, resistance was found only to *Fusarium* wilt. This disease has a long and destructive history in the banana-producing regions of the world (Stover 1962, 1990; Ploetz et al. 1990). The disease, known also as Panama disease, is caused by the soilborne fungus *Fusarium oxysporum* f. sp. *cubense* (FOC).

Four FOC races have been identified; three of them are primary pathogens of *Musa*, while race 3 is a pathogen of *Heliconia*. The major damage by FOC - race 1 was avoided by replacing the susceptible cultivar Gros Michel (AAA) by a group of cultivars belonging to the Cavendish subgroup (AAA). The Cavendish cultivars have remained resistant in tropical America despite their monoculture. In contrast, in the Canary Islands, Queensland, South Africa, the Philippines, and Taiwan, bananas are attacked by race 4 (Stover 1990).

Since 1922 the major objective of various banana breeding projects has been to find new cultivars resistant to FOC (Stover and Buddenhagen 1986). Despite this long period, no acceptable replacements for the AAA dessert cultivars have been developed by hybridization in the breeding programs. One of the most significant contributions of somaclonal variation to disease resistance is the identification of a race 4 FOC-resistant variant in Taiwan (Hwang and Ko 1987). Another case of selecting somaclonal variants resistant to FOC was reported from Australia (Smith et al. 1993).

## 2.3 Somaclonal Variation for Improved Performance

An attempt to select improved Grand Nain and Williams was made in Israel. The selection started among conventionally propagated bananas which were then micropropagated. Data on the vitroplants were recorded over several years. This selection was done several times. In one of the experiments the annual variation in yield was between 15 and 100 kg per stool. The best selections in each experiment were transferred to further in vitro propagation and then to additional horticultural evaluation. All the commercially grown in vitro bananas in Israel are the offspring of these selections.

In Australia, two somaclonal variants with beneficial characters were identified: J.D. Special, selected from Mons Mari, has fingers 2–3 cm longer than usual; J.D. Dwarf, selected from Williams, has a very sturdy pseudostem (Daniells and Smith 1993).

In Israel, stable clones which produce very low rates of undesired dwarf variants during in vitro culture were selected from vitroplants of Williams and Grand Nain (see Sect. 5.1 and Israeli et al. 1995).

# 3 Factors Affecting Somaclonal Variation

The information about variation among banana vitroplants was obtained mainly from plants regenerated from shoot-tip cultures. Most of the described variations are transmissible by vegetative multiplication cycles (Côte et al. 1993b). In sterile plants like bananas this is a reliable way to confirm that a certain phenotypic change is a result of a genetic change, somaclonal variation. Most of the somaclonal variants are stable and maintained after several conventional propagation cycles (Israeli et al. 1991; Vuylsteke et al. 1991). Also unstable variants were found, like some of the mosaic-like plants which reverted to the original phenotype after a few multiplication cycles (Israeli et al. 1991) and other variants (Vuylsteke et al. 1991; Daniells and Smith 1993). Generally, most of the somaclonal variants are morphologically similar to existing cultivar phenotypes but occur at a much higher rate in vitroplants than in conventionally propagated plants. The nature of this genetic change is unclear because in vitro-proliferated meristem tips and axillary buds are considered as structures with high stability due to a regular cell cycle and a continuous cell division which eliminates karyological and genetic irregularities (Bajaj 1990; D'Amato 1990).

There are limited experimental data to explain the origin and causes of variation in bananas. In other plants (Swartz 1991) as well as in bananas, variation among vitroplants may be explained by three major mechanisms: (1) genetic changes that were already present, or the heterogeneity of the primary explants; (2) genomic modifications in the vitroplants; and (3) the in vitro culture environment which induces nontransmissible epigenetic modifications. Other factors suggested to change the level of somaclonal variation include culture medium composition, types of budding, and duration in culture (Scowcroft 1984).

## 3.1 Culture Medium Composition

Reuveni et al. (1993) compared the occurrence of variants among Williams banana vitroplants multiplied on 16 culture media differing mainly in their auxin and cytokinin contents. The rate of variants obtained was not affected by medium composition or by the rate of multiplication. When seven cultivars of plantain were multiplied under identical in vitro conditions characterized by a high concentration of benzyladenine, three cultivars showed little phenotypic variation. It was concluded that the mutation rate was not affected by the medium composition (Vuylsteke et al. 1991). These results still do not indicate that culture media are not the cause of variation, but they do show that variants appear at comparable rates on a wide range of media (Côte et al. 1993b).

## 3.2 Types of Budding

The formation of adventitious shoots in conjunction with axillary shoots is another commonly proposed cause of variation (George and Sherrington 1984). Such a possibility was reported to occur in banana plants where axillary and adventive budding was found (Banerjee et al. 1986; Novak et al. 1990). In the case of Red banana, it is presumed that the cultivar is a periclinal chimera where a red outer layer covers a green core (Stover and Simmonds 1987). This chimera is dissociated in vitro and green-red plants are regenerated (Israeli et al. 1991). A significantly higher rate of dwarf variants was found among Williams banana vitroplants regenerated from adventive budding (Reuveni et al. 1993). Adventive budding therefore can be an important cause for the regeneration of variants when heterogeneity exists in the primary explants.

### 3.3 Duration in Culture

The general opinion is that somaclonal variation is enhanced by prolongation of the culture period. Only small differences in the rate of variants were found when cultures of Agbagba plantain kept for a long period were compared to short-period cultures (Vuylsteke et al. 1988, 1991). In the case of Williams bananas, in general the length of time in culture affected the rate of variants (Reuveni and Israeli 1990). On the other hand, it was concluded that the duration in culture was not a mutation-inducing factor, since no variants were found among plants derived from certain primary explants (Reuveni et al. 1993). It was found that variation can appear spontaneously at any time during successive multiplication cycles (Table 3). Therefore, the percentage of variation is likely to increase as an exponential function of frequency of variation and the number of multiplication cycles (Côte et al. 1993a). The lack of relation sometimes observed between the number of multiplication cycles and the variation percentages was explained by relatively high early variant occurrence that could conceal the effect of the multiplication cycle number on the percentage (Côte et al. 1993a). There might be other explanations, such as instability of the variants and their in vitro reversion to the original phenotype (Vuylsteke et al. 1991), or a lower multiplication rate of the variants. In cases where the multiplication rates of the variants are similar to

**Table 3.** Number of plants evaluated of different families in which dwarfism (DW) and mosaic-like (MO) variants were found (+) in different multiplication cycles. (Reuveni et al. 1993)

| Family no. | Plants evaluated (no.) | Variants found | | | | | | |
|---|---|---|---|---|---|---|---|---|
| | | Kind | | Multiplication cycle | | | | |
| | | DW | MO | 0 | 1 | 2 | 3 | 4 |
| 10/3 | 43 | + | | + | | | | |
| 10/7 | 59 | + | | | + | | | |
| 10/29 | 56 | + | | | | + | | |
| 10/20 | 40 | + | | | | + | + | |
| 10/30 | 63 | + | + | | | + | + | |
| 10/4 | 53 | + | | | | + | | |
| 10/17 | 52 | + | + | + | | + | | |
| 10/9 | 106 | + | + | | + | + | | |
| 10/10 | 45 | + | + | | | | | + |

normal plants as in Williams bananas (Reuveni and Israeli 1990), and with the appearance of new variants in successive multiplication cycles, a higher percentage of variants is expected with an increase in the number of multiplication cycles.

### 3.4 Heterogeneity of the Primary Explant

The term "heterogeneity of the primary explant" might refer to differences between primary explants at different levels of organization, starting with differences between cultivars, through differences between clones and different explants from the same mother plant. In plantain, the percentage of variants in seven cultivars ranged from 0.5 to 69% (Vuylsteke et al. 1991). The differences in percentage of variants for the same banana cultivar (Table 1) can be attributed only partially to differences in techniques. Differences in variant percentages were reported for different clones of the same cultivar (Table 4). Krikorian et al. (1993) compared the rate of variation among vitroplants of French-type and False Horn-type plantains. The French type Superplatano was found to be stable whether primary explants were derived from vegetative corm or floral bud apex. The False Horn-type Maricongo was not stable and presented considerable reversion to French-type bunch (Table 2). The frequency of bunch reversion varied from 0.4 to 100% and was confined to individual stem tips rather than clones. Evidence was presented of the existence of heterogeneity among primary explants derived from the same mother stock plant. It was suggested (Reuveni and Israeli 1990; Reuveni et al. 1993) that heterogeneity of primary explants is the cause of differences in percentages of dwarf variants among different families of Williams. Each family corresponds to the entire in vitro progeny of one primary explant.

In the case of both bananas and plantains, the chimeric heterogeneity of the primary explants was suggested as the reason for large differences in variation. Since chimerism is very common in plants propagated vegetatively for a long period (Tilney-Basset 1986; Swartz 1991), its role in *Musa* somaclonal variation should be considered. Such plants are easily dissociated in vitro and the regenerated plants could be somaclonal variants (Swartz 1991). This dissociation is promoted by adventive budding, which was found to occur in in vitro culture of bananas (Banerjee et al. 1986; Novak et al. 1990). As mentioned above, a significantly higher rate of dwarf variants was found among Williams banana vitroplants regenerated from adventive buds than from axillary buds (Reuveni et al. 1993).

**Table 4.** Percentage of somaclonal variants among vitroplants of different cultivars and clones (232 plants per clone) of Cavendish bananas. (Johns 1994)

| Cultivar and clone | Dwarf | Mosaic-like | Total |
|---|---|---|---|
| Chinese Cavendish | 8.6 | 2.6 | 11.2 |
| Williams-BSH | 6.0 | 1.3 | 7.3 |
| Williams-IA | 10.3 | 2.6 | 12.9 |
| Mons Mari-WR | 0.9 | 0.4 | 1.3 |
| Mons Mari-GL | 28.0 | 35.0 | 63.0 |

### 3.5 Genomic Modification in the Vitroplant

Another causal mechanism involved with genetic modification expressed as somaclonal variation is associated with chromosomal variability of vitroplants (Lee and Phillips 1988). It was reported (Reuveni 1990) that the mosaic-like somaclonal variants possess aneuploid cells. Recent studies by Sandoval et al. (1993) showed that the mosaic somaclonal variant of Grand Nain have 50% aneuploid cells, as compared with 14% in normal true-to-type and dwarf somaclonal vitroplants. Whether the high mixoploidy of the mosaic-like variants was induced by the in vitro environment, or if it preexisted as cytochimeras in the primary explants, requires further study.

Karyological analysis of regenerated *Musa* plants is still lacking, but some morphological features, such as drooping leaves associated with slower growth, reported among plantain variants, may indicate a change in ploidy (Vuylsteke et al. 1991). In addition to numerical chromosome variation, structural and point mutations are also expected. There is not enough information about their occurrence and their role in banana somaclonal variation.

### 3.6 Epigenetic Modification

It is difficult sometimes to distinguish between genetic and epigenetic variation in bananas, because of their sterility. However, changes that are induced by the in vitro culture and are not transmissible through several vegetative multiplication cycles, are considered to be of epigenetic origin.

Vitroplants, as compared to conventionally propagated plants, were reported to be more vigorous, taller, have larger bunches, a shorter flowering-to-harvest cycle, and significantly greater production of suckers (Mascarenhas et al. 1983; Hwang et al. 1984; Drew and Smith 1990; Perez 1991; Robinson et al. 1993). These characters are typical of the first cycle of vitroplants and are less expressed in the following cycles. Therefore, they can be defined as epigenetic modifications.

## 4 Detection of Somaclonal Variation

There is a need to detect the in vitro variants as soon as they occur. Since there are conceptional and operational differences between eliminating and keeping the somaclonal variants, they are discussed below separately.

### 4.1 Detection of Somaclonal Variants in Micropropagation, Conservation, and Germplasm Transfer

Although shoot-tip culture is widely used for micropropagation, short-term conservation and germplasm transfer carry a risk that a certain level of instability will counterbalance the advantages of the technique. The development of tests by which variants could be detected, and especially when they occur, would be highly valuable. Basically, there are different potential marker systems that could be developed and used (Withers 1993). In addition to their fidelity, simplicity, and cost, the strategy of incorporating them into propagation and conservation protocols has to be considered. The development of such tests for all possible variants will require much effort. Therefore attention was given to the more important variations.

### *4.1.1 Detection of Dwarf Variants*

Dwarf variants are the most common form of variant found in the Cavendish subgroup, except in Dwarf Cavendish (Table 1). The plants are very similar to Dwarf Cavendish but vary significantly from the original cultivar. Their height, leaf size, distance between petioles, bunch length, bunch mass, and finger dimensions are smaller than those of the original cultivar (Smith and Drew 1990; Israeli et al. 1991). Some of the flowers have a persistent perianth and the male flowers and bracts are persistent on the male axis, all of which are typical characteristics of Dwarf Cavendish.

Dwarfs may be detected when a few parameters are compared at the same time. When only one parameter is measured, some of the plants may not differ from the original cultivar (Israeli et al. 1991). The earliest stage when dwarfs could be detected was at the end of their growth in the nursery and prior to transplanting to the field. Differences in height, distance between petioles, petiole length, and leaf size and index can be distinguished if plants are grown under uniform conditions, with the dwarf having significantly smaller values (Israeli et al. 1991; Côte et al. 1993b; Smith and Hamill 1993). Although growth regimes and facilities vary in commercial nurseries dealing with tens of thousands of plants, most dwarf off-types are culled visually.

When explants taken from dwarf Williams were established in culture and compared with normal plants, no differences in visual appearance or multiplication rates were found on various media (Reuveni and Israeli 1990). Only after subculturing them on $GA_3$-enriched medium did normal plants differ from dwarf ones by suppression of leaf production and a larger internode elongation, while dwarf variants were less sensitive (Reuveni 1990). A similar response was obtained when in vitro-derived dwarfs of Grand Nain and Nathan (a variant of Dwarf Cavendish) were compared with their normal counterparts.

The application of $GA_3$ to Grand Nain vitroplants during acclimatization assisted in discriminating dwarfs from normal plants expressed by faster petiole extension in the normal plants (Côte et al. 1993b). $GA_3$ concentration in Grand Nain normal and giant variant plants was found to be 3.5 and 4.5 times higher, respectively, than in dwarf variants. The $GA_{20}$ content was similar for dwarf and normal plants but was three times higher in the giant variant. It is suggested that the somaclonal variation affecting banana plant height is associated with gibberellin metabolism (Sandoval et al. 1994). Studies are in progress to develop a simple immunological method to assay gibberellins in bananas (Côte et al.

1993b). If successful, it will enable detection of dwarf variants at the test-tube stage much faster and earlier than with the usage of a GA-enriched medium.

Many attempts were made to use isozyme polymorphism as genetic markers, as reviewed by Novak (1992) and studied in detail by Lebot et al. (1993). Isozymes were found to be reliable markers for identifying *Musa* clones differing in their genomic group A or B. Reuveni et al. (1986) were not able to identify distinct differences in isozyme patterns and in electrophoretic and two-dimensional total protein analysis among different banana cultivars of the Cavendish group (AAA) and their dwarf and mosaic-like variants. There is one report (Rodriguez et al. 1991) where a dwarf somaclonal variant and other variants of the cooking banana Burro Cemsa (ABB) were distinguished from the true-to-type plants by isoenzyme analysis of polyphenol oxidase.

Kaemmer et al. (1992) reported that DNA oligonucleotide and amplification fingerprinting have been used successfully to detect genetic polymorphism in representative cultivars of the genus *Musa*. Using the Random Amplified Polymorphic DNA (RAPD) technique, it was possible to differentiate among several cultivars of the Cavendish banana group and an induced mutant of Grand Nain of the same group. Preliminary results of using the RAPD technique for detecting dwarf somaclonal variants were reported by Ford-Lloyd et al. (1993) with Uunapope and by Damasco et al. (1994) with New Guinea Cavendish and Williams. Shoseyov et al. (1995) used RAPD markers in normal Williams, Grand Nain, and Nathan bananas for comparison with somaclonal variants. One out of ten primers generated a polymorphic RAPD marker for each of the cultivars compared with their dwarf variants. The calculated degree of polymorphism determined by comparing each of the three normal cultivars with each other revealed lower values than those observed for each cultivar and its in vitro somaclonal dwarf variant. The RAPD technique and other PCR methods (Jarret et al. 1993) seem, therefore, to be very promising for detecting dwarf somaclonal variants and potentially other somaclonal variants.

The only dwarf variant, and one of the few variants which can be detected visually at the test tube stage, is the extra dwarf variant. A high incidence of an extra dwarf variant was found among vitroplants of Nathan. This variant has a typical compact appearance of rosette-like leaves and can be identified easily at the acclimatization and nursery stages (Israeli et al. 1991).

### *4.1.2 Detection of Mosaic-like Variants*

The second most common off-type among banana vitroplants is mosaic-like variants (Table 1). In contrast to dwarf variants which phenotypically are very similar to each other, the mosaic-like variants are a group of variants which can differ in their stature and their typical phenotypic characteristics. Almost all mosaic-like variants can be detected during the end of the nursery stage (Israeli et al. 1991). A significantly lower number of stomata per leaf area, and larger stomatal cells were found in mosaic-like variants than in normal banana plants of Nathan, Grand Nain, and Williams (Reuveni 1990). These differences were found at the test-tube and later stages. The explanation for these differences is the fact that mosaic-like variants possess aneuploid cells (Reuveni 1990; Sandoval et al. 1993).

More studies are required to identify the involvement of ploidy level in somaclonal variation. In practice, determination of ploidy level in *Musa* by counting chromosomes in meristematic cells has a few obstacles: (1) a very low frequency of dividing cells; (2) a rare occurrence of well-spread metaphase plates for chromosome counting; and (3) the procedure is time-consuming (Novak 1992). To overcome these obstacles, a nuclear DNA flow-cytometry technique was calibrated for detecting DNA content in many thousands of isolated cell nuclei of bananas (Afza et al. 1993). The method is simple and robust and has a great potential for identification of genomically unstable plants (aneuploids, polyploids) among vitroplants at the early test-tube stage.

## 4.2 Detection of Beneficial Variants

All important commercial cultivars of bananas have arisen by mutation in the field (Stover and Simmonds 1987). Since a higher rate of somaclonal variation was reported to occur among banana vitroplants, there is also an expectation of obtaining a higher rate of beneficial variants among these plants. The term "beneficial variant" has a broad meaning: it may refer to a variant with better general agronomical characters expressed in a higher yield or a more specific character, like *Fusarium* wilt resistance (Hwang and Ko 1987). Basically, the methodology applied is different when selection is done to find better general agronomic characters or resistance to specific pests and diseases. The two approaches are described below for variants selected among spontaneous or induced somaclonal variants.

### *4.2.1 Detection of Somaclonal Variants for Improved Agronomical Performance*

It is relatively easy to make a selection for a certain character, but extremely difficult to identify a variant which excels in general performance. Some specific characters which may improve and are important in banana production include dwarf plants with large fingers and bunches. The combination of both characters in the same plant is important, since size of the bunch and length of the fingers tend to become shorter with dwarfism. Each of the characters within the same clone was reported to occur among vitroplants; the two in combination have not been reported yet. Since no screening methods for use at an early stage for detecting long fingers and large bunches have been developed, the detection of such plants will have to be done in the field among dwarf vitroplants.

Early-flowering plants with a shorter cycle will result in higher yields under both tropical and subtropical conditions. Novak et al. (1990) reported on an early-flowering putative mutant plant of Grand Nain obtained by gamma irradiation of shoot apices. This mutant showed differences in the zymograms of soluble proteins and esterase isozymes and was recognized from its original variety by the RAPD technique (Kaemmer et al. 1992).

A less sophisticated method is to select in the field the best-performing plants and take from them suckers to be used as a source of primary explant for new batches of in vitro propagation. By applying this procedure, a gradual improvement in the vitroplants is made. This is a procedure that has been applied in Israel for more than a decade, which may explain the better performance of Israeli Grand Nain in South Africa (Morse 1994, pers. comm.).

### *4.2.2 Detection of Somaclonal Variants for Resistance to Diseases*

Much effort has been expended on two extremely important diseases: leaf spot diseases and the soil borne FOC. The most important leaf spot diseases are: Sigatoka, caused by *Mycosphaerella musicola*; black leaf streak, caused by *M. fijiensis*; and black Sigatoka, caused by *M. fijiensis* var. *difformis*. The simplest way by which cell or tissue cultures have been screened is by direct exposure to the pathogen. However, these cells or tissues are usually very sensitive to this treatment and it may not be possible to distinguish resistant and susceptible selections in vitro, and both resistant and susceptible variants die after the innoculation (Ploetz 1993b).

The role of toxic metabolites obtained from *Mycosphaerella* cultures was reviewed by Novak (1992). Cercosporin culture extracts and isolated toxins were applied to susceptible and resistant in vitro young vitroplants. In certain assays no correlation was observed between susceptible and resistant cultivars (Strobel et al. 1993). In other assays there was a correlation between resistance of several cultivars to a toxin and their resistance to the infection, but this is no proof of the involvement of the toxin in pathogenicity. The toxin can be applied to detached leaves of plantlets or to in vitro-cultured plantlets if it is confirmed that toxin tolerance is expressed also in the regenerated plants. The validity of such screening techniques depends on the actual involvement of toxins as primary or secondary determinants of symptom development (Lepoivre et al. 1993). Further studies of host-pathogen relations are needed in order to develop reliable screening techniques for the early stage of the test tube and the nursery.

Recent investigations have revealed the existence of two kinds of host-pathogen interactions (Beveraggi et al. 1993; Fouré 1993): high resistance, which is characterized by hypersensitivity, and different levels of partial resistance. These two types of resistance were found to be governed by distinct mechanisms. It was also found that toxins secreted by *M. fijiensis* play no role in the early infection stages, but only later during disease development. Other potential resistance markers based on increase in chitinase and glucanase activity were suggested (Lepoivre et al. 1993).

Su and Su (1984) used small meristem-derived plantlets for screening resistance to race 1 and race 4 of FOC. Roots were placed in conidial spore suspensions for 1 min and then incubated at 28 °C in a growth chamber. Most plantlets showed yellowish leaf symptoms and died within 4 weeks. Different reactions of Cocos (dwarf mutant of Gros Michel) and Cavendish varieties to race 4 were observed. Resistance to race 4 of FOC was found among vitroplants planted in an infected area. Six somaclones with tolerance were identified among approximately 20 000 plants planted without prior selection. The healthy plants were micropropagated and tested in the field (Hwang and Ko 1987; Hwang 1990).

From a selection standpoint, host-specific toxins are, unfortunately, not always involved in pathogenesis (Ploetz 1993a). On the other hand, nonspecific toxins produced by the pathogen may be useful even when the relationship between tolerance to a toxin and resistance to the disease is not clear. Fusaric acid is one such nonspecific toxin. It is produced in vitro by many forms of *F. oxysporum*, and has been detected in bananas (Beckman 1987). It has been used to select *Fusarium* wilt-resistant variants of bananas, but the results were discouraging (Epp 1987; Krikorian 1990). Although crude filtrates of FOC races 1 and 4 reduced growth of isolated shoot tips proportionally to their concentration, no significant differences were detected between susceptible (Pisang Mas) and immune (SH-3362) cultivars. Since fusaric acid was unable to induce leaf wilt symptoms, and no sublethal concentrations were identified, it may not be a useful tool for selecting *Fusarium* wilt-resistant somaclones for bananas.

Non-natural (abiotic) elicitors have been tested as means of mimicking pathogen (biotic) effects, and in some cases the study of regulation of defense gene(s) is possible. Such a method was employed by Krikorian (1990), who used potassium vanadate as an elicitor in cell culture of Cardaba (ABB) cooking banana. A small percentage of cells survived the treatment, and organized propagules were recovered but have not been evaluated in the field.

Much attention has been given to classifying the apparent great variation that exists among and within populations of FOC races. With the advent of new technologies, it has become clear that members of a given race are not always related and there is evidence for geographic and pathogenic variations (Ploetz 1990, 1993a). Although the effect of edaphic factors on disease development in different areas cannot be discounted, the differential responses of these FOC strains are probably caused by different virulence genes in a given race (Stover and Buddenhagen 1986). It was noted that soils with poor drainage were most affected by FOC (Stover and Malo 1972). Crop rotation has been shown in Taiwan to be an effective method for controlling the disease for 1 to 2 years (Hwang 1990). It is possible, therefore, that the somaclonal resistant variants selected in Taiwan are primarily tolerant to a certain stress existing in long-term cultivated land (e.g., lack of aeration). Under subtropical conditions of Australia, cold stress could be linked with resistance to race 4. Resistance of the extra-dwarf Dwarf Parfitt to cold stress may be the key factor in its resistance to FOC (Moore et al. 1993). It is possible, therefore, to look for other indirect factors in screening for FOC resistance in bananas.

## 5 Somaclonal Variation and Banana Improvement

The use of spontaneous and induced regeneration of somaclonal variants was studied in banana improvement programs, and can be divided into two main groups: (1) programs aimed at achieving clonal fidelity and genetic stability by eliminating undesired somaclonal variants; and (2) programs aimed at selecting spontaneous or induced variants with beneficial traits. The methodology of detecting both kinds of variants was reviewed in Section 4. This section reviews the results obtained in the two main lines of study.

### 5.1 Reduction of Undesired Somaclonal Variation in Micropropagation

Reuveni and Israeli (1990) suggested a few measures to reduce somaclonal variation among Williams vitroplants: (1) the source plant used for obtaining primary explants has to match the known description of the cultivar, must have the desired agronomic characteristics, and be relatively stable in in vitro propagation; (2) the duration time in culture should be shortened by reducing the number of plants produced from a primary explant; (3) many primary explants have to be used per batch in order to minimize the risk of having a high percentage of variants originated from a single explant that had mutated at an early stage of propagation (Table 3); and (4) off-type plants should be screened out and discarded at all stages from the in vitro regeneration stage to the final stage in the nursery. By following these suggestions, the rate of somaclonal variants among banana vitroplants was reduced dramatically (Daniells and Smith 1993; Robinson 1993).

Another suggestion was to study the possibility of selecting stable clones for in vitro propagation (Reuveni and Israeli 1990; Reuveni et al. 1993). In a recent study, clones of Williams and Grand Nain which restore a very low rate of dwarf variants with time in situ and in vitro were selected (Israeli et al. 1995). The selection was done first by evaluating families derived from different primary explants that were exceptional because of having a very low rate of dwarf variants. At the second stage only families with no dwarf variants were multiplied further, and thousands of plants from each family were evaluated in the field. It was possible to demonstrate that the source of the primary explant determines the stability of the vitroplants.

Similar results were obtained by Krikorian et al. (1993) with plantain vitroplants. Vitroplants of Superplatano, which is a French-type plantain, were stable whether primary explants were derived from vegetative corm or from a floral-bud apex. Considerable variation in bunch phenotype was found among vitroplants of Maricongo, which is a False Horn-type plantain. The variation was observed independently of whether the primary explants were derived from shoot or floral meristems, and was confined to individual originating stem tips rather than to clones.

It may be concluded that in cases where variations are associated with source material heterogeneity, selection of stable clones is feasible in order to overcome a high rate of undesired somaclonal variation in micropropagation, germplasm transfer, and conservation.

### 5.2 Beneficial Somaclonal Variants

In vitro shoot-tip cultures were used as means of inducing mutation in banana and plantain by gamma irradiation and mutagenic agents and the literature on induced mutation breeding has been reviewed (Omar et al. 1989; Novak et al. 1990; Novak 1992). It was expected to contribute to the broadening of genetic variation among *Musa* clones with obligate vegetative reproduction. In practice, these methods have failed to generate the expected variability, especially in regard to disease-resistant variants. Heterogeneity of the shoot tips was suggested as a source of undesirable hidden traits that cannot be eliminated because of competition between the few modified cells and the unchanged tissue (De Langhe 1993). Another obstacle seems to be the lack of efficient screening methods for detecting variants, especially at the test-tube stage (see Sect. 4).

Considerable phenotypic variation was observed among plants regenerated from shoot tips after mutagenic treatment. An early-flowering putative mutant plant of Grand Nain (GN-60-Gy/A) was selected in the greenhouse in the $M_1V_4$ generation and has now been evaluated in the field (Novak 1992). In a recent report, a high degree of variation was observed among Grand Nain vitroplants regenerated from shoot tips treated with EMS and evaluated in the field (Navarro et al. 1994). After 11 months, 13 clones tolerant to black Sigatoka were selected. Further evaluation is needed to verify these promising findings.

Phenotypic variation is common among banana vitroplants. Except dwarfism, other phenotypic variation was observed among Williams vitroplants, and sometimes occurred at high rates (Daniells and Smith 1993). Most somaclonal variants are inferior for commercial cultivation. However, a variant with longer fingers selected in Australia has been evaluated in commercial groves. Recent data on this variant indicate an increase of 15% in finger length associated with a 24% increase in bunch mass (Daniells 1993). The most significant case of a positive variation is the reported race 4 *Fusarium* wilt-resistant variant in Taiwan. This variant was agronomically inferior, but after it was put back into culture for additional variation, its performance was improved without losing resistance. The resistant clone GCTCV-215-1 is now being planted in Taiwan on a commercial scale (Hwang et al. 1993).

## 6 Summary and Conclusions

Much information has accumulated on a range of in vitro techniques for *Musa* improvement. With the increased use of shoot-tip culture for *Musa* micropropagation a high frequency of nonbeneficial somaclonal variation among vitroplants was reported. Most of the variation may be grouped into changes in plant stature, leaf structure, pseudostem pigmentation or inflorescence, and fruit morphology. Dwarfism is the most common variation among Cavendish banana vitroplants, while variations in type of inflorescence are the most common variants among Horn or False Horn plantain vitroplants.

Of many banana diseases, resistance was found among vitroplant variants only to *Fusarium* wilt. Two somaclonal variants with other beneficial characters were identified: one which has longer fingers and one with a very sturdy pseudostem.

The rate of variants obtained was not affected by medium composition. Variation can appear spontaneously at any time during successive multiplication cycles. The heterogeneity of the primary explants was suggested as the main reason for large differences in variation between cultivars, clones, and different explants of the same plant. The chimeric heterogeneity of the primary explants was suggested as one of the reasons for the large differences in variation. Another causal mechanism is associated with chromosomal variability where somaclonal variants were found to possess aneuploid cells.

Different methods suggested for early detection of nonbeneficial variants among micropropagated vitroplants are reviewed. They are based on morphological, physiological, and chemical parameters. Early detection of beneficial variants with improved agronomical performance or diseases resistance did not achieve great success, so selection has to be done in the field.

Measures to reduce nonbeneficial somaclonal variants among vitroplants are described, including the selection of stable clones which restore a very low rate of variation. Other examples of selecting beneficial somaclonal variants are brought.

## References


Afza R, Kämmer D, Dolezel J, Köing J, Van Duren M, Kahl G, Novak FJ (1993) The potential of nuclear DNA flow-cytometry and DNA fingerprinting for *Musa* improvement programs. In: Ganry J (ed) Breeding bananas and plantains for resistance to diseases and pests. CIRAD, INIBAP, Montpellier, France, pp 65–75

Arias O (1993) Commercial micropropagation of banana. In: Biotechnology applications for banana and plantain improvement. INIBAP, Montpellier, pp 139–142

Arias O, Valverde M (1987) Producción y variación somaclonal de plantas de banano, variedad Grand Nain producidas por cultivos de tejidos. ASBANA (Costa Rica) 28: 6–11

Bajaj YPS (ed) (1990) Biotechnology in agriculture and forestry, vol 11. Somaclonal variation in crop improvement I. Springer, Berlin Heidelberg New York

Bakry F, Haicour R, Horry JP, Megia R, Rossignal L (1993) Applications of biotechnologies to banana breeding: haplogenesis, plant regeneration from protoplasts, and transformation. In: Biotechnology applications for banana and plantain improvement. INIBAP, Montpellier, pp 52–62

Banerjee N, Vuylsteke D, De Langhe E (1986) Meristem tip culture of *Musa*, histomorphological studies of shoot bud proliferation. In: Withers LA, Alderson PG (eds) Plant tissue culture and its agricultural applications. Butterworth, London, pp 139–147

Beckman CH (1987) The nature of wilt diseases of plants. APS Press, St Paul, 175 pp

Berg LA, Bustamante M (1974) Heat treatment and meristem culture for the production of virus free bananas. Phytopathology 64: 320–322

Beveraggi A, Mourichon X, Salle G (1993) Study of host-parasite interactions in susceptible and resistant bananas inoculated with *Cercospora fijiensis*, pathogen of black leaf streak disease. In: Ganry J (ed) Breeding banana and plantain for resistance to diseases and pests. CIRAD, INIBAP, Montpellier, pp 171–192

Buddenhagen IW (1993) Whence and whither banana research and development. In: Biotechnology applications for banana and plantain improvement. INIBAP, Montpellier, pp 12–26

Carrieres JL (1991) Bananiers vitroculture et vitroplants. In: Centre de Coop en Recherche Agron pour le Dev Rap d'Act. IRFA-Guadeloupe, pp 11–14

Côte FX, Perrier X, Teisson C (1993a) Somaclonal variation in *Musa* sp. : theoretical risks, risk management, future research prospects. In: Biotechnology applications for banana and plantain improvement. INIBAP, Montpellier, pp 192–199

Côte FX, Sandoval JA, Marie Ph, Auboiron E (1993b) Variations in micropropagated bananas and plantains: literature survey. Fruits 48: 15–22

Cox EA, Stotzky G, Goos RD (1960) In vitro culture of *Musa balbisiana* Colla embryos. Nature 185: 403–404

Cronauer SS, Krikorian AD (1986) Banana (*Musa* spp.) In: Bajaj YPS (ed) Biotechnology in agriculture and forestry, vol 1. Trees I. Springer, Berlin Heidelberg New York, pp 233–252

Damasco OP, Godwin ID, Henry RJ, Adkins SW, Smith MK (1994) RAPD detection and characterization of a DNA polymorphism associated with dwarf banana somaclonal variants. Plant Genome II, 31 p (Abstr)

D'Amato F (1990) Somatic nuclear mutations in vivo and in vitro in higher plants. Caryologia 43: 191–204

Daniells JW (1988) Comparison of growth and yield of bananas derived from tissue culture and conventional planting material. Banana Newsl Int Group Hort Physiol Banana, University of Western Australia, Nedlands, 11: 2

Daniells J (1993) Breeding bananas for Australian conditions. In: Ganry J (ed) Breeding banana and plantain for resistance to diseases and pests. CIRAD, INIBAP, Montpellier, pp 283–292

Daniells J, Bryde N (1993) Results of survey of offtypes in tissue culture plantings 1992. Bananatopics 19: 4

Daniells JW, Smith MK (1991) Post-flask management of tissue-cultured bananas. ACIAR Canberra. Tech Rep 18, 8 p

Daniells JW, Smith MK (1993) Somatic mutations of bananas – their stability and potential. In: Valmayor RV, Hwang SC, Ploetz R, Lee SW, Roa NV (eds) Proc Int Symp Recent developments in banana cultivation technology. INIBAP-ASPNET Los Banos, Philippines, pp 162–171

De Langhe E (1993) Genetic improvement of banana and plantain: the new era. In: Ganry J (ed) Breeding banana and plantain for resistance to diseases and pests. CIRAD, INIBAP, Montpellier, pp 1–9

De Langhe EA (1984) The role of in vitro techniques in germplasm conservation. In: Holden JHW, Williams JT (eds) Crop genetic resources: conservation and evaluation. Allen and Unwin, London, pp 131–137

De Smet K, Van den Houwen I (1991) The banana germplasm collection at the INIBAP transit center. Annu Rep 1991 INIBAP, Montpellier, pp 35–37

Drew RA, Smith MK (1990) Field evaluation of tissue-culture bananas in south-eastern Queensland. Aust J Exp Agric 30: 569–574

Epp MD (1987) Somaclonal variation in bananas: a case study with *Fusarium* wilt. In: Persley GJ, De Langhe EA (eds) Banana and plantain breeding strategies. ACIAR Proc No 21 Canberra, pp 140–150

Escalant JV, Paduscheck C, Babeau J, Teisson C (1993) Somatic embryogenesis in triploid banana cultivars. In: Ganry J (ed) Breeding banana and plantain for resistance to diseases and pests. CIRAD, INIBAP, Montpellier, pp 313–316

Ford-Lloyd BV, Howell E, Newbury HJ (1993) An evaluation of random amplified polymorphic DNA (RAPD) as a tool for detecting genetic instability in *Musa* germplasm stored in vitro. In: Ganry J (ed) Breeding banana and plantain for resistance to diseases and pests. CIRAD, INIBAP, Montpellier, 375 pp (Abstr)

Fouré E (1993) Characterization of the reactions of banana cultivars to *Mycosphaerella fijiensis* Morelet in Cameroon and genetics resistance. In: Ganry J (ed) Breeding banana and plantain for resistance to diseases and pests. CIRAD, INIBAP, Montpellier, pp 159–170

George EF, Sherrington PD (1984) Plant propagation by tissue culture—handbook and directory of commercial operations. Exegetics, Eversley Basingstoke

Haicour R, Rossignol L, Megia R, Sihachakr D, Bui Trang V, Schwendiman J (1993) Use of cell and protoplast cultures for banana improvement biotechnologies. In: Ganry J (ed) Breeding banana and plantain for resistance to diseases and pests. CIRAD, INIBAP, Montpellier, pp 327–338

Hamilton KS (1965) Reproduction from banana adventitious buds. Trop Agric (Trinidad) 42: 69–73

Hwang SC (1990) Somaclonal resistance in Cavendish banana to *Fusarium* wilt. In: Ploetz RC (ed) *Fusarium* wilt of banana. APS Press, St Paul, pp 121–125

Hwang SC, Ko WH (1987) Somaclonal variation of bananas and screening for resistance to *Fusarium*

wilt. In: Persley GJ, De Lange EA (eds) Banana and plantain breeding strategies. ACIAR Proc No 21 Canberra, pp 151–156
Hwang SC, Ko WH, Chao CP (1993) GCTCV-215-1: a promising Cavendish clone resistant to Race 4 of *Fusarium oxysporum* f. sp. *cubense*. In: Valmayor RV, Hwang SC, Ploetz R, Lee SW, Roa NV (eds) Proc Int Symp Recent developments in banana cultivation technology. INIBAP-ASPENT, Los Banos, Philippines, pp 62–74
Hwang SC, Chen CL, Lin JC, Lin HL (1984) Cultivation of banana using plantlets from meristem culture. HortScience 19: 231–233
Israeli Y, Reuveni O, Lahav E (1991) Qualitative aspects of somaclonal variations in banana propagated by in vitro techniques. Sci Hortic 48: 71–88
Israeli Y, Ben-Basat D, Reuveni O (1995) Selection of stable banana clones which do not produce undesired dwarf somaclonal variants during in vitro culture (submitted)
Jarret RL, Vuylsteke DR, Gawel NJ, Pimentel RB, Dunbar LJ (1993) Detecting genetic diversity in diploid banana using PCR and primers from a highly repetitive DNA sequence. Euphytica 68: 69–76
Johns GG (1994) Field evaluation of five clones of tissue cultured bananas in northern New South Wales. Aust J Exp Agric 34: 521–528
Jones DR, Tezenas du Montcel H (1993) Safe movement of *Musa* germplasm. Infomusa 2: 3–4
Kaemmer D, Afza R, Weising K, Gunter K, Novak FJ (1992) Oligonucleotide and amplification fingerprinting of wild species and cultivars of banana (*Musa* spp.). Bio/Technology 10: 1031–1035
Krikorian AD (1989) In vitro culture of bananas and plantains: background, update and call for information. Trop Agric (Trinidad) 66: 194–200
Krikorian AD (1990) Baseline studies and cell culture studies for use in banana improvement schemes. In: Ploetz RC (ed) *Fusarium* wilt of banana. APS Press, St Paul, pp 127–133
Krikorian AD, Cronauer SS (1984) Banana. In: Sharp WR, Evans DA, Ammirato P, Yamada Y (eds) Handbook of plant cell culture, vol 2. Crop Species. Macmillan. New York, pp 327–348
Krikorian AD, Irizarry H, Cronauer-Mitras, Rivera E (1993) Clonal fidelity and variation in plantain (*Musa* AAB) regenerated from vegetative stem and floral axis tips in vitro. Ann Bot 71: 519–535
Lebot V, Aradhya KM, Manshardt R, Meilleur B (1993) Genetic relationship among cultivated bananas and plantains from Asia and the Pacific. Euphytica 67: 163–175
Lee M, Phillips RL (1988) The chromosomal basis of somaclonal variation. Annu Rev Plant Physiol 39: 413–437
Lepoivre P, Acuna CP, Riveros AS (1993) Screening procedures for improving resistance to banana black leaf streak disease. In: Ganry J (ed) Breeding banana and plantain for resistance to diseases and pests. CIRAD, INIBAP, Montpellier, pp 214–220
Ma SS, Shii CT (1972) In vitro formation of adventitious buds in banana shoot apex following decapitation. J Chin Soc Hortic Sci 18: 135–142
Ma SS, Shii CT (1974) Growing banana plantlet from adventitious buds. J Chin Soc Hortic Sci 20: 6–12
Mascarenhas AF, Nadgauda RS, Kulkarni VM (1983) Tissue culture of banana. Banana Newsl Int Group Hortic Physiol of Banana 6: 11
Mendez T, Loor FH (1979) Recent advances in vegetative propagation and their application to banana breeding. Proc 4th ACORBAT Conf, Panama City, 21–24 May 1979, pp 211–222
Moore NY, Pegg KG, Langdon PW, Smith MK, Whiley AW (1993) Current research on *Fusarium* wilt of banana in Australia. In: Valmayor RV, Hwang SC, Ploetz R, Lee SW, Roa NV (eds) Proc Int Symp Recent development in banana cultivation technology. INIBAP-ASPNET, Los Banos, Philippines, pp 270–284
Navarro W, Valerin AT, Salazar R, Radriz J (1994) Genetic improvement of bananas through in vitro mutation breeding in Costa Rica. ACORBAT XI, San-José, Costa Rica, 13–18 Feb (Abstr)
Novak FJ (1992) *Musa* (Bananas and Plantains). In: Hammerschlag FA, Litz RE (eds) Biotechnology of perennial fruit crops. CAB International, Wallingford, pp 449–488
Novak FJ, Afza R, Van Duren M, Perea-Dallos M, Conger BV, Tang Xiaolang (1989) Somatic embryogenesis and plant regeneration in suspension cultures of dessert (AA and AAA) and cooking (ABB) bananas (*Musa* spp.). Bio/Technology 7: 154–159
Novak FJ, Afza R, Van Duren M, Omar MS (1990) Mutation induction by gamma irradiation of in vitro culture shoot-tips of banana and plantain (*Musa* cvs.) Trop Agric (Trinidad) 67: 21–28

Omar MS, Novak FJ, Brunner H (1989) In vitro action of ethyl-methanesulphonate on banana shoot tips. Sci Hortic 40: 283–295

Panis B, Swennen R (1995) Cryopreservation of germplasm of banana and plantain (*Musa* species). In: Bajaj YPS (ed) Biotechnology in agriculture and forestry, vol 32. Cryopreservation of plant germplasm I. Springer, Berlin Heidelberg New York, pp 381–397

Panis B, Sagi L, Swennen R (1994) Regeneration of plants from protoplasts of *Musa* species (Banana). In: Bajaj YPS (ed) Biotechnology in agriculture and forestry, vol 29. Plant protoplasts and genetic engineering V. Springer, Berlin Heidelberg New York, pp 102–114

Perez L (1991) Comparison de varios metodes de propagacion, en cuanto a algunas variables de produccion y crecimiento en el cv. Gran Enano (*Musa* AAA) durante los tres primeros ciclos de cosecha. In: Proc 9th Meeting ACORBAT, Merida, Venezuela, pp 65–75

Ploetz RC (1990) Population biology of *Fusarium oxysporum* f. sp. *cubense*. In: Ploetz RC (ed) *Fusarium* wilt of banana. APS Press, St Paul, pp 63–76

Ploetz RC (1993a) Molecular approaches to identifying *Fusarium* wilt resistance. In: Biotechnology for banana and plantain improvement, INIBAP, Montpellier, pp 104–115

Ploetz RC (1993b) *Fusarium* wilt (Panama disease) In: Ganry J (ed) Breeding banana and plantain for resistance to diseases and pests. CIRAD, INIBAP, Montpellier, pp 149–158

Ploetz RC, Herbert J, Sebasigari K, Hernandez JH, Pegg KG, Ventura JA, Mayato LS (1990) Importance of *Fusarium* in different banana-growing regions. In: Ploetz RC (ed) *Fusarium* wilt of banana. APS Press, St Paul, pp 9–26

Ramcharan C, Gonzalez A, Knausenberger WI (1987) Performance of plantains produced from tissue-cultured plantlets in St. Croix, U.S. Virgin Islands. In: Int Assoc for Res on Plantain and Bananas Proc 3rd Meet of IARPB, Abidjan, Cote D'Ivoire. INIBAP, Montpellier, pp 36–39

Reuveni O (1990) Methods for detecting somaclonal variants in Williams bananas. In: Jarret RL (ed) Identification of genetic diversity in the genus *Musa*. INIBAP, Montpellier, pp 108–113

Reuveni O, Israeli Y (1990) Measures to reduce somaclonal variation in in vitro-propagated bananas. Acta Hortic 275: 307–313

Reuveni O, Israeli Y, Degani H, Eshdat Y (1986) Genetic variability in banana plants multiplied via in vitro techniques. Res Rep AGPG IBPGR/85/216, Rome, 36 pp

Reuveni O, Israeli Y, Golubowicz S (1993) Factors influencing the occurrence of somaclonal variations in micropropagated bananas. Acta Hortic 336: 358–364

Robinson JC (1993) Tissue culture planting material for bananas in South Africa. Subtropica 14: 14–18

Robinson JC, Frazer C, Eckstein K (1993) A field comparison of conventional suckers with tissue culture banana planting material over three crop cycles. J Hortic Sci 68: 831–836

Rodriguez A, Ventura JC, Rodriguez R, Lopez J, Pino J, Roman M (1991) Banana genetic improvement program at INIVIT in Cuba: a progress report. Infomusa 1: 3–5

Sagi L, Remy S, Panis B, Swennen R, Volckaert G (1994) Transient gene expression in electroporated banana (*Musa* spp., cv. Bluggoe ABB group) protoplasts isolated from regenerated embryogenic cell suspensions. Plant Cell Rep 13: 262–266

Sandoval J, Côte F, Esconte J, Soleymani A, Teisson C (1993) Chromosome counts in true-to-type and off-type banana plantlets (AAA, cv. Grand Nain). In: Ganry J (ed) Breeding banana and plantain for resistance to diseases and pests. CIRAD, INIBAP, Montpellier, 380 pp (Abstr)

Sandoval J, Doumas P, Teisson C, Côte F (1994) Identification and quantification of gibberellins in *Musa* (cv. Grand Nain AAA) somaclonal variants and true-to-type plants utilizing HPLC and mass spectrometry systems. ACORBAT XI Meeting, San José, Costa Rica, 13–18 Feb. (Abstr)

Sandoval JA, Tapia FAC, Muller L, Villalobos AV (1991) Observaciones sobre le variabilidad encontrada en plantas micropropagadas de *Musa* cv. Falso Cuerno AAB. Fruits 46: 535–539

Scowcroft WR (1984) Genetic variability in tissue culture: impact on germplasm conservation and utilization. Tech Rep Int Board of Plant Genetic Resources (IBPGR) Rome, (84/152), 41 pp

Shoseyov O, Zabari G, Reuveni O (1995) Detection of in vitro somaclonal dwarf variants of banana cultivars (*Musa*) by RAPD markers (submitted)

Smith MK (1988) A review of factors influencing the genetic stability of micropropagated bananas. Fruits 43: 219–223

Smith MK, Drew RA (1990) Growth and yield characteristics of dwarf off-types recovered from tissue-cultured bananas. Aust J Exp Agric 30: 557–578

Smith MK, Hamill SD (1993) Early detection of dwarf off-types from micropropagated Cavendish bananas. Aust J Exp Agric 33: 639–644

Smith MK, Hamill SD, Thomas JE, Pegg KG, Peterson RA (1993) The role of tissue culture in banana disease research: an Australian perspective. In: Valmayor RV, Hwang SC, Ploetz R, Lee SW, Roa NV (eds) Proc Int Symp Recent developments in banana cultivation technology. INIBAP-ASPNET, Los Banos, Philippines, pp 148–161

Stover RH (1962) Fusarium wilt (Panama disease) of bananas and other *Musa* species. Commonwealth Mycological Institute (CMI) Kew, Surrey Phytopathology No 4, 117 pp

Stover RH (1987) Somaclonal variation in Grand Nain and Saba bananas in the nursery and field. In: Persley GJ, De Lange EA (eds) Banana and plantain breeding strategies. ACIAR Proc No 21 Canberra, pp 136–139

Stover RH (1988) Variation and cultivar nomenclature in *Musa* AAA group, Cavendish subgroups. Fruits 43: 353–356

Stover RH (1990) *Fusarium* wilt of banana: some history and current status of the disease. In: Ploetz RC (ed) *Fusarium* wilt of banana. APS Press, St Paul, pp 1–7

Stover RH, Buddenhagen IW (1986) Banana breeding; polyploidy, disease resistance and productivity. Fruits 41: 175–191

Stover RH, Malo SE (1972) The occurrence of fusarial wilt in normally resistant Dwarf Cavendish banana. Plant Dis Rep 56: 1000–1003

Stover RH, Simmonds N (1987) Bananas. 3rd edn. Longman, Harlow

Strobel GA, Stierle AA, Upadhyay R, Hershenhorn J, Molina G (1993) The phytotoxins of *Mycosphaerella fijiensis*, the causative agent of black sigatoka disease, and their potential use in screening for disease resistance. In: Biotechnology applications for banana and plantain improvement. INIBAP Montpellier, pp 93–103

Su EJ, Su HJ (1984) Rapid method for determining differential pathogenicity of *Fusarium oxysporum* f. sp. *cubense* using banana plantlets. Trop Agric (Trinidad) 61: 7–8

Swartz HJ (1991) Post culture behavior: genetic and epigenetic effects and related problems. In: Debergh PC, Zimmerman RH (eds) Micropropagation: technology and application. Kluwer, Dordrecht, pp 95–122

Tapia VS, Pons S (1989) Cultivo in vitro de embriones de musacaceas. Scientia (Panama) 4: 51–57

Tilney-Basset RAE (1986) Plant chimeras. Edward Arnold, London

Turner DW (1972) Banana plant growth. Gross morphology. Aust J Exp Agric Anim Husb 12: 209–215

Vuylsteke DR (1989) Shoot-tip culture for the propagation, conservation and exchange of *Musa* germplasm. IBPGR, Rome, 56 pp

Vuylsteke D, Swennen R, De Langhe E (1991) Somaclonal variation in plantains (*Musa* spp., AAB group) derived from shoot-tip culture. Fruits 46: 429–439

Vuylsteke D, Swennen R, Wilson GF, De Langhe E (1988) Phenotypic variation among in vitro-propagated plantain (*Musa* spp. cultivar AAB). Sci Hortic 36: 79–88

Walduck GD, Daniells JW, Gall EN (1988) Results of survey of off-types in tissue cultured Cavendish banana in north Queensland 1987. Bananatopics 8: 11–12

Withers LA (1993) Early detection of somaclonal variation. In: Biotechnology application for banana and plantain improvement. INIBAP, Montpellier, pp 200–208

# I.13 Somaclonal Variation for Resistance to *Fusarium*- and *Glomerella*-Caused Diseases in Strawberry (*Fragaria* x *ananassa*)

H. TOYODA[1]

## 1 Introduction

### 1.1 Importance and Distribution

The large fruited garden strawberry, *Fragaria* x *ananassa* is one of the most popular fruits. Because of the high commercial value of this plant, some cultivars have been bred and intensively cultivated all over Japan. Octoploid-cultivated strawberries were first introduced to Japan from Holland in the early 19th century. After World War II, Fairfax and Dorsett were introduced from the US, and used as foundation stock to breed Japanese cultivars. Consequently, many cultivars have been bred by national and prefectural experimental stations, private companies, and some individuals (Oda 1991). The cultivar Hokowase, which was released in 1960 at Hyogo Agricultural Experimental Station, spread over the country and became the most important cultivar, representing more than 50% of the total acreage in 1982. Since 1984, the new cultivars Nyoho and Toyonoka have spread rapidly, replacing most of the earlier cultivars, including Hokowase. Total world production of strawberries has increased considerably in the last decade by the opening of new area for cultivation and the introduction of new cultivars as well as the intensification of the growing process (Jungnickel 1988). In contrast, in Japan, the acreage of land cultivated for strawberry has been decreasing over this period. Despite this decline, the highest level of strawberry production has been maintained by refining growing techniques and enhancing strawberry yield per ha. With the current growing system, strawberry plants are intensively cultivated by means of vegetative propagation. Once a mother plant is infected by viruses or fungi, further vegetative propagation means a spread of disease in a strawberry cultivation area because of its uniformity. Thus, the major breeding objectives are the production of lines resistant to various pathogens, especially to phytopathogenic fungi.

### 1.2 Breeding Objectives and Necessity of Inducing Somaclonal Variation

In spite of its high fruit productivity and established cultivation techniques, the cultivar Hokowase was completely replaced by the substitute cultivars. The

[1] Laboratory of Plant Pathology, Faculty of Agriculture, Kinki University, 3327-204 Nakamachi, Nara 631, Japan

Biotechnology in Agriculture and Forestry, Vol. 36
Somaclonal Variation in Crop Improvement II (ed. by Y.P.S. Bajaj)

major reason for such a drastic change was due to the rapid softening of harvested fruits and the high susceptibility of this cultivar to *Fusarium* wilt. Nowadays, only a few strawberry cultivars have been extensively cultivated in Japan, and most of the yields of strawberry fruits have been obtained from the cultivars Nyoho and Toyonoka (Oda 1991). However, these two cultivars are highly susceptible to crown rot and powdery mildew, respectively, and the yield loss due to these diseases is a serious problem in strawberry cultivation. Therefore, the production of the resistant lines has been a major breeding objective in Japan. To solve such phytopathological and commercial problems, the following strategies could be used for improving strawberry; (1) classical breeding by crossing between resistant and susceptible cultivars, (2) genetic engineering of susceptible host plants by introducing genes involved in resistance, and (3) isolation of disease-resistant genetic variation following tissue culture of the susceptible cultivars. Recently, we succeeded in genetically transforming Japanese strawberry cultivars with *Agrobacterium rhizogenes* (Toyoda et al. 1993), and detected the expression of introduced genes in fruits of transgenic plants. Such a gene manipulation approach may enable us to delay or suppress the progress of fruit softening by transforming strawberry plants with antisense sequences of the polygalacturonase gene and the ACC synthase gene which were involved in ripening or softening of tomato fruits. In addition, the feasibility of a chitinase gene was substantially elucidated with the aim of protecting strawberry plants from powdery mildew (*Sphaerotheca humuli*) (Ikeda et al. 1992). However, the molecular strategy is not always easy and effective, especially for protecting strawberry from certain fungal diseases such as *Fusarium* wilt or crown rot. The tissue culture approach was therefore tried to select valuable genetic variation and to produce disease-resistant lines. This chapter describes an efficient system for tissue culture of Japanese strawberry cultivars (Toyoda et al. 1990), and focuses on possible applications of tissue culture technology for the selection of strawberry lines resistant to *Fusarium* wilt (Toyoda et al. 1991) and crown rot.

### 1.3 Biology of the Pathogens and Mode of Infection, Symptoms, and Control

The causal pathogen of a wilt of strawberry plants is *Fusarium oxysporum* f. sp. *fragariae* (Winks and Williams 1965). The fungus spreads along stolons from infected parent plants into runners. Although these runners appear to be healthy, the vascular tissue is usually discolored, and the fungus can be isolated. Infected plants occur at random on the basis of fungal spread by root contact or cultural operations. The disease causes sudden wilting of all the leaves of the infected plants during the summer months. In the cooler months, however, the disease is less obvious and the affected plants are generally rolled, chlorotic, and often stunted (Winks and Williams 1965).

All species of *Glomerella* have the same type of imperfect stage (*Gloeosporium* or *Colletotrichum*) producing hyaline, ovoid, cylindrical, or somewhat dumbbell-shaped conidia in acervuli. *Glomerella* produce beaked perithecia, either in groups in a stroma or separately. The ascospores are hyaline, one-celled, and curved. *G. cingulata* is the cause of bitter rot in apple and of anthracnose in a large number of other plants (Alexopoulos 1962), including strawberry (Okayama 1988). The pathogen infected into strawberry plants can survive in the crown in an inactive state during the winter season and spread rapidly in the hot season. The important roles of high temperature and humidity on disease development of strawberry anthracnose have been well documented in the field and in the greenhouse. Brown lesions first appear on petioles of strawberry plants in the naturally infested field. Symptoms further develop on mother plants, then on runners and daughter plants. Control of the disease has been carried out by fumigating soil with a fungicide (Horn and Carver 1962) and by improving the watering system to prevent spores from dispersing (Okayama 1993).

## 2 Brief Review of Somaclonal Studies on Strawberry

The main objectives in strawberry breeding have been focused on improvement of cultivars with respect to high yield, favorable fruiting habit, and disease resistance. Although the classical outcrossing and inbreeding techniques have been mainly employed as a major breeding method for these purposes, it is worthwhile mentioning that plants regenerated from meristem- or anther-derived callus can undergo various phenotypical and morphological changes such as chlorosis and dwarfing, fruit size, multi-apexing, fasciate flower bearing, and varying plant size (Niemirowicz-Szczytt 1990). Such variations are shown to originate from medium composition, duration of culture, and number of subcultures (Shaeffer et al. 1980). In commercial micropropagation of strawberry, these variants may be undesirable. However, the selection of varieties showing valuable somaclonal variation could be of interest. Cytological aspects of somaclonal variation were noted in relation to types of tissue cultures and its utilization was discussed for improving genetic traits of strawberry (Niemirowicz-Szczytt 1990). Recently, attempts have been made to isolate disease-resistant lines from strawberry tissue cultures (Toyoda et al. 1991; Takahashi et al. 1992), where *Fusarium* wilt resistance or *Alternaria* black spot resistance was successfully isolated by combination of direct inoculation with the pathogen and the devised selection method, respectively. Especially, the successful selection of black spot-resistant strawberry demonstrated that the direct application of the pathogen to callus made it possible to efficiently isolate resistant calliclones by evading the technical difficulties and time-consuming process of isolation of the host-specific toxin (HST) produced by the pathogen (*A. alternata*) (Takahashi et al. 1992).

## 3 Somaclonal Variation for Disease Resistance

### 3.1 Callus Induction and Plant Regeneration

In tissue culture of strawberry, meristem tip culture has been utilized to produce virus-free strawberry plantlets in vitro. Because viruses are scarcely translocated to meristem tissues, and because of the rare occurrence of somaclonal variations, this culture is useful for maintaining the commercial strawberry cultivars (Niemirowicz-Szczytt 1990). However, stability is not always advantageous, especially when we wish to isolate genetic variation from tissue cultures and apply it to breeding. For this purpose, it is necessary for a different culture system, which supplies abundant variations, to be developed. In a series of studies, the culture conditions for leaf explants of various plant species were pointed out, and somaclonal variations, including disease resistance, were frequently induced in leaf explant-derived callus tissues, and successfully propagated to plants regenerated from these callus tissues (Toyoda and Ouchi 1991). Similar reports have been made by many workers in a large number of plant species (Larkin and Scowcroft 1981; Bajaj 1990). Also in Japanese cultivars of strawberry, such culture systems have been examined and elaborated as a primary step for obtaining useful variants (Toyoda et al. 1990). In this section, an efficient culture system for callus induction and plant regeneration in strawberry leaf explants is described.

Young leaves newly developing from runners of the cvs Hokowase and Nyoho were harvested and surface-sterilized with 70% EtOH and 2% sodium hypochlorite. After washing several times with sterilized water, excised leaf segments (leaf explants) were put onto MS medium (Murashige and Skoog 1962) supplemented with growth regulators, solidified with 0.2% Gellan gum, and adjusted to pH6 with NaOH before autoclaving. The growth regulators used were α-naphthaleneacetic acid (NAA), 2,4-dichlorophenoxyacetic acid (2,4-D), and 6-benzylaminopurine (BA). NAA or 2,4-D was mixed with BA at various concentrations and added to MS medium. Culture bottles were tightly sealed with aluminum sheets and Parafilm, and cultured at 26 °C under a constant illumination of 4,000 lx until some morphological changes in explants or induced calli were observed (for 30–40 days). Induction of actively growing callus tissues was observed in many combinations of NAA and BA, and was most effective especially when leaf explants were cultured with 1 mg/l of NAA and BA. When much higher concentrations (2.5–3 mg/l) of NAA were used, both leaf explants and callus tissues frequently turned brown and the subsequent growth of callus tissues ceased. With much lower concentrations of the growth regulators, callus tissues became brownish, but their growth was vigorous, and many adventitious roots developed from callus tissues 20–30 days after incubation. In any combinations of NAA and BA, however, shoot formation from callus tissues was not observed even when the tissues were excised from explants and subcultured for several passages at intervals of 2 weeks. Successful shoot formation was observed when NAA was replaced by 2,4-D. The shoots developed from callus tissues surrounding leaf explants in the presence of 0.1 mg/l 2,4-D and 1 mg/l BA for Hokowase and 0.1 mg/l 2,4-D and 1.5 mg/l BA for Nyoho. Figure 1A shows shoot developed from callus tissues of Hokowase. These callus tissues were greenish and their growth was much slower than that in callus tissues induced by NAA and BA. In callus tissues which were first induced at the edge of leaf explants, shoot formation was not synchronous. However, the formation was well synchronized after the tissues were separated from the explants and subcultured for two to three passages with the fresh (same) medium (Fig. 1B). In the present experiments, 80–90% of transplanted callus clumps (5–10 mm in diameter) formed shoots and each callus clump developed three to five shoots 10–15 days after transfer. The

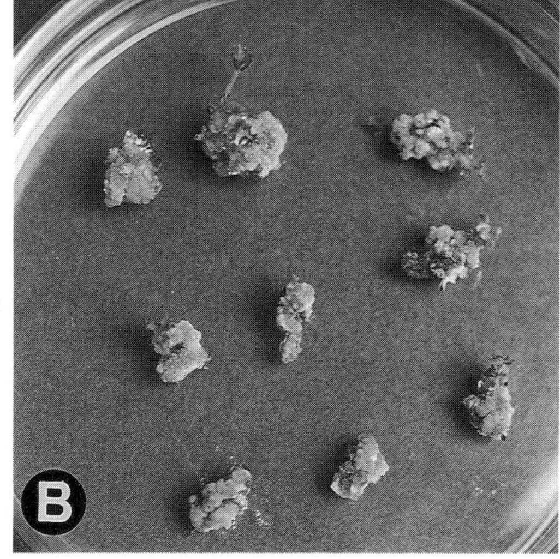


**Fig. 1A, B.** Effects of 2,4-D and BA on callus culture and redifferentiation of strawberry leaf explants. **A** Shoots developed from callus tissues surrounding leaf explants (0.1 mg/l 2,4-D and 1 mg/l BA, 30 days after incubation). **B** Synchronized formation of shoots in callus clumps transferred to the same fresh medium as **A** (10 days after transfer) (Toyoda et al. 1990)

capability of callus tissues to redifferentiate was maintained for at least ten passages when the tissues were subcultured at a 2-week-interval. After leaflets were well developed, shoots were transferred to growth regulator-free MS medium for root initiation. Roots were initiated for 3–4 days and vigorously elongated for 10–15 days after transfer. Regenerated plantlets were transplanted to soil and acclimated for 3–4 days in a moist chamber. These plants grew normally under field conditions. Thus, the present system enables an efficient supply of plants regenerated from leaf-derived callus tissues and analysis of the occurrence of variations in these regenerants.

## 3.2 Selection for *Fusarium* Wilt Resistance

### *3.2.1 Pathogen and Selection System*

Using in vitro systems, some workers effectively selected variants resistant to host-specific toxins and produced lines resistant to the diseases caused by fungal toxin producers. However, it has not been easy to apply such a system to the selection for resistance in diseases caused by soil-borne pathogens, where the mechanisms of pathogenicity or host resistance have not been clarified. Heath-Pagliuso et al. (1988) and Toyoda et al. (1989) reported the successful isolation of plants resistant to soil-borne diseases caused by *F. oxysporum* f. sp. *apii* and *Pseudomonas solanacearum*, respectively, by directly inoculating regenerants with the pathogens. Also in strawberry, the disease caused by the soil-borne pathogen, *F. oxysporum* f. sp. *fragariae*, is a serious problem and the production of resistant lines has been an urgent matter.

Three-week-old cultures of *F. oxysporum* f. sp. *fragariae* grown on potato sucrose agar were macerated in water and used as the inoculum. A pathogen-infested soil was produced by mixing the inoculum with soil in a bed ($1 \times 2$ m, 30 cm depth). This treatment was repeated five times at 1 week intervals. The pathogen was isolated from the test soil using Komada's selective medium for *F. oxysporum* (Komada 1975). The populations of the pathogen were between $5 \times 10^5$ and $10^6$ CFU (colony-forming units)/g soil, indicating that the levels of the pathogen in the present soil were high enough to induce the infection in plants (Komada 1975). For examining the infectivity of the pathogen applied to soil, control strawberry plants from which callus tissues were originally induced were planted in this bed, and the appearance of disease symptoms was examined. In order to ensure the active infection by the pathogen, all experiments were carried out in a greenhouse during the summer months when the soil temperature was between 27 and 32 °C (Winks and Williams 1965).

In the pathogen-infested soil prepared in the present study, approximately 80% of the control plants tested (38 of 50 plants) showed typical primary symptoms of the disease (wilting of leaves, reddish brown vascular discoloration) within 2–3 weeks, and plants withered 1 month after planting. The remaining plants which escaped infection developed runners and produced secondary daughter plants (Table 1). In these daughter plants, however, when they were transplanted to infested soil, symptoms appeared more quickly and all of the plants wilted within 2 weeks. These results clearly indicated that the present method would enable us to obtain resistant lines by isolating nondiseased plants normally multiplied in this soil over two generations of daughter plants obtained through runner formation.

### 3.2.2 *Selection of Resistant Lines*

Plant regenerants obtained from culture were grown for 2 weeks in soil (noninfested with the pathogen) in pots in a greenhouse (Fig. 2A) and then the plants were transplanted to the test soil. Figure 2B shows the regenerated strawberry plants normally growing in this soil. Two regenerated plants developed runners and produced daughter plants throughout three generations (Table 1). Thus, two putative resistant lines (KHFR-1126 and −0228) from 1225 regenerants were isolated. In order to further confirm the successful propagation of resistance acquired in the original plant regenerants (KHFR-1126 and −0228), the daughter plants (third generation since original regenerants) were tested in the summer

**Table 1.** Selection of resistant strawberry lines to *F. oxysporum* f. sp. *fragariae* using plants regenerated from leaf-derived callus tissues. (Toyoda et al. 1991)

| Selection[a] | No. of plants tested | No. of plants | | No. of plants forming daughter plants | |
|---|---|---|---|---|---|
| | | Wilting | Forming daughter plants | Wilting | Resistant |
| 1st selection | | | | | |
| Control | 50 | 38 | 12 | 12 (1)[b] | 0 |
| Regenerants | 1225 | 1042 | 183 | 181 (1)[b] | 2 (3)[c] |
| 2nd selection | | | | | |
| Control | 50 | 50 | 0 | – | – |
| KHFR-1126 | 5 | 0 | 5 | 0 | 5 (4)[c] |
| KHFR-0228 | 5 | 0 | 5 | 0 | 5 (4)[c] |

[a] Regenerated and control strawberry plants were transplanted to a pathogen-infested soil, and normally growing plants which could produce daughter plants through runner formation were selected as putative resistant lines in the first selection. In the second selection, daughter plants of both isolated lines (third generation since the original regenerants) were directly inoculated with this pathogen and then transplanted to the pathogen-infested soil.
[b] Generation of daughter plants in which wilting symptoms were detected.
[c] Generation of daughter plants vegetatively produced from plants transplanted to a pathogen-infested soil.

season of the following year. In this experiment, the roots of plants were dipped in a spore suspension ($10^6$ spores/ml) of the pathogen and then transplanted to a pathogen-infested soil. With this inoculation, control plants (50 plants) were infected and all the plants died within 3 weeks after planting. These results indicated that the present inoculation was effective enough to eliminate the infection plant escapes. With this method, the disease resistance of the selected strawberry lines was examined. The data showed that the present lines grew normally and developed runners even after direct inoculation and produced daughter plants through four generations in a pathogen-infested soil (Fig. 2C and Table 1). Thus, this study demonstrated that the plant disease resistance acquired through tissue culture could be stably passed on to daughter plants vegetatively, though the mechanism for disease resistance acquired is not obvious.

## 3.3 Selection for Crown Rot Resistance

### *3.3.1 Pathogen and Selection System*

The crown-rot pathogen, *Glomerella cingulata* (*Colletotrichum gloeosporioides*), which was isolated and kindly donated by Dr. K. Okayama, (Nara Agricultural Experiment Station, Japan; Okayama 1988), was cultured in liquid Czapek medium (without agar) at 26 °C for 3–4 weeks. Newly produced conidiospores were washed with sterilized water and collected by centrifugation. The conidia were suspended in sterilized water to give $10^6$ spores/ml, and used as inoculum. Leaves were harvested from 8–15 petioles per plant, and one of trifoliolate leaves on each petiole was used for inoculation. Excised leaves were placed in

**Fig. 2A–C.** Selection for resistance to wilting disease caused by the fungal pathogen, *F. oxysporum* f. sp. *fragariae* using regenerated strawberry plants from leaf-derived callus tissues. **A** Regenerated plants acclimated for 2 weeks in a greenhouse. These plants were transplanted to soil infested with the pathogen for selection. **B** Normally growing regenerant (KHFR-1126) in a pathogen-infested soil (1 month after planting). Withered or heavily stunted plants by field infection of the pathogen are indicated by *arrow*. Note the vigorous growth of the isolated line during the period of 1 month. Daughter plants vegetatively produced through runner formation of three generations were used for an additional inoculation. **C** Test of the stable propagation of disease resistance acquired in the line, KHFR-1126, to daughter plants. Daughter plant (*0*) of KHFR-1126 (third generation since the original regenerant) was directly inoculated by dipping roots in a pathogen spore suspension and transplanting to a pathogen-infested soil. Daughter plants in each generation from the first (*1*) to fourth (*4*) normally grew and developed runners and new daughter plants, while all of control plants died by the present dual inoculations. (Toyoda et al. 1991)

a moistened Petri dish and punctured with a bundle of sterilized sewing needles ("needle-prick inoculation method"). A drop (100 μl) of the conidial suspension was placed onto needle-pricked portions of leaves, and inoculated leaves were incubated at 28 °C for 6 days under a continuous illumination of 4000 lx. The progress of necrotic lesion formation around the inoculation sites was examined for determining the degree of resistance or susceptibility of tested plants.

### *3.3.2 Selection of Resistant Lines*

Experiments were designed to select the crown-rot-resistant cells which preexisted in strawberry leaf tissues, and to produce resistant strawberry plants through tissue culture. For this purpose, the following two steps were considered essential; (1) easy and effective inoculation procedure for precise evaluation of the resistant or susceptible responses, and (2) highly efficient system for callus induction and plant regeneration. Since the tissue culture system for strawberry had been established in our laboratory, the present study first focused on the responses of the resistant (Hokowase) and susceptible cultivars (Nyoho and Toyonoka) to inoculation with *G. cingulata*. As shown in Fig. 3, inoculated leaves exhibited three different types of necrotic lesion formation. The first type of lesion was a small necrotic spot (LEL; Fig. 3A) in which the pathogen was strictly limited around the inoculation sites. The second type was a necrotic lesion (MEL; Fig. 3B) showing moderate expansion, and the last type was a rapidly and extensively expanded lesion (REL; Fig. 3C) which, in most cases, fused with other lesions 5 or 6 days after inoculation. On the other hand, no necrotic lesion was formed when water was dropped onto the punctured sites. Thus, the present results clearly indicated that the necrotic lesion formation was due to the pathogen infection, but not to the mechanical injury.

The types of lesions were detected in both the susceptible and resistant cultivars, but their frequencies were considerably different among these cultivars (Fig. 4). In the resistant cultivar, the LEL was formed at approximately 70% of the inoculation sites, whereas in the susceptible cultivars, this type of lesion was found in approximately 40 and 25% of the inoculation sites in Toyonoka and Nyoho, respectively. There were no significant differences in the appearance of MEL among the three cultivars. These results indicated that the frequencies of LEL reflect the degree of resistance or susceptibility of the cultivars. Judging from these results, we deduced that a resistant cultivar could be produced by enhancing the frequency of LEL in the susceptible cultivar through tissue culture. To effectively select cells that lead to the formation of LEL, it is important to demonstrate the existence of these cells in leaves of the cultivar to be improved. Therefore, inoculated leaves were classified on the basis of the type of the necrotic lesion. More than 50% of inoculated leaves of the cultivar Nyoho exhibited two or three types of lesions on the same inoculated leaf, although approximately 40% of inoculated leaves responded only with the REL at all of the inoculation sites. These results indicate that types of cells different in respect to responses to the pathogen exist in the same leaf tissues, hence in the same plants. To produce the strawberry plants exhibiting predominantly a LEL response to the pathogen,

**Fig. 3A–C.** Formation of necrotic lesions in strawberry leaves inoculated with the crown-rot disease pathogen (*G. cingulata*) by a needle-pricked inoculation method. The present method produced three types of necrotic lesions. **A** Small necrotic spots (LEL; cv. Hokowase). **B** Moderately expanded lesions (MEL; cv. Nyoho). **C** Rapidly expanded and fused lesions (REL; Nyoho) formed around the inoculation sites (*arrow heads*) 6 days after inoculation. (H. Toyoda, unpubl.)

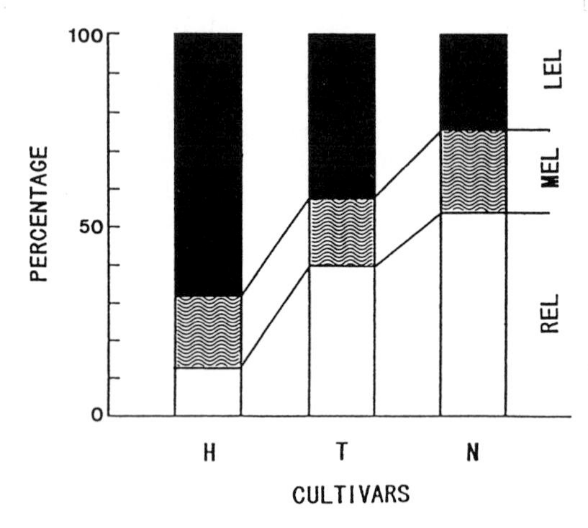


**Fig. 4.** Frequencies of three types of necrotic lesions on leaves of strawberry cultivars Hokowase (*H*), Toyonoka (*T*), and Nyoho (*N*) after inoculation with the crown-rot disease pathogen. (H. Toyoda, unpubl.)

we tried to concentrate these cells by repeating the tissue culture and selection by inoculation, using the leaves showing different types of necrotic lesion. However, this approach was unfeasible, because inoculated leaves become unsuitable for tissue culture due to deterioration of leaf tissues and contamination with the pathogen inoculated. Therefore, in the present study, callus tissues were induced from leaf explants harvested at random from uninoculated leaves, and the regenerated plants obtained were inoculated with the pathogen as above. In this experiment, 1806 leaves of 336 regenerants were used for inoculation, and the results are summarized in Fig. 5, where the regenerant numbers are plotted on the basis of the frequencies of the LEL on inoculated leaves. The strawberry plants regenerated from leaf-callus tissues were diverse in their responses; 228 regenerants showed lower values and the remaining regenerants values higher than those of the parental cultivar (23.1%). Moreover, scores higher than that of the resistant cultivar Hokowase (69.9%) were recorded in some regenerants. These results

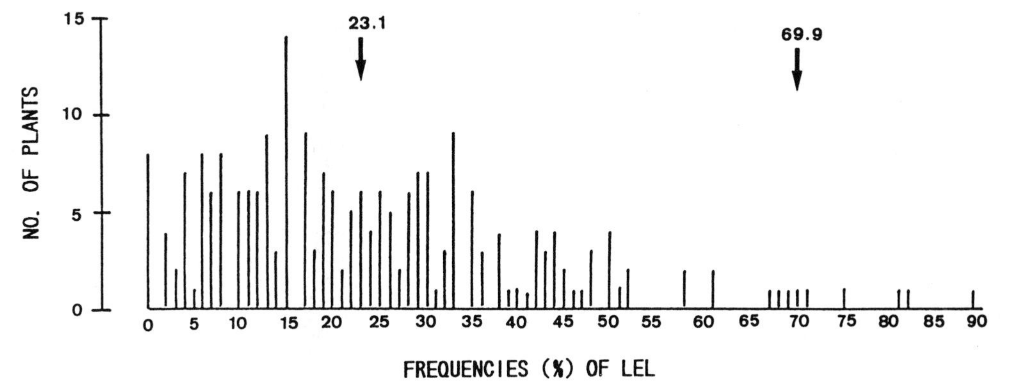


**Fig. 5.** Distribution of regenerated plants derived from leaf explants of strawberry cultivar Nyoho on the basis of frequencies of LEL after inoculation with the crown-rot pathogen. (H. Toyoda, unpubl.)

suggest that the "resistant-type" cells (cells respond with LEL formation) were highly and efficiently concentrated in these high-score regenerants. In a subsequent experiment, therefore, callus induction and plant regeneration were further conducted using leaf explants obtained from the high-score regenerants. As a result, six regenerant candidates were examined for lesion formation after inoculation. Table 2 summarizes the types of lesion formation in each of the regenerant leaves. The data clearly indicate that almost all the leaves of these secondary regenerants showed the LEL or the LEL + MEL response, and no secondary regenerant leaves responded with only the REL response after inoculation. These regenerated plants exhibited increased levels of resistance to infection with this pathogen. Thus, these data suggest that the "resistant-type" cells can be preferentially concentrated in callus-derived regenerants produced by the present culture system.

## 4 Summary and Conclusions

Lines of strawberry resistant to fungal wilt disease were selected from plants regenerated from leaf-derived callus tissues. Regenerants were transplanted to a field heavily infested with this pathogen, and normally growing plants were selected as the putative resistant lines. Daughter plants produced vegetatively through runner formation of the lines were similarly tested in the pathogen-infested field over an additional three generations. Finally, two resistant lines were obtained from a total of 1225 regenerants. The stable propagation of disease resistance in these lines was confirmed by directly inoculating the daughter plants with the pathogen and planting them in a pathogen-infested soil. All the control plants were efficiently infected and died within 1 month. The isolated plant lines grew and developed runners even after direct inoculation, and produced daughter plants in this soil. The present strawberry lines were directly selected from regenerated plants using the dual inoculation method (field infection and root

**Table 2.** Formation of necrotic lesions on leaves of secondary regenerants of strawberry after inoculation with *G. cingulata*. (H. Toyoda, unpubl.)

| Secondary | Lesion types formed on leaves[a] | | | | | | | | |
|---|---|---|---|---|---|---|---|---|---|
| regenerants | 1 | 2 | 3 | 4 | 5 | 6 | 7 | 8 | 9[b] |
| N2-132 | L | LMR | L | L | L | L | L | L | L |
| N2-105 | LMR | LM | L | L | LM | L | L | L | L |
| N2-096 | L | L | L | LM | L | L | | | |
| N2-101 | L | L | L | L | LMR | L | L | | |
| N2-157 | LM | LM | LM | L | L | L | | | |
| N2-171 | LMR | LM | L | L | L | LMR | | | |

[a] L, M, and R represent LEL, MEL, and REL, respectively. The dual and triple notations, LM or LMR, shows the simultaneous occurrence of different lesion types at the inoculation sites of the same leaves. Refer to the legend of Fig. 1 for LEL, MEL and REL.
[b] The secondary regenerants were obtained from callus tissues of the original regenerants with high frequencies of the LEL. Leaves were harvested from 6–9 petioles of these secondary regenerants and inoculated with the pathogen by a needle-prick inoculation method.

inoculation with the pathogen). Although selection pressure was not made with tissue cultures in the present study, utilization of the pathogen-infested soil reflected the natural infection by the pathogen in the farm and moreover promoted the primary selection for resistant lines. Thus, the present selection system may be applicable to other soil diseases in which the mechanisms for resistance or pathogenicity have not been clarified.

To establish an efficient system for producing crown-rot-resistant lines of strawberry, a tissue culture system was used for selecting and enhancing resistant variant cells which exist indigenously in leaf tissues of the susceptible cultivar. Leaves of the resistant and susceptible cultivars were inoculated with the crown-rot pathogen by a needle-prick inoculation method. Although three types of lesions (small necrotic spots, moderate, and extensively expanding lesions) were formed on leaves of both the resistant and susceptible cultivars, the former was characterized by a higher percentage of the small necrotic spots. Cells from the resistant-type lesion were multiplied by inducing callus tissues from leaf explants of the susceptible cultivar, and regenerants showing higher frequencies of the resistant-type lesion were selected. The multiplication of these cells to the level of a resistant cultivar was achieved by repeating callus induction and plant regeneration from the selected regenerants. Although the mechanisms for lesion formation have not been elucidated, the present study suggests that this tissue culture approach is effective in selection and multiplication of variant cells for important traits. At present, however, this study provides a useful and practical technique for improving a strawberry species with respect to the varieties under cultivation.

## References

Alexopoulos CJ (1962) Pyrenomycetes the perithecial fungi. In: Introductory mycology. John Wiley, New York, pp 292–332

Bajaj YPS (1990) Somaclonal variation – origin, induction, cryopreservation, and implications in plant breeding. In: Bajaj YPS (ed) Biotechnology in agriculture and forestry, vol 11. Somaclonal variation in crop improvement I. Springer, Berlin Heidelberg New York, pp 3–48

Heath-Pagliuso S, Pullman J, Rappaport L (1988) Somaclonal variation in celery: screening for resistance to *Fusarium oxysporum* f. sp. *apii*. Theor Appl Genet 75: 446–451

Horn NI, Carver RG (1962) A new crown rot of strawberry plants caused by *Colletotrichum fragariae*. Phytopathology 53: 768–770

Ikeda S, Toyoda H, Yoshida K, Koreeda K, Chatani K, Ouchi S (1992) Digestion of haustoria of *Sphaerotheca* powdery mildew fungi by an endogenous chitinase. Ann Phytopathol Soc Jpn 58: 780–783

Jungnickel F (1988) Strawberries (*Fragaria* spp. and hybrids). In: Bajaj YPS (ed) Biotechnology in agriculture and forestry, vol 6. Crops II. Springer, Berlin Heidelberg New York, pp 38–103

Komada H (1975) Development of a selective medium for quantitative isolation of *Fusarium oxysporum* from natural soil. Rev Plant Prot Res 8: 114–125

Larkin PJ, Scowcroft WR (1981) Somaclonal variation – a novel source of variability from cell cultures for plant improvement. Theor Appl Genet 60: 197–214

Murashige T, Skoog F (1962) A revised medium for rapid growth and bioassays with tobacco tissue cultures. Physiol Plant 15: 473–497

Niemirowicz-Szczytt K (1990) Somaclonal variation in strawberry. In: Bajaj YPS (ed) Biotechnology in agriculture and forestry, vol 11. Somaclonal variation in crop improvement I. Springer, Berlin Heidelberg New York, pp 511–528

Oda Y (1991) The strawberry in Japan. In: Dale A, Luby JJ (eds) The strawberry into the 21st century. Timber Press, Portland, pp 36–46

Okayama K (1988) Occurrence and control of strawberry anthracnose. Plant Prot 42: 559–563

Okayama K (1993) Effects of rain shelter and capillary watering on disease development of symptomless strawberry plants infected with *Glomerella cingulata* (*Colletotrichum gloeosporioides*). Ann Phytopathol Soc Jpn 59: 514–519

Shaeffer GW, Damiano C, Scott DH, McGrew JR, Krul WR, Zimmerman RH (1980) Transcription of a panel discussion on the genetic stability of tissue culture propagated plants. In: Proc Conf Nursery production of fruit plant through tissue culture – application and feasibility. USDA-ARS, pp. 64–79

Takahashi H, Takai T, Matsumoto T (1992) Resistant plants to *Alternaria alternata* strawberry pathotype selected from calliclones of strawberry cultivar Morioka-16 and their characteristics. J Jpn Soc Hortic Sci 61: 323–329

Toyoda H, Ouchi S (1991) The use of somaclonal variation for the breeding of disease-resistant plants. In: Patil SS, Ouchi S, Mills D, Vance C (eds) Molecular strategies of pathogens and host plants. Springer, Berlin Hiedelberg New York, pp 229–239

Toyoda H, Shimizu K, Chatani K, Kita N, Matsuda Y, Ouchi S (1989) Selection of bacterial wilt-resistant tomato through tissue culture. Plant Cell Rep 8: 317–320

Toyoda H, Horikoshi K, Inaba K, Ouchi S (1990) Plant regeneration of callus tissues induced from leaf explants of strawberry. Plant Tissue Cult Lett 7: 38–41

Toyoda H, Horikoshi K, Yamamoto Y, Ouchi S (1991) Selection for *Fusarium* wilt disease resistance from regenerants derived from leaf callus of strawberry. Plant Cell Rep 10: 167–170

Toyoda H, Kami C, Shumitani K, Zheng SJ, Hosoi Y, Ouchi S (1993) Transformation of Japanese cultivars of strawberry with *Agrobacterium rhizogenes*. Plant Tissue Cult Lett 10: 92–94

Winks BL, Williams YN (1965) A wilt of strawberry caused by a new form of *Fusarium oxysporum*. Queensland J Agric Anim Sci 22: 475–479

# I.14 In Vitro Selection for Salt Tolerance in *Citrus* Rootstocks

J. Bouharmont[1] and N. Beloualy[1]

## 1 Introduction

*Citrus* is one of the most important groups of fruit trees, and grows in tropical and subtropical regions. The genus *Citrus* (Rutaceae) includes several species, but the taxonomy is very confused because of the variability of many species and the possibility of interspecific hybridization: the number of species proposed varies from 16 to 157 (Cameron and Soost 1976). The center of origin of the genus is southeast Asia. Domestication started very early in that region, and propagation of hybrids and mutants increased the complexity of the group; wild relatives have not been found.

Many *Citrus* species are apomictic and produce adventitious (nucellar) embryos. This mode of reproduction by seeds is useful for clonal propagation of new varieties and rootstocks. Grafting is commonly used and rootstocks are selected firstly for their resistance to soil-borne diseases. *Poncirus trifoliata* is one of the most important rootstocks, and it is disease-resistant and cold-tolerant.

Many citrus varieties derive from spontaneous mutations; interspecific and sometimes intergeneric hybridizations are also applied in breeding. The objectives are varied: precocity, yield, quality of the fruits and juices, disease resistance, cold hardiness, and parthenocarpy are traits that pertain to the scion, while the most important characters of the rootstocks are pest and disease resistance, and also salt tolerance. The main commercial areas are in the subtropical, more or less arid regions, and the cultures are frequently irrigated. According to Furr and Ream (1968), *Citrus* is among the crops most sensitive to salinity. It seems that water of bad quality has a more negative effect than saline soil.

Traditional breeding using crosses and recombination has not been applied in citrus rootstocks because of the long generation time of the plants and of the frequent nucellar embryony. On the contrary, the use of interspecific and intergeneric hybridization is frequent, since the incompatibility barriers are rare. All rootstocks are salt-sensitive, but some of them are more affected by salinity than others. In 1989, Gallasch and Dalton reported a comparison of 28 imported rootstocks and 3 others used in Southern Australia; after irrigation with salt water, 15 genotypes were not affected in growth and were better than the 3 older rootstocks. Furr and Ream (1968) showed that when a highly salt-tolerant rootstock, such as Rangpour lime, was crossed with one of low tolerance, such as

[1] Laboratoire de Cytogénétique, Université Catholique de Louvain, Place Croix-du-Sud 4, B-1348, Louvain-la-Neuve, Belgium

Biotechnology in Agriculture and Forestry, Vol. 36
Somaclonal Variation in Crop Improvement II (ed. by Y.P.S. Bajaj)

Rubidoux trifoliate orange, the progeny exhibited varying degrees of salt tolerance, most seedlings being intermediate. In crosses of tolerant parents, a large proportion of seedlings were tolerant. Interspecific hybridizations were also used in Florida for improving rootstocks for various traits: a hybrid (citrumelo Swingle) between Duncan grapefruit and *Poncirus trifoliata* was tolerant to tristeza virus and *Phytophthora parasitica*, and moderately tolerant to soil salinity (Hutchinson 1974). Little is known about the behavior of subsequent generations: nucellar embryony is an obstacle to citrus breeding, but is valuable for the reproduction of genetically uniform plants for rootstocks (Cameron and Soost 1976).

Mutagenesis appears to be the best way for inducing characters which are not present in the available rootstocks and related species. It seems that mutagenic treatments have not been applied to plants or seeds for improving salt tolerance of citrus rootstocks. Nevertheless, mutagenesis has been used at the cell level and in vitro selection was successful. In vitro culture of citrus cells and plant regeneration have been frequently reported, and thus it was realistic to experiment with in vitro selection and to look for somaclonal variation in these plants.

## 2 In Vitro Culture in *Citrus*

*Citrus* represents one of the few groups of woody plants where in vitro cultures have been successful for plant regeneration, micropropagation, micrografting, protoplast fusion, and cell transformation, and the subject has been reviewed (see Button and Kochba 1977; Navarro 1992; Moore et al. 1993; Louzada and Grosser 1994).

The applications of in vitro cultures in *Citrus* species are not new: Bové and Morel (1957) reported callus induction from plantlets of *C. limon*, and Rangan et al. (1968) described in vitro initiation of nucellar embryos in monoembryonic *Citrus*. A medium was proposed by Murashige and Tucker (1969) for citrus tissue culture and was then used in many laboratories. A number of experiments were later reported concerning the influence of auxins, cytokinins, and other substances on cell proliferation, embryogenesis, organ induction, and differentiation.

Somatic embryogenesis was achieved in several species starting from different explants. Since many *Citrus* species are polyembryonic, ovule and nucellus are very convenient explants, giving embryos directly or after callus proliferation (Kochba et al. 1972). Bajaj (1984), and Marin and Duran-Vila (1988) described cryopreservation of orange explants in liquid nitrogen.

Anther culture has been successful in several *Citrus* species: Drira and Benbadis (1975) reported the proliferation of androgenic calli for *C. limon*, while Hidaka et al. (1979) recovered haploid, aneuploid, diploid, and mixoploid embryos and plants from anther culture of *Poncirus trifoliata*. Callus proliferation has also been obtained starting from cellular endosperm of several species excised 12–14 weeks after anthesis, and triploid plantlets of *Citrus sinensis* were regenerated (Gmitter et al. 1990). Embryogenic cultures of *C. sinensis* on a medium supplemented with colchicine was also used for inducing polyploidy: diploid and tetraploid plantlets were regenerated, while untreated cultures yielded only diploid regenerants (Gmitter et al. 1991).

Protoplasts have been isolated from ovular calli of several species by Vardi et al. (1975, 1982); embryos and normal plants were regenerated after plating. Using protoplast fusion, some hybrids obtainable by sexual fertilization have been recovered, but some new interspecific or intergeneric combinations have also been described and their hybrid origin was confirmed by morphological, cytological, and biochemical evidence. The first somatic hybrid was obtained by fusion of protoplasts isolated from embryogenic cells of *Citrus sinensis* and from leaves of *Poncirus trifoliata* (Ohgawara et al. 1985). Other hybrids were later reported between Navel orange and *C. unshiu* and between orange and grapefruit (Kobayashi et al. 1988; Ohgawara et al. 1989), *C. sinensis* and *Severinia disticha*, *C. reticulata*, *Citropsis gilletiana*, and *Fortunella crassifolia* (Grosser et al. 1988, 1990; Deng et al. 1992). The cold, salt, and boron tolerance of *Severinia* and its resistance to *Phytophthora* and nematodes could be used for rootstock improvement. Saito et al. (1991), using electroporation, recovered eight somatic hybrids between *C. sudachi* and *C. aurantiifolia*. More recently, Louzada et al. (1993) obtained somatic hybrids between two sexually incompatible species, *C. sinensis* and *Atalantia ceylanica*.

Vardi et al. (1990) succeeded in transferring cytoplasmic organelles of *Microcitrus* into *Citrus jambhiri* and *C. aurantium* protoplasts. The colonies and regenerated plants exhibited the morphological features of the recipient species. These plants were homoplastomic, containing either *Microcitrus* or *Citrus* chloroplasts.

Several experiments indicate the possibility of obtaining transgenic plants in citrus. Hidaka et al. (1990) used cocultivation of callus and pollen embryoid cells of *C. sinensis* and *C. reticulata* with an *Agrobacterium tumefaciens* strain which harbored a vector containing a gene for phosphotransferase II or hygromycine phosphotransferase. Transformed colonies were kanamycin or hygromycin-resistant; they yielded embryos and plantlets. Vardi et al. (1990) used polyethyleneglycol for direct transformation of *C. jambhiri* protoplasts and regenerated transgenic plants after selection for antibiotic resistance. Moore et al. (1992, 1993) obtained two transgenic plants of Carrizo citrange after coculture of internodal stem segments with *Agrobacterium*. The transformation was proved by the expression of the gene for $\beta$-glucuronidase (GUS) and it was confirmed in the plants by Southern analysis.

## 3 Somaclonal Variation and In Vitro Selection

### 3.1 Previous Works

The importance of somaclonal variation in citrus is not clearly established. Kobayashi (1987) reported no modifications in floral or leaf characters, oil composition, isozyme patterns, and chromosome numbers in 27 protoclones derived from orange protoplasts. Nevertheless, Nadel and Spiegel-Roy (1987) succeeded in selecting *C. limon* cell variants resistant to a toxin of *Phoma tracheiphila*. According to Vardi et al. (1986), the sensitivity of nucellus calli to a *Phytophthora citriophora* culture filtrate is related to the tolerance of the plants to the pathogen, indicating that the filtrate could be used for in vitro selection. A few experiments of in vitro selection have actually been reported, but selection came after application of mutagenic treatments.

In vitro selection for salt tolerance has been the subject of experiments at the Volcani Center (Israel). Callus cells of *C. sinensis* capable of growing in the presence of 0.2 M NaCl were obtained after $\gamma$-irradiation and exposure to a medium containing salt. The increased tolerance for salt was retained after four consecutive transfers in medium without salt (Ben-Hayyim and Kochba 1982). Ben-Hayyim and Kochba (1983) confirmed the stability of the induced variation; they reported observations on ion uptake by normal and selected cells and their growth when cultured in the presence of different salts, indicating that salt tolerance is due to a partial avoidance. Embryogenesis occurred in both the presence and absence of NaCl, but the globular embryos were white on saline medium. The embryos regenerated from salt-tolerant calli were also salt-tolerant, as well as calli derived from regenerated plantlets, indicating acquisition of salt tolerance at the whole-plant level (Ben-Hayyim and Goffer 1989). NaCl-tolerant cell lines were later selected in sour orange (*Citrus aurantium*): the mechanism operating in these cells seemed to differ from the one operating in *C. sinensis* (Ben-Hayyim et al. 1985).

Deng et al. (1989) applied mutagenic treatments ($\gamma$-rays or chemical mutagens) before five to seven passages in the presence of NaCl for selection. In this way, mutant cell lines able to survive in 0.8% NaCl (137 mM) were obtained in four rootstocks and four scion varieties. Salt tolerance was maintained after three passages on a nonselective medium. The mutant lines exhibited a 30–70% higher proline accumulation and a lower peroxidase activity; $Na^+$ and $Cl^-$ absorption were also altered. Plantlets were regenerated.

Garcia-Agustin and Primo-Millo (1995) treated unfertilized ovaries of Troyer citrange with EMS for inducing mutations in the nucellar tissues in order to improve the tolerance to high NaCl levels in irrigation water. After in vitro culture, embryos developed from the nucellus and the plantlets were tested for NaCl tolerance. The selected plants were characterized by a higher concentration of $Na^+$ in roots and shoots and a lower accumulation of that ion in leaves with increasing NaCl levels in the nutrient solution.

Table 1 gives a summary of the applications of in vitro cultures for improving NaCl tolerance in citrus.

### 3.2 Callus Selection and Plant Regeneration

In our study, mature embryos of three rootstocks were used. *Poncirus trifoliata*, *Citrus aurantium*, and Carrizo citrange (*C. sinensis* × *P. trifoliata*). After removing the cotyledons, the explants were grown in the medium described by Murashige and Tucker (1969) for *Citrus* tissues (Beloualy 1991). Induced calli were sub-

**Table 1.** Summary of the work done on the in vitro tolerance of citrus to NaCl

| Plant species | Explant used | Observations | Reference |
|---|---|---|---|
| *Citrus sinensis* | Ovules | γ-Ray mutagenesis | Ben-Hayyim and Kochba (1982) |
| *Citrus sinensis* | Ovules | Cell selection | Ben-Hayyim and Kochba (1983) |
| *Citrus aurantium* | Ovules | | Ben-Hayyim et al. (1985) |
| *Citrus sinensis* *Citrus aurantium* | Ovules | Cell selection | Spiegel-Roy and Ben-Hayyim (1985) |
| *Citrus sinensis* | Ovules | Plant regeneration | Ben-Hayyim and Goffer (1989) |
| *Citrus* and rootstocks | Ovules | γ-Ray and chemical mutagenesis | Deng et al. (1989) |
| *Poncirus trifoliata* | Mature embryos | Plant regeneration | Beloualy and Bouharmont (1992) |
| *Poncirus trifoliata* Citrange | Mature embryos | | Bouharmont et al. (1993) |
| Citrange | Ovaries | EMS treatment | Garcia-Agustin and Primo-Millo (1995) |

cultured every month by transfer to fresh medium. Shoots and embryos were regenerated after transfer of calli to culture media supplemented with naphthalene acetic acid and benzylaminopurine. Rooted plantlets were later obtained, propagated in vitro and acclimatized to soil.

Mutagenic agents were not applied to cultured cells. For selection, callus fragments of the three rootstocks of about 50 mg were grown on media containing different concentrations of NaCl. In the presence of 85 and 171 mM NaCl, cell proliferation was strongly limited, but most calli survived after 4 weeks. These concentrations were then used for a large-scale selection of somaclonal variants: for each rootstock, more than 6000 calli were grown for 4 weeks on these two media. Only six calli of *Poncirus* survived and showed a growth which was better than that of the original calli cultured on a normal medium. Four calli derived from a single explant were selected on 85 mM NaCl, and two others, from another explant, were recovered from the medium containing 171 mM NaCl. Two calli derived from a citrange embryo were also selected on 85 mM NaCl: their weight on the selective medium was nearly as high as that of original calli cultured on the medium without salt. The selection was not successful for *C. aurantium*.

The selected cell lines were subcultured for 5 months on saline medium. After their transfer to a nonsaline medium, the behavior of the cell populations derived from the eight selected calli varied markedly: some subcultures grew normally, the growth of others was slow or irregular, while many cell lines did not grow, and died. Only cell lines growing normally on both saline and nonsaline media were retained and their salt tolerance was confirmed during a new culture on a medium containing 171 mM NaCl. These calli were then transferred to the regeneration

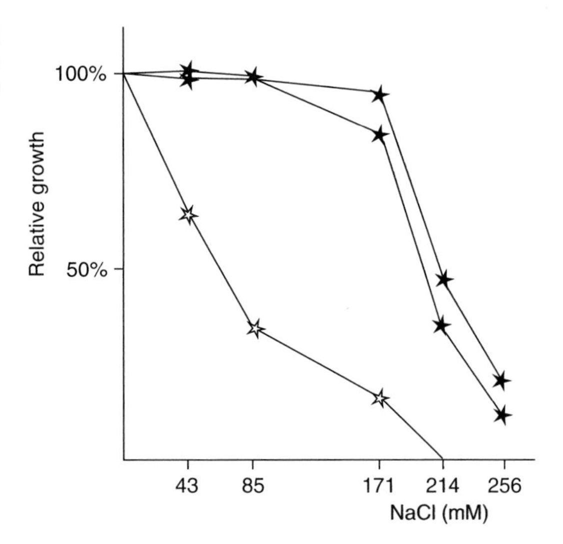


**Fig. 1.** Relative growth of unselected (☆) and selected (★) cells of citrange grown on different salt concentrations

medium supplemented with 0, 43, 85, or 171 mM NaCl. In the absence of NaCl in the medium, the regeneration rates were similar in selected and unselected cells of *Poncirus*, but they were improved in salt-tolerant lines of citrange. The addition of NaCl in the medium inhibited regeneration from unselected calli, while regeneration and plantlet growth were not affected or were improved for selected cell lines of both rootstocks. Finally, the persistence of the salt tolerance was controlled by induction of calli from the regenerated plants and their comparison with the previously selected cell lines.

### 3.3 Characterization of the Selected Cells

The citrus cells are very salt-susceptible: the addition of 43 mM of NaCl to the culture medium drastically reduces the growth of *Poncirus* calli. On a nonsaline culture medium, the growth of the selected calli was generally similar to that of nonselected cells, and it was often improved. It remained high for salt concentrations as high as 171 mM; the growth inhibition was very clear for higher concentrations (Beloualy and Bouharmont 1992). The results were very similar for citrange: the growth of selected cell lines remained stable for salt concentrations up to 171 mM and decreased for higher concentrations (Fig. 1).

In both rootstocks, the proline content rose rapidly in the original cells with increasing salt concentrations in the culture medium, while it was not much affected in the salt-tolerant cell lines (Bouharmont et al. 1993). Figure 2 shows the behavior of two salt-tolerant cell lines of citrange compared to the original cells. According to these observations, the salt tolerance of the selected cells did not seem to be due to a proline accumulation in the cells.

When unselected calli of *Poncirus* and citrange were grown on media containing NaCl (43 or 171 mM), the ion content ($Na^+$ and $Cl^-$) increased mainly during the first days, particularly in citrange, and then remained rather constant.

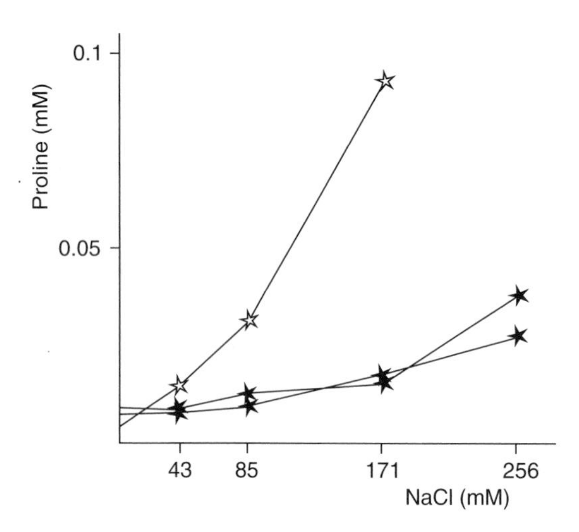


**Fig. 2.** Proline content in unselected (☆) and selected (★) cells of citrange for increasing salt concentrations

In comparison, the uptake of $Na^+$ and $Cl^-$ was much slower in the selected cell lines (Fig. 3). Nevertheless, after a month, the final concentrations were a little higher in the salt-tolerant cells of *Poncirus* and the difference was significant for sodium when NaCl concentration in the culture medium was at least 85 mM. For citrange, on the contrary, $Na^+$ and $Cl^-$ level remained much lower in the salt-tolerant cells than in the unselected ones. The slower ion penetration in selected calli could be explained by modifications of the cell membrane properties, but some consequences of these modifications should differ in *Poncirus* and citrange.

Increasing NaCl concentration in the culture medium reduced progressively $K^+$ and $Ca^{2+}$ uptake by the unselected cells. In selected cells, on the contrary, these cations rose for moderate NaCl concentrations and decreased for high salt levels. Consequently, the availability of $K^+$ and $Ca^{2+}$ remained higher in selected cells.

In a last experiment, the influence of NaCl, KCl, $Na_2SO_4$, sea water, and mannitol on callus growth was compared for equimolar concentrations. All treatments reduced more drastically the growth of unselected calli than that of NaCl-tolerant cells of *Poncirus* and citrange, particularly for the highest concentrations (171 and 200 mM), but the responses were different according to the nature of the chemical added to the medium (Fig. 4). The influence of KCl was particularly strong. The growth of the salt-tolerant calli was rather similar on media supplemented with NaCl and seawater, but it was better with seawater, for the highest concentrations, probably because of a broader ionic diversity. The salt-tolerant cells were moderately affected by mannitol, compared to the original calli: thus it seems that tolerance to water stress plays a part in the adaptation of these cells.

### 3.4 Characters of Regenerated Plants

Plantlets regenerated from selected and unselected cell lines of both rootstocks were micropropagated and the clones were compared on different culture media

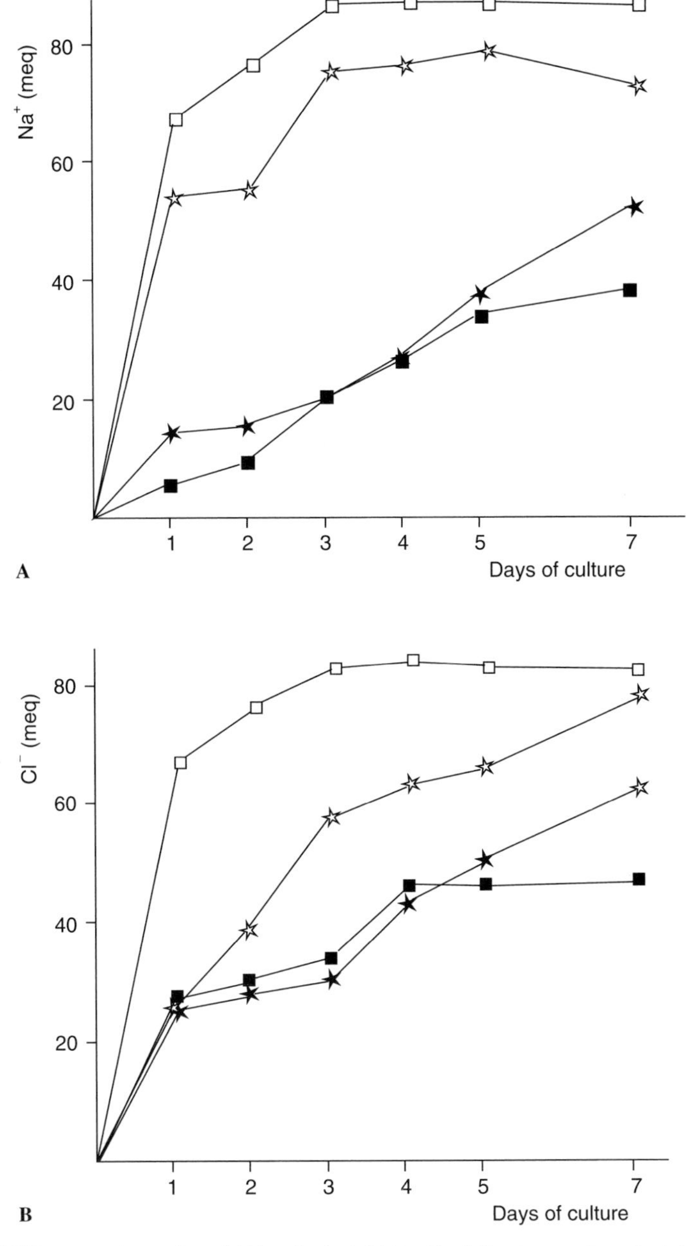


**Fig. 3. A** $Na^+$ and **B** $Cl^-$ content in unselected (☆) and selected (★) cells of *Poncirus*, and unselected (□) and selected (■) cells of citrange, cultured for 1–7 days on a medium containing 171 mM NaCl

for their survival and growth. The first important observation was the improved tallness of the plantlets recovered from selected calli after culture on different regeneration media (Fig. 5). On an average, the length of the cuttings of *Poncirus* and citrange grown for 7 weeks on a nonsaline medium was respectively 66 and

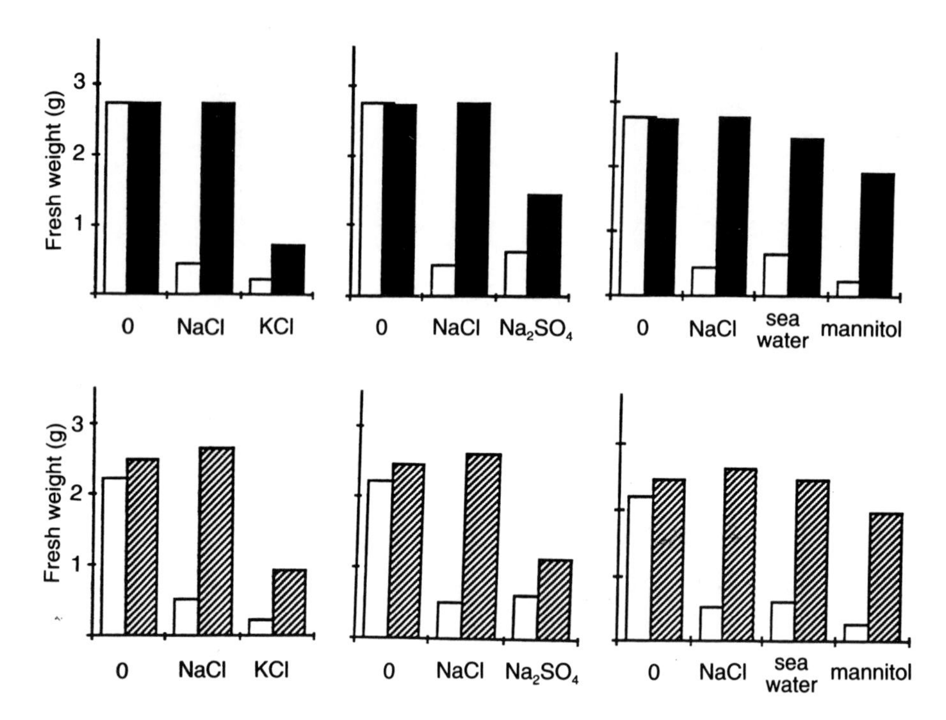


**Fig. 4.** Effect of NaCl, KCl, $Na_2SO_4$, sea water, and mannitol (171 mM) on cell growth of salt-tolerant lines of *Poncirus* (■) and citrange (▨) compared to unselected cells (□)

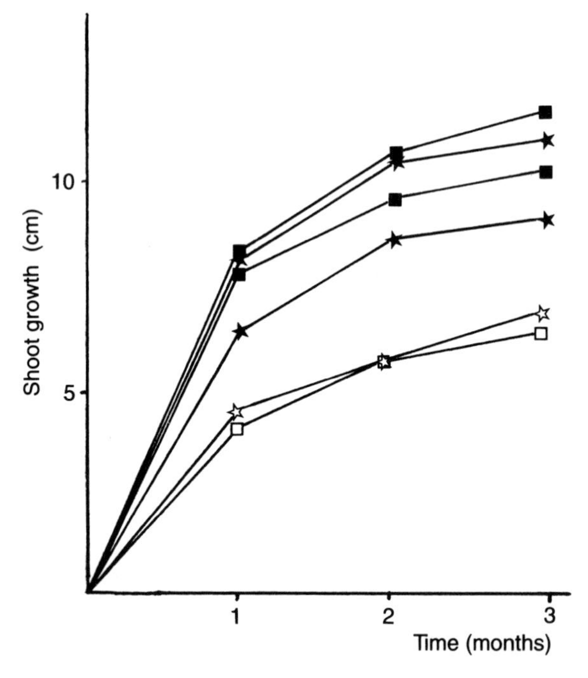


**Fig. 5.** Shoot growth of unselected (☆) and selected (★) clones of *Poncirus*, unselected (□) and selected (■) clones of citrange

56% higher than that of plantlets originating from unselected cells. The number of leaves was also nearly twice as high in the plantlets grown from tolerant cells, while the size of the roots developed in the agar medium was not modified.

On the other hand, the salt tolerance of the plantlets was assessed by culture on media containing different NaCl concentrations. The salt sensitivity of the nonselected cuttings was similar to that of the seedlings: their growth was drastically affected by low NaCl concentrations (43 or 85 mM), while higher doses were lethal. The growth of plantlets derived from salt-tolerant calli remained stable and much better than that of standard ones for salt concentrations of 43 and 85 mM (Bouharmont et al. 1993). A slightly lower increase in length appeared for 171 mM NaCl in a few plant clones. Nevertheless, the shoot growth was not strongly affected by concentrations lower than 171 mM, and plantlets survived in the presence of 256 mM NaCl. On such a medium, the plantlets were similar to the unselected ones growing on a medium containing 43 mM NaCl.

Thus it seems that the better growth of selected plantlets of *Poncirus* and citrange results from two origins: an increased vigor of the plants and the expression of the salt tolerance observed at the cell level. McHughen (1987) obtained similar results in flax after in vitro selection: the salt-tolerant line was characterized by an improved vigor in the plants grown on saline and nonsaline soil, but other characters, like seed weight, amount of oil, and earliness were also modified.

## 4 Summary and Conclusions

Salt tolerance was much improved in cell lines of two rootstocks (*Poncirus trifoliata* and Carrizo citrange) after callus culture in the presence of lethal concentrations of NaCl. The number of selected calli was very low, suggesting that somaclonal variation is not frequent in these plants. Thus application of in vitro selection is necessary, since it would be difficult to apply screening tests on a large number of plants regenerated from unselected cell cultures. On the other hand, the low frequency of mutations induced by cell culture could justify mutagenic treatments. The salt tolerance was kept during a number of subcultures in the presence or absence of salt, it was also expressed in the plantlets derived from the cells and remained unchanged in cell lines induced from these plantlets. Although the transmission of the character through seed has not been proved, it is highly probable that the new cell lines and plants result from stable mutations (Beloualy and Bouharmont, 1993).

The mechanisms involved in salt tolerance in *Poncirus* and citrange are not clear and the modifications are probably complex. Alterations of the cell membranes are probably frequent, resulting in higher or lower, sometimes slower uptake of some ions. Better tolerance to water stress is probably involved, but it is not due to an accumulation of proline: the proline content does not differ in unstressed selected and nonselected cells, and increases slowly in salt-tolerant cells when the NaCl goes up. Finally, the improvement of the salt tolerance at the plantlet level is also very important; it is partly due to the improved cell tolerance, but also to a higher growth rate of the shoots. The simultaneous selection, in independent cell lines, of modifications of the cell physiology, and of the plant development suggests the occurrence of a single fundamental event.

## 5 Protocol

In vitro selection and regeneration of salt-tolerant citrus rootstocks

**1. Callus culture and Selection**

1. Dissect mature embryos and eliminate the cotyledons.
2. Culture explants in Petri dishes on Murashige and Tucker medium in the dark.
3. Subculture the calli every 4 weeks.
4. Transfer callus fragments (±50 mg) on a medium containing NaCl (85 or 171 mM).
5. Isolate calli characterized by a normal weight after a month.
6. Transfer the selected calli on fresh selective media: monthly for 5 months.
7. Culture the selected calli on a nonsaline medium (three passages).
8. Select calli growing at least as well as the original cells.
9. Control the salt tolerance of the cell lines on a selective medium.

**2. Plant regeneration**

1. Transfer calli on regeneration media (1 mg/l NAA, 5 mg/l BAP), with or without NaCl, culture at 25 °C under 16/8 h (L/D) photoperiod, with subcultures every 4 weeks.
2. Embryoids and shoots develop after 16 to 20 weeks.
3. Stems elongate and root on saline medium (1 mg/l NAA or 1 mg/l $GA_3$).
4. Micropropagate shoot segments (±1 cm).
5. Verify the salt tolerance by culture of shoots on selective media.
6. Acclimatize rooted shoots in vermiculite and soil.

## References


Bajaj YPS (1984) Induction of growth in frozen embryos of coconut and ovules of citrus. Curr Sci 53: 1215–1216

Beloualy N (1991) Plant regeneration from callus culture of three *Citrus* rootstocks. Plant Cell Tissue Organ Cult 24: 29–34

Beloualy N, Bouharmont J (1992) NaCl-tolerant plants of *Poncirus trifoliata* regenerated from tolerant cell lines. Theor Appl Genet 83: 509–514

Beloualy N, Bouharmont J (1993) Amélioration de la tolérance à la salinité par sélection in vitro chez deux porte-greffes de citrus. In: Chlyah H, Demarly Y (eds) Le progrès génétique passe-t-il par le repérage et l'inventaire des gènes? Libley, Paris, pp 301–304

Ben-Hayyim G, Goffer Y (1989) Plantlet regeneration from a NaCl-selected salt-tolerant callus culture of shamouti orange [*Citrus sinensis* (L.) Osbeck]. Plant Cell Rep 7: 680–683

Ben-Hayyim G, Kochba J (1982) Growth characteristics and stability of tolerance of citrus callus cells subjected to NaCl stress. Plant Sci Lett 27: 87–94

Ben-Hayyim G, Kochba J (1983) Aspects of salt-tolerance in a NaCl-selected stable cell line of *Citrus sinensis*. Plant Physiol 72: 685–690

Ben-Hayyim G, Spiegel-Roy P, Neumann H (1985) Relation between ion accumulation of salt-sensitive and isolated stable salt-tolerant cell lines of *Citrus aurantium*. Plant Physiol 78: 144–148

Bouharmont J, Beloualy N, Van Sint Jan V (1993) Improvement of salt tolerance in plants by in vitro selection at the cellular level. In: Lieth H, Masoom A (eds) Towards the rational use of high salinity-tolerant plant, vol 2. Kluwer, Dordrecht, pp 83–88

Bové J, Morel G (1957) La culture de tissus de *Citrus*. Rev Gen Bot 64: 34–39

Button J, Kochba J (1977) Tissue culture in the citrus industry. In: Reinert J, Bajaj YPS (eds) Applied and fundamental aspects of plant cell, tissue, and organ culture. Springer, Berlin Heidelberg New York, pp 70–92

Cameron JW, Soost RK (1976) Citrus. In: Simmonds NW (ed) Evolution of crop plants. Longman Edinburg, pp 261–265

Deng XX, Grosser JW, Gmitter FG Jr (1992) Intergeneric somatic hybrid plants from protoplast fusion of *Fortunella crassifolia* cultivar Meina with *Citrus sinensis* cultivar Valencia. Sci Hortic 49: 55–62

Deng ZN, Zhang WC, Wan SY (1989) In vitro mutation breeding for salinity tolerance in *Citrus*. Mutat Breed Newsl 33: 12–14

Drira N, Benbadis A (1975) Analyse, par culture d'anthères in vitro, des potentialités androgénétiques de deux espèces de citrus (*Citrus medica* L. et *Citrus limon* (L.) Burm.). C R Acad Sci Paris 281: 1321–1324

Furr JR, Ream CL (1968) Breeding and testing rootstocks for salt tolerance. Calif Citrogr 54: 34–35

Gallasch PT, Dalton G (1989) Selecting salt-tolerant citrus rootstocks. Aust J Agric Res 40: 137–144

Garcia-Agustin P, Primo-Millo E (1995) Selection of NaCl-tolerant Citrus plant. Plant Cell Rep 14: 314–318

Gmitter FG Jr, Ling XB, Deng XX (1990) Induction of triploid *Citrus* plants from endosperm calli in vitro. Theor Appl Genet 80: 785–790

Gmitter FG Jr, Ling XB, Cai CY, Grosser JW (1991) Colchicine-induced polyploidy in *Citrus* embryogenic cultures, somatic embryos, and regenerated plantlets. Plant Sci 74: 135–141

Grosser JW, Gmitter FG Jr, Chandler JL (1988) Intergeneric somatic hybrid plants from sexually incompatible woody species: *Citrus sinensis* and *Severinia disticha*. Theor Appl Genet 5: 397–401

Grosser JW, Gmitter FG Jr, Tusa N, Chandler JL (1990) Somatic hybrid plants from sexually incompatible woody species: *Citrus reticulata* and *Citropsis gilletiana*. Plant Cell Rep 8: 656–659

Hidaka T, Yamada Y, Shichijo T (1979) In vitro differentiation of haploid plants by anther culture in *Poncirus trifoliata* (L.) Raf. Jpn J Breed 29: 248–254

Hidaka T, Omura M, Ugaki M, Tomiyama M, Kato A, Oshima M, Motoyoshi F (1990) *Agrobacterium*-mediated transformation and regeneration of *Citrus* spp. from suspension cells. Jpn J Breed 40: 199–207

Hutchinson D (1974) Swingle citrumelo–a promising rootstock hybrid. Proc Fla State Hortic Soc 87: 89–91

Kobayashi S (1987) Uniformity of plants regenerated from orange (*Citrus sinensis* Osb.) protoplasts. Theor Appl Genet 74: 10–14

Kobayashi S, Ohgawara T, Ohgawara E, Oiyama I, Ishii S (1988) A somatic hybrid plant obtained by protoplast fusion between navel orange (*Citrus sinensis*) and satsuma mandarin (*C. unshiu*). Plant Cell Tissue Organ Cult 14: 63–69

Kochba J, Spiegel-Roy P, Safran H (1972) Adventive plants from ovules and nucelli in citrus. Planta 106: 237–247

Louzada ES, Grosser JW (1994) Somatic hybridization in *Citrus* with sexually incompatible wild relatives. In: Bajaj YPS (ed) Biotechnology in agriculture and forestry, vol 27. Somatic hybridization in crop improvement I. Springer, Berlin Heidelberg New York, pp 427–438

Louzada ES, Grosser JW, Gmitter FG Jr (1993) Intergeneric somatic hybridization of sexually incompatible parents: *Citrus sinensis* and *Atalantia ceylanica*. Plant Cell Rep 12: 687–690

Marin ML, Duran-Vila N (1988) Survival of somatic embryos and recovery of plants of sweet orange [*Citrus sinensis* (L.) Osb.] after immersion in liquid nitrogen. Plant Cell Tissue Organ Cult 14: 51–57

McHughen AG (1987) Salt tolerance through increased vigor in a flax line (STS-II) selected for salt tolerance in vitro. Theor Appl Genet 74: 727–732

Moore GA, Jacono CC, Neidigh JL, Lawrence SD, Cline K (1992) *Agrobacterium*-mediated transformation of *Citrus* stem segments and regeneration of transgenic plants. Plant Cell Rep 11: 238–242

Moore GA, Jacono CC, Neidigh JL, Lawrence SD, Cline K (1993) Transformation in *Citrus*. In: Bajaj YPS (ed) Biotechnology in agriculture and forestry, vol 23. Plant protoplasts and genetic engineering IV. Springer, Berlin Heidelberg New York, pp 194–208

Murashige T, Tucker DPH (1969) Growth factor requirements of *Citrus* tissue culture. Univ Calif Riverside Symp 3: 1155–1161

Nadel B, Spiegel-Roy P (1987) Selection of *Citrus limon* cell culture variants resistant to the malsecco toxin. Plant Sci 53: 177–182

Navarro L (1992) Citrus shoot tip grafting in vitro. In: Bajaj YPS (ed) Biotechnology in agriculture and forestry, vol 18. High-Tech and micropropagation II. Springer, Berlin Heidelberg New York, pp 327–338

Ohgawara T, Kobayashi S, Ohgawara E, Uchimiya H, Ishii S (1985) Somatic hybrid plants obtained by protoplast fusion between *Citrus sinensis* and *Poncirus trifoliata*. Theor Appl Genet 71: 1–4

Ohgawara T, Kobayashi S, Ishii S, Yoshinaga K, Oiyama I (1989) Somatic hybridization in *Citrus*: navel orange (*C. sinensis* Osb.) and grapefruit (*C. paradisi* Maef), Theor Appl Genet 78: 609–612

Rangan TS, Murashige T, Bitters WP (1968) In vitro initiation of nucellar embryos of monoembryonic citrus. HortScience 3: 226–227

Saito W, Ohgawara T, Shimizu J, Ishi S (1991) Acid citrus somatic hybrids between sudachi (*Citrus sudachi* Hort. ex Shirai) and lime (*C. aurantiifolia* Swing) produced by electrofusion. Plant Sci 77: 125–130

Spiegel-Roy P, Ben-Hayyim G (1985) Selection and breeding for salinity tolerance in vitro. Plant Soil 89: 243–252

Vardi A, Spiegel-Roy P, Galun E (1975) *Citrus* cell culture: isolation of protoplasts, plating densities, effect of mutagens and regeneration of embryos. Plant Sci Lett 4: 231–236

Vardi A, Spiegel-Roy P, Galun E (1982) Plant regeneration from *Citrus* protoplasts: variability in methodological requirements among cultivars and species. Theor Appl Genet 62: 171–176

Vardi A, Epstein E, Breiman A (1986) Is the *Phytophthora citrophthora* culture filtrate a reliable tool for the in vitro selection of resistant *Citrus* variants? Theor Appl Genet 72: 569–574

Vardi A, Frydman-Shani A, Galun E, Gonen P, Bliechman S (1990) *Citrus* cybrids-transfer of *Microcitrus* organelles into *Citrus* cultivar. Acta Hortic 280: 239–245

# I.15 In Vitro Selection for Salt/Drought Tolerance in Colt Cherry (*Prunus avium x pseudocerasus*)

S.J. Ochatt[1]

## 1 Introduction

Top-fruit trees are routinely exposed to several environmental abiotic stresses, amongst which salinity and drought probably stand as the most important factors, especially where irrigation water is scarce and/or contains high levels of salts. In this context, *Prunus* species were recently classified as being sensitive to both drought and salt stress (Maas 1985) and it would, therefore, be of special interest for fruit growers to have rootstocks of a wider environmental adaptation. Of the various possible *Prunus* genotypes, it was felt that a cherry rootstock would be the best genotype for the induction of tolerance to such stresses, since marginal soils are increasingly being used as sites for cherry orchards around the world, and also because there is no natural source of salt/drought tolerance among cultivated cherries (Fogle 1975). Over the last decades, selecting plants with resistance to salinity and drought has been one of the main objectives of many traditional breeding programs (Tal 1990). More recently, biotechnological approaches based on the exploitation of somaclonal variation have also been adopted for this purpose (Chandler and Thorpe 1986; Tal 1992).

In order to ensure the stability of novel traits eventually incorporated, a direct (rather than stepwise) recurrent selection strategy was adopted for Colt cherry tissues, so as to eliminate nonheritable characters progressively within the selected cell population (Ochatt and Power 1989a). This approach also simultaneously enriches the population with potentially stable rather than just physiologically adapted cells. Experiments in the past have shown a need to assess the heritability of newly acquired traits from one generation to the next (Tal 1990) or, alternatively, the persistence of a new trait from in vitro-cultured cells to whole plants (Tyagi et al. 1981). Given the long life cycle (3–4 years to flowering) of fruit trees, the latter approach was adopted for Colt cherry.

In vitro selection for salt or water stress tolerance has been reviewed (Chandler and Thorpe 1986; Tal 1990). Protoplasts (Rosen and Tal 1981) or protoplast-derived tissues (Rossi et al. 1988) were rarely used for this purpose, and protoplast-derived salt/drought-tolerant plants were not obtained from either study. In addition, with the sole exception of *Citrus* (Ben-Hayyim 1987 and references therein), woody plant species have not been assessed with respect to salt or water stress tolerance in vitro, probably due to the scarcity of reliable

[1] INRA, Station d'Amélioration des Espéces Fruitières et Ornementales, B.P. 57, F-49071 Beaucouzé Cedex, France

Biotechnology in Agriculture and Forestry, Vol. 36
Somaclonal Variation in Crop Improvement II (ed. by Y.P.S. Bajaj)

techniques designed to regenerate trees from true callus or protoplasts. Such procedures have recently been reviewed for several rosaceous fruit trees (Ochatt et al. 1992) including stone fruits (Ochatt 1993), and against this background, therefore, in vitro selection for salt/drought tolerance was undertaken using explant and protoplast-derived tissues of Colt cherry.

## 2 In Vitro Studies for Salt Tolerance

### 2.1 Identification of Tolerant Tissues Within the Original Cell Populations

Protoplasts were isolated from leaves of in vitro-grown axenic shoots and from root cell suspensions of Colt cherry, and were cultured to give callus following protocols described previously (Ochatt et al. 1987). The two callus lines were designated PL and PR calli, respectively. Callus cultures were also established from leaf disks (L calli) and root segments (R calli) on a basal semisolid proliferation medium (BPM) of the same composition as used for maintenance of the cell suspensions above, i.e., MS (Murashige and Skoog 1962) medium + 2mg/l NAA and 0.5 mg/l BAP, pH 5.8, with an osmolality of 312 mOsm/kg. For all four sources, cultures were kept at 25 °C with a continuous illumination of 1000 lx (cool white fluorescent tubes) and subcultured onto the appropriate medium every week. Thereafter, callus (lines PL, PR, L, and R) were transferred to BPM medium, which was further supplemented with NaCl, KCl, or $Na_2SO_4$ at concentrations to give solutions with a normality of 0, 25, 50, 100, or 200 mN (i.e., low to high salinity), or iso-osmotic (of NaCl) concentrations (0, 8.7, 17.1, 33.3, or 66.6 g/l) of mannitol (i.e., mild to severe water stress). Therefore, the respective final osmolality of the various culture media was 312, 360, 407, 499, and 682 mOsm/kg, respectively, as determined by freezing point depression.

For the isolation of tolerant cells, an initial population of 100 nonselected callus pieces (200 mg fr.wt.) per treatment per tissue source (thereby involving an overall total of 1700 calli) was subcultured (three times, every 3 weeks) onto the same medium, and one surviving callus was then randomly selected, from each treatment (PL, PR, L, and R callus), for subsequent cloning and for the selection of tolerant lines. Results for this first screen are shown in Fig. 1. Thus, only rarely were cells within the control, nonselected callus populations capable of surviving in the presence of stress-inducing agents. In this respect, tissues of root origin (R and PR calli) had a larger natural stress tolerance than those of leaf origin (L and PL calli). The extent of tissue tolerance was inversely correlated with the concentrations of stress-inducing agents added to the BPM medium. Accordingly, the cell lines subsequently subjected to direct recurrent selection, for each of the stress conditions assessed, were initiated from one individual piece of original callus (of 50 mg fr.wt.) of each source that had shown an ability to withstand a particular stress level for at least three successive subculture passages on selection medium (Ochatt and Power 1989a).

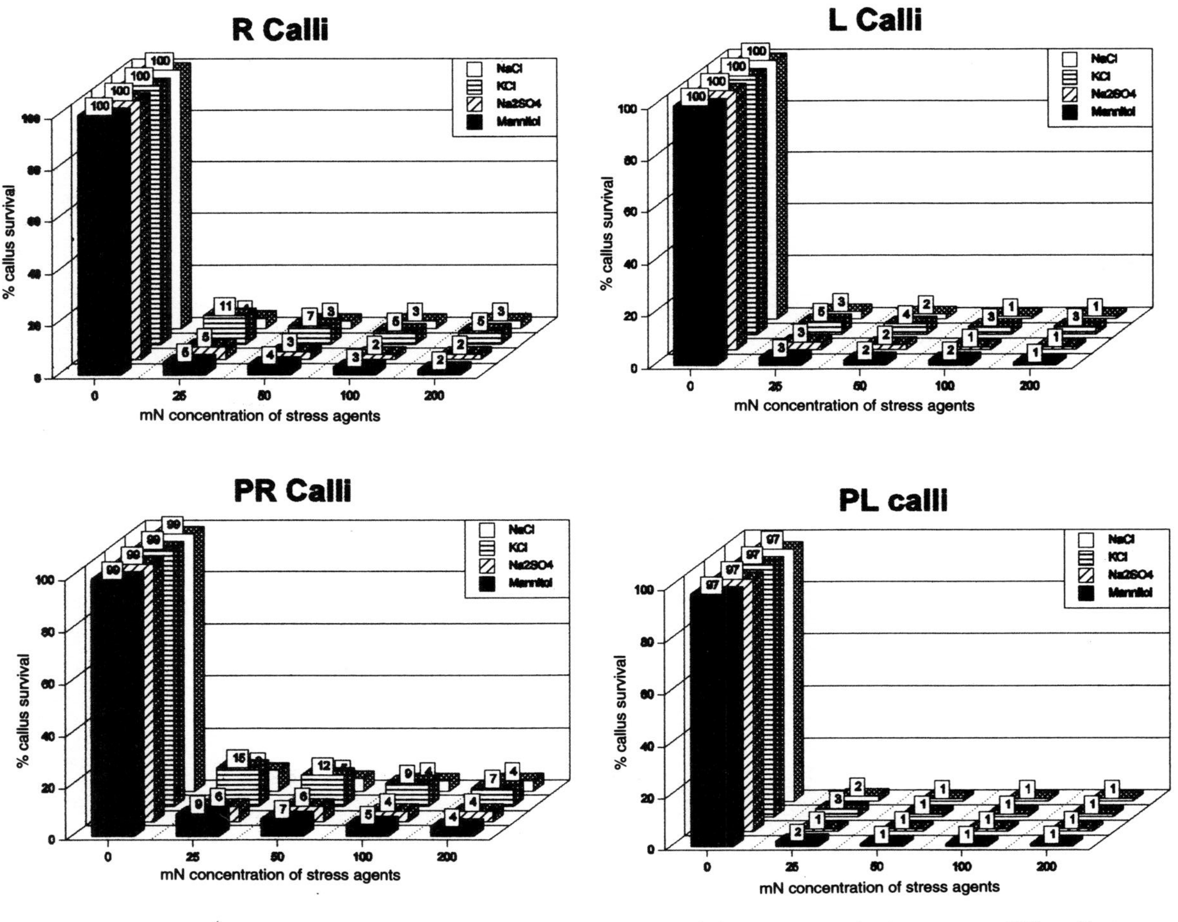


**Fig. 1.** Percentage callus survival in stress media before the application of the recurrent selection strategy (100 replicates per treatment per callus origin)

## 2.2 Design and Implementation of a Direct Recurrent Selection Strategy for the Induction of Salt/Drought Tolerance in Colt Cherry

A cell line which developed at the end of the initial screening from an individual inculum was subjected to a direct recurrent selection which was implemented after a minimum of six passages on the appropriate selection medium, and only

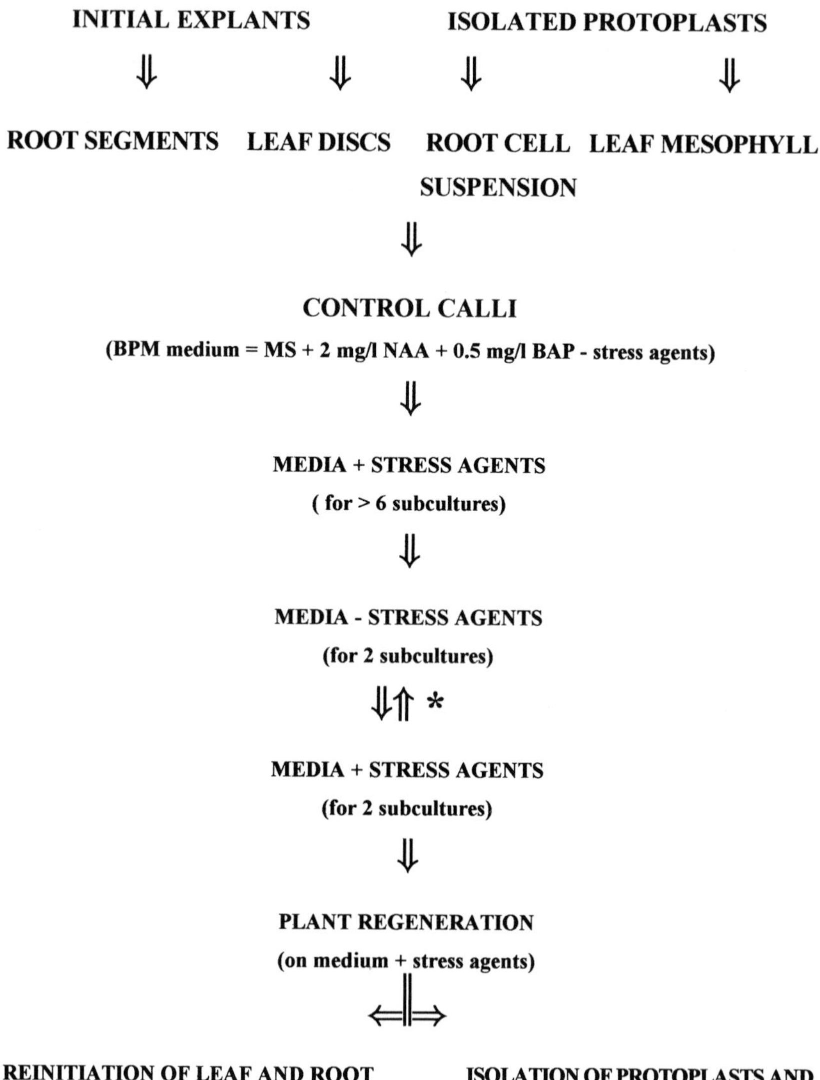


**Fig. 2.** Scheme of the direct recurrent selection strategy.* Recurrent selection passages repeated at least 3 times. All assessments including at least 25 replicates per treatment

when the original callus line had been firmly established under stress conditions. This direct recurrent selection strategy consisted of recurrent cycles, each involving a sequence of alternate, two successive subculture passages (3 weeks each) on a medium with the stress-inducing agent, followed by two successive subculture passages on stress-free medium. At least three such cycles of recurrent selection were applied prior to any attempt at plant regeneration from the putatively tolerant calli that survived this strategy. Figure 2 schematically depicts the selection strategy as adopted (Ochatt and Power 1989a).

The rationale behind the adoption of this approach was that nonstable, physiologically adapted cells within the selected cell population would outgrow stably tolerant cells during the stress-free subculture periods, only to be no longer able to survive after being transferred back to stress conditions. Selected cell lines would, therefore, be progressively enriched with stable and tolerant cells.

Cultural responses of cell lines were measured in terms of the extent of callus growth (fresh and dry weights) at the end of each subculture, overall callus survival (the percentage of surviving calli after each recurrent selection cycle), and percentage cell viability. This latter was carried out with 10 mg fr.wt. callus portions mechanically disassociated into individual cells or small cell clumps, suspended in 5 ml of an appropriate liquid medium with stress agents counterpart, and then mixed with an equal volume of medium with FDA and observed under U.V. light. In turn, cultural responses of the selected cell lines were subsequently assessed with respect to control, nontolerant tissues (original calli that were not subjected to recurrent selection), which were rarely able to survive or grow on media with a normality higher than 25 mN of either salt or when the osmolality exceeded 360 mOsm/kg.

### *2.2.1 Direct Recurrent Selection and the Responses of Cell Lines to Salinity*

Salt-tolerant cell lines were established at all salt concentrations tested and for all three salts. With increasing salt concentration, all parameters, callus growth, percentage cell viability, and callus survival, had declined after applying the recurrent selection methed three times.

As was the case for the original cell population, the tissue source affected the responses of cells to salt stress, with root-derived cultures (R and PR calli) appearing to be more tolerant than those of leaf origin (L and PL calli). In turn, PR calli were more tolerant than R calli, whereas ontogenetic drift (i.e., the ontogenetic effect of sources on the responses of cells to salinity) determined the reverse situation for leaf-derived cultures, with PL calli showing a higher sensitivity to salts than L calli (Fig. 3).

Sodium ions, as compared to potassium ions, were primarily responsible for growth arrest and cell damage, and no differences were detected when sodium was applied as chloride or sulphate. The anionic affect on Colt cherry was, therefore, negligible. However, cross-tolerance to stress, as induced by a salt other than the one for which tolerance was selected initially, was not observed (Ochatt and Power 1989a).

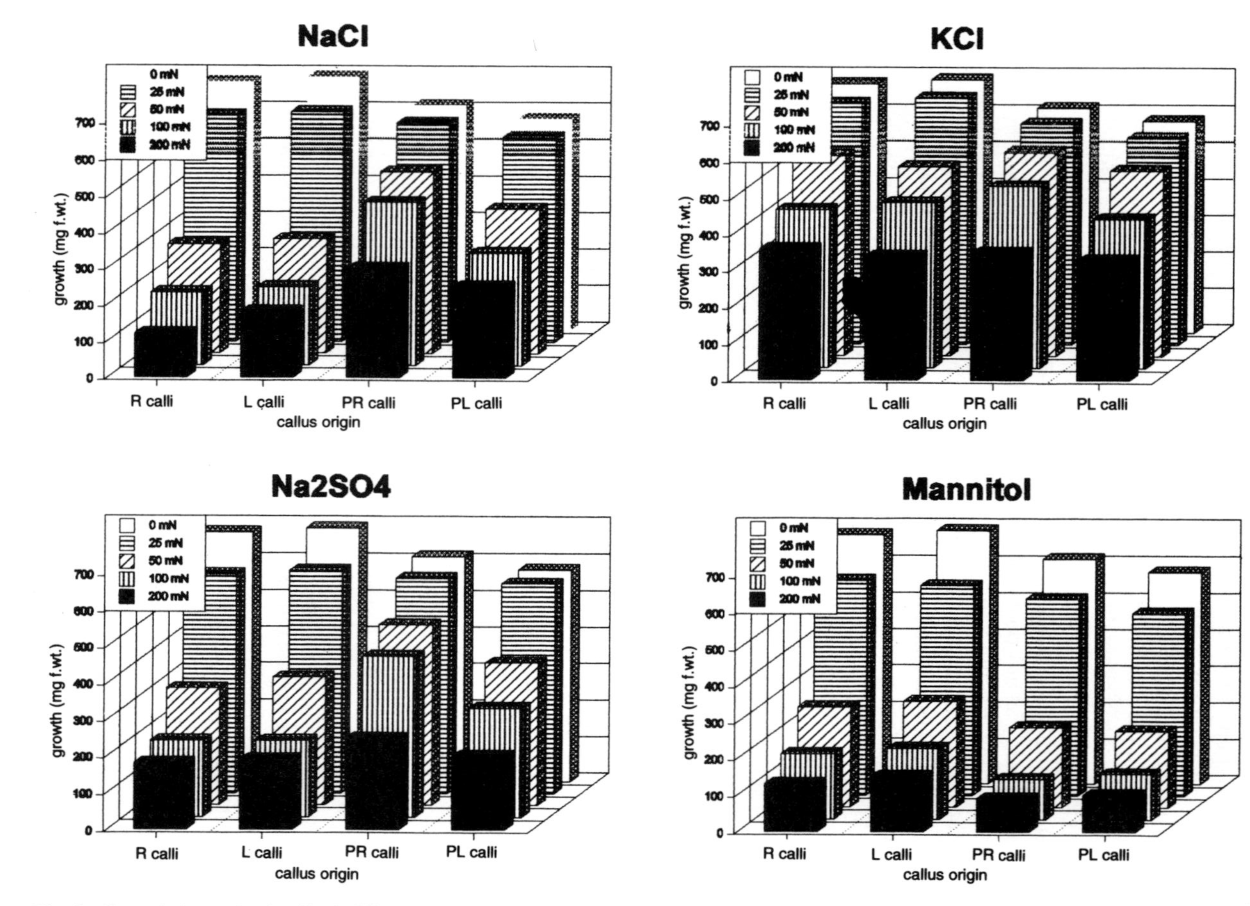


**Fig. 3.** Growth (mg.wt) of calli of different origins in response to salt and drought stress after three cycles of recurrent selection (25 replicates per treatment)

### *2.2.2 Direct Recurrent Selection and the Responses of Cell Lines to Drought*

In relation to the effects of mannitol-induced water stress on the cultured tissues, growth again was progressively inhibited with increased mannitol concentration, with protoplast-derived cultures (both PL and PR calli) exhibiting a greater inhibition of growth than those of explant origin (L and R calli; Fig. 3). Likewise, cell viability and percentage callus survival declined with an increased mannitol concentration added to BPM medium. Growth inhibition, induced with mannitol, was always more pronounced than with ion (i.e., $Na^+$) toxicity when responses on isoosmotic saline media were compared. It was possible, however, to recover cell lines tolerant of up to 66.6 g/l mannitol (iso-osmotically equivalent to 200 mN of NaCl, medium osmolality = 682 mOsm/kg). It was noteworthy that cell lines capable of withstanding a given osmolality of the culture medium, as provided by the various mannitol concentrations, also showed cross-tolerance to media with the same osmolality provided by either of the three salts assessed.

To date, there are few reports which have described the maintenance of osmotically stressed cell cultures and the interactions of osmotic and ionic stress in cultures maintained in the presence of elevated salt levels (Harms and Oertli 1985; Ben-Hayyim 1987). Although for Colt cherry cells, culture leads to ion toxicity and cell damage, a reduced water potential seemed more harmful to these cells. These observations, coupled with the apparent sensitivity of nonselected cells to water stress and the demonstrable existence of a cross-tolerance between salt and water-stress-adapted cells suggested that one of the mechanisms accounting for salt tolerance was ion compartmentation followed by osmotic adjustment, as was suggested for Shamouti orange (Ben-Hayyim 1987). This would also appear to confirm the widely accepted model for an effective cellular response to salt stress (Tal 1990).

## 2.3 Plant Regeneration from Salt- and Drought-Tolerant Cell Lines

For all sources, shoot regeneration from tolerant cell lines was undertaken using the same media and cultural conditions as for control, nontolerant protoplast-derived tissues (Ochatt et al. 1987, 1988). The regeneration media, however, were supplemented with the same concentration of stress-inducing agent as tolerated and selected for at the callus level, using 25 replicates per treatment per callus origin.

Subsequent responses of tolerant lines were assessed in terms of the percentage of calli capable of regeneration and the number of shoot buds per regenerating callus portion, as compared to the same parameters for control, nontolerant calli. Regenerated shoot buds were then detached from the callus and propagated and rooted as for control Colt cherry shoot cultures. The regenerated plants were finally transferred to pots with a commercial compost and grown in a glasshouse, alongside (normal) regenerants for further agronomical observation.

Plants were regenerated from all the salt and mannitol-induced water-stress tolerant cell lines established after three recurrent selection cycles (Ochatt and Power 1989a). There was a marked reduction in plant regeneration capacity in

**Table 1.** Plant regeneration from salt and water-stress-tolerant cell lines of Colt cherry. Data are the mean ± SD, from 25 replicates per treatment per callus origin. Control cell line data were RC = 72 ± 6%; BC = 5 ± 2

| Stress | Concentration (mN) | | | | | | | |
|---|---|---|---|---|---|---|---|---|
| Agent | 25 | | 50 | | 100 | | 200 | |
| | RC[a] | BC[b] | RC | BC | RC | BC | RC | BC |
| NaCl | 68 ± 10 | 5 ± 2 | 36 ± 7 | 2 ± 2 | 12 ± 3 | 1 ± 1 | 8 ± 4 | 1 ± 1 |
| KCl | 72 ± 9 | 5 ± 1 | 56 ± 6 | 2 ± 1 | 48 ± 7 | 2 ± 1 | 32 ± 5 | 2 ± 1 |
| $Na_2SO_4$ | 64 ± 7 | 4 ± 1 | 44 ± 5 | 2 ± 2 | 20 ± 3 | 1 ± 1 | 12 ± 2 | 1 ± 1 |
| Mannitol[c] | 72 ± 7 | 4 ± 1 | 68 ± 5 | 4 ± 1 | 68 ± 4 | 2 ± 1 | 64 ± 5 | 3 ± 2 |

[a] RC = Percentage of regenerating calli.
[b] BC = Number of shoot buds regenerated per callus portion.
[c] Iso-osmotically equivalent concentrations (8.7, 17.1, 33.6, and 66.6 g/l mannitol) to those of salts were used.

those callus lines maintained at the highest concentrations of salts, whereas for mannitol-tolerant cell lines, mannitol concentration had little effect on regeneration responses (Table 1). Such responses remained unaltered even for plant regeneration media that were not supplemented with the respective concentrations and types of stress-inducing agents as tolerated at the callus level. This concurs with a previous report whereby selection, at the cell level, simultaneously resulted, perhaps through a close genetic, biochemical or metabolic linkage, in modifications to the expression of other characters (Hasegawa et al. 1986); in the case of Colt cherry, the inability of cells to differentiate to plants.

A previous report (Rosen and Tal 1981) on in vitro salt tolerance in tomato protoplasts compared the cultural responses of leaf protoplasts of a wild, naturally salt-tolerant species, *Lycopersicon peruvianum*, with the cultivated tomato (*L esculentum*). Protoplasts of the wild species had a higher plating efficiency in saline medium. Cutures were, however not taken beyond the cell colony stage, unlike in Colt cherry, but the results with tomato generally concurred with those for Colt cherry. Later, protoplast-derived microcalli of a tomato variety with a high natural tolerance to salinity were also challenged with salts; some callus clones were able to grow in saline media (Rossi et al. 1988). Salt-tolerant plants were not regenerated in this case, nor was the stability of the acquired stress tolerance examined in either of these reports.

Interestingly, the requirements for weaning and acclimatization to autotrophism for the selected, stress-tolerant regenerated plants of Colt cherry differed markedly from those for control, nonstress-tolerant regenerants. Thus, for the former, the whole process took only 1 week under plastic bags after transfer to pots and with little watering, following planting of the in vitro-rooted stress-tolerant regenerated plants. Moreover, when a weaning strategy as prescribed for control regenerants was adopted, the stems of stress-tolerant plants turned brown after 10 days, and collapsed by 15 days, with ultimate plant death.

Conversely, when weaned under appropriate conditions, all the stress-tolerant Colt cherry plants were successfully transferred to soil, and samples of each treatment could be grown further, under both glasshouse and field conditions.

## 2.4 Assessments of the Stability of the Induced Stress Tolerance

Three different strategies were followed to examine the stability of the newly acquired stress tolerances in the regenerated Colt cherry plants:

1. Callus lines of leaf and root origins were reestablished, using explants taken from the first generation of (tolerant) regenerants, and were grown under the appropriate stress conditions (BPM medium + salts or mannitol).
2. The cross-tolerance to stress as induced by a salt other than the one for which tolerance was initially selected was also examined, using callus lines reestablished from leaf and root explants of selected tolerant regenerants (as in point 1. above).
3. Mesophyll protoplasts were isolated from the leaves of tolerant regenerants, and were cultured in the presence of the same concentration of stress-inducing agent as tolerated by the source calli for the regenerated plants.

The culture of leaf disks and root segments, taken from the regenerated plants of tolerant cell lines (of all treatments), on the corresponding stress-induced media, produced friable fast-growing calli under stress conditions that would normally lead to tissue browning and death of explants of nontolerant plants. These callus lines showed a cross-tolerance to salt stress, a response that was never observed in the parental callus lines. Similarly, cross-tolerance was also observed between salt and water stress in the callus tissues derived from explants taken from regenerated plants that had shown tolerance to mannitol-induced water stress. As an example, Table 2 shows the percentage of cell viability for root-derived calli of both these generations, on media with an osmolality of

**Table 2.** Percentage cell viability for R callus (taken from roots of normal plants and subjected to selection), and for callus derived from roots taken from putatively tolerant regenerated plants (rR callus) after three cycles of recurrent selection on stress-inducing media with an osmolality of 499 mOsm/kg

| Stress Agent | Original stress agent for which tolerance was selected | | | | | | | |
|---|---|---|---|---|---|---|---|---|
| | NaCl | | KCl | | $Na_2SO_4$ | | Mannitol | |
| | R calli | rR calli | R calli | rR calli | R calli | rR calli | R calli | rR calli |
| NaCl | $54 \pm 6$ | $58 \pm 6$ | $2 \pm 2$ | $76 \pm 4$ | $4 \pm 3$ | $56 \pm 7$ | $76 \pm 7$ | $72 \pm 4$ |
| KCl | $16 \pm 4$ | $65 \pm 7$ | $76 \pm 3$ | $81 \pm 2$ | $18 \pm 2$ | $61 \pm 4$ | $85 \pm 5$ | $76 \pm 2$ |
| $Na_2SO_4$ | $5 \pm 4$ | $56 \pm 11$ | $3 \pm 3$ | $75 \pm 5$ | $57 \pm 3$ | $52 \pm 5$ | $79 \pm 3$ | $70 \pm 5$ |
| Mannitol | $59 \pm 5$ | $63 \pm 5$ | $78 \pm 5$ | $79 \pm 3$ | $59 \pm 5$ | $60 \pm 2$ | $88 \pm 4$ | $77 \pm 3$ |

All stress-inducing agents added at a concentration to give a normality of 100 mN (499 mOsm/kg), as tolerated by original selected regenerants. All data (mean $\pm$ SD) from 500 cells per count; 10 replicates per treatment per callus origin.

499 mOsm/kg (iso-osmotically equivalent to 100 mN NaCl) and after applying three successive cycles of recurrent selection (Ochatt and Power 1989a). Similar responses were detected for callus cultures derived from the respective leaf origins and for the other osmolalities tested, whilst the results for callus growth and the percentage of callus survival followed a similar trend.

Salt tolerance in cultured plant cells of other systems (Hasegawa et al. 1986) has been associated with a reduction in size of callus cells. In this respect, the mesophyll protoplasts isolated from the tolerant regenerants of Colt cherry appeared to be nearly one third smaller in diameter ($13 \pm 5$ μm) than those of the original axenic shoot cultures ($19 \pm 4$ μm). These smaller protoplasts, however, showed no significant difference in cell viability or plating efficiency after 14 days on media with the respective salt concentrations as tolerated by the donor tissues, as compared to control protoplasts on BPM medium without salts. Such a reduction in cell size was also observed in cells of tolerant callus cultures and for all four sources (L, R, PL, and PR calli) after applying recurrent selection, and was associated with an increased membrane area: volume ratio which could, in turn, facilitate ion transport and the maintenance of ion compartmentation in the salt-adapted cells (Hasegawa et al. 1986; Funkhouser et al. 1993; Wink 1993). Plant regeneration could not be achieved for a second time in callus derived from such salt-tolerant protoplasts.

## 2.5 Responses to Salt Stress at the Protoplast Level

Protoplasts were isolated from leaves of nonstress-tolerant (normal) axenic shoot cultures and root cell suspensions of Colt cherry and cultured in their optimum media and growth conditions, but with NaCl, KCl, or $Na_2SO_4$ added to the medium at the concentrations as used for selection of tolerant cells. In addition, the effect of the presence or absence in the culture medium of the inhibitor of cell wall synthesis, 2, 6-dichlorobenzonitrile (DBN) at a concentration of 11.8 μM on the response of cells to salt stress conditions was also examined.

Results were expressed in terms of the protoplast/cell viability after 14 days, using FDA, and by the plating efficiency (PE), assessed only for media which lacked DBN, and expressed as the percentage of the initially cultivated protoplasts that had regenerated a cell well and divided at least once.

### *2.5.1 Examination of Protoplast Responses to Salt Stress*

The culture of normal leaf and root cell suspension protoplasts on the respective stress media confirmed a detrimental effect of all stress agents on cell survival. Sodium ions appeared as more effective, in terms of depressing protoplast viability, than those of potassium (Fig. 4). This was particularly evident for the highest concentrations (200 mN, 682 mOsm/kg; Ochatt and Power 1989b).

A marked decline in cell viability was concomitant with the onset of cell wall regeneration for both mesophyll and root cell suspension protoplasts, after 7–9 days of culture. Thereafter a progressive decline (1–2% per day) was noted

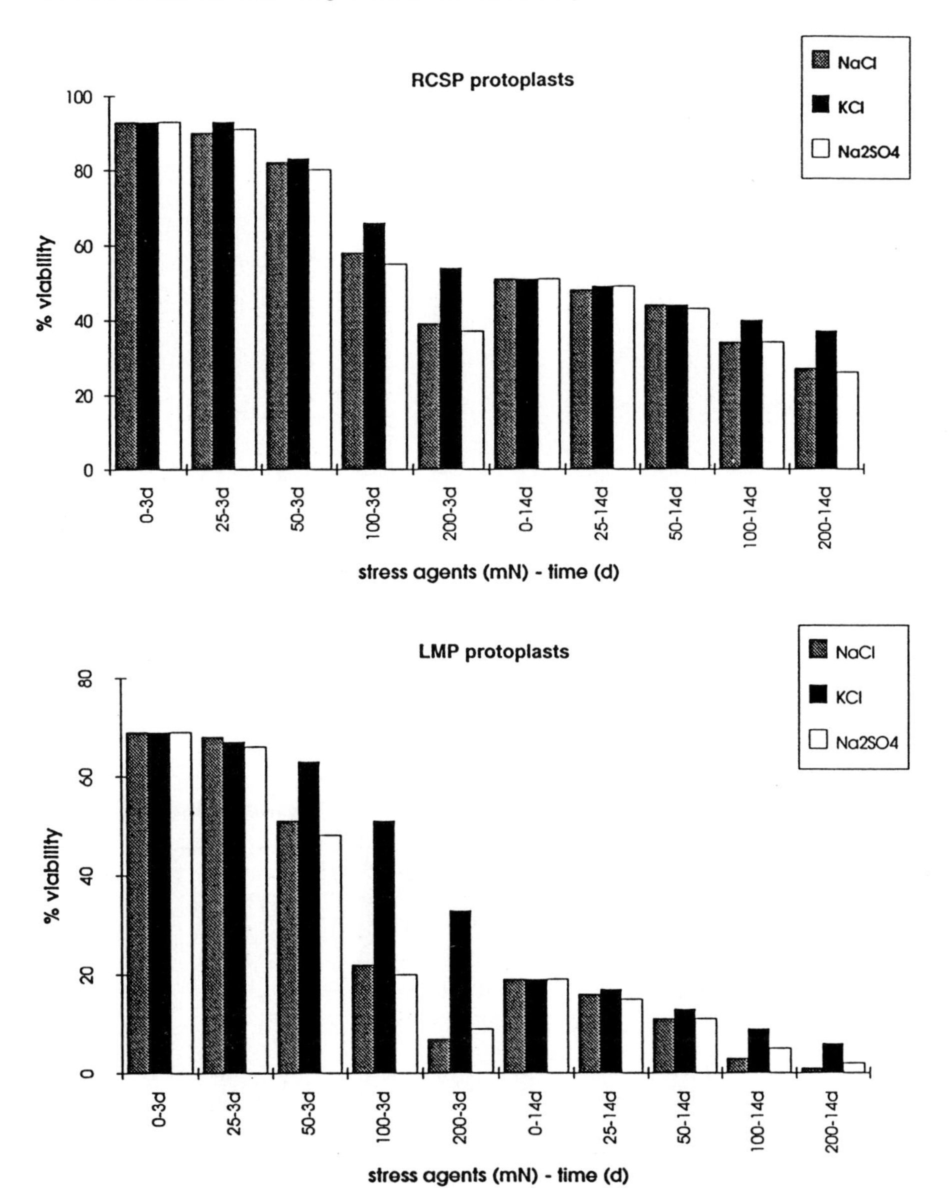


**Fig. 4.** Percentage viability at days 3 and 14, for root cell suspension (RCSP) and leaf mesophyll (LMP) protoplasts of nontolerant (normal) tissues cultured in BPM medium plus salts

(Fig. 4). These responses suggested that a relationship might exist between the responses of cultured protoplasts to salinity and the presence or absence of a cell wall. A new series of experiments was, therefore, initiated in order to investigate such hypothesis.

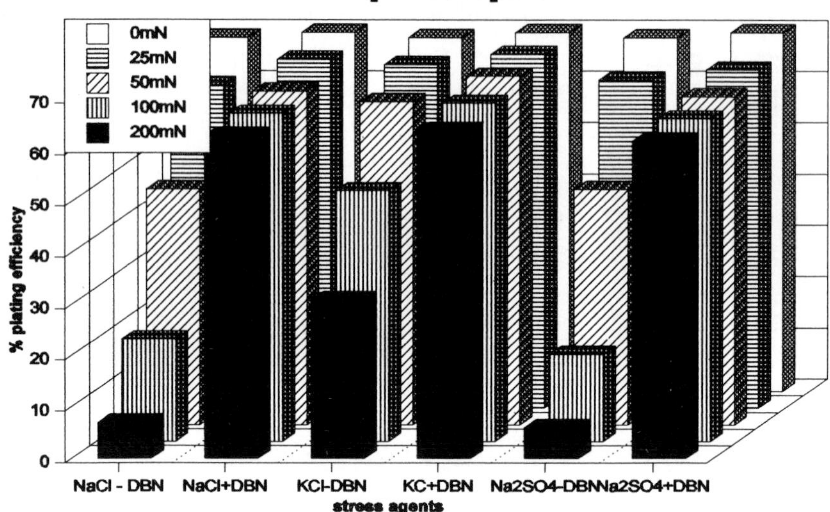


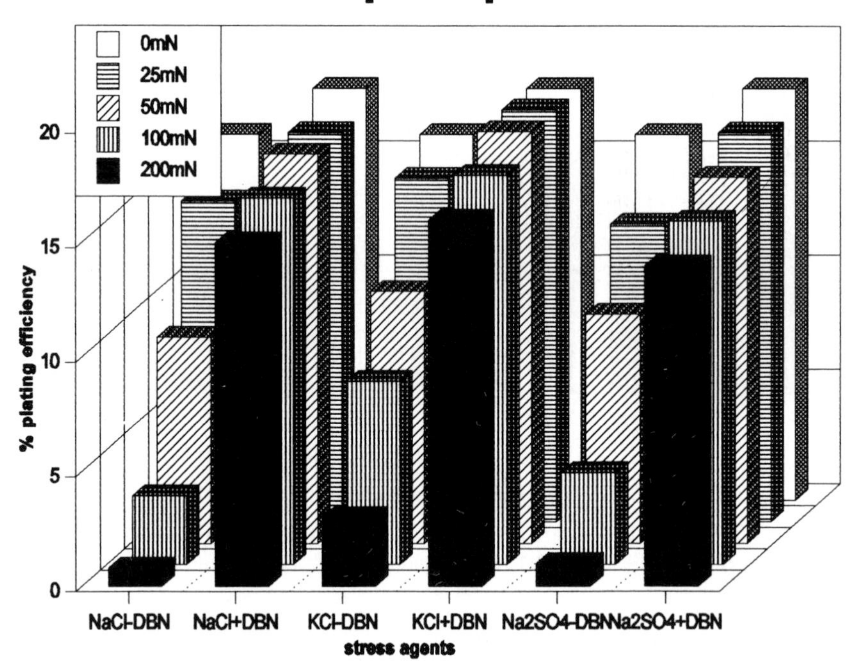


**Fig. 5.** Percentage plating efficiency (day 14) for Colt cherry protoplasts of different origins in media supplemented with salts and with or without DBN

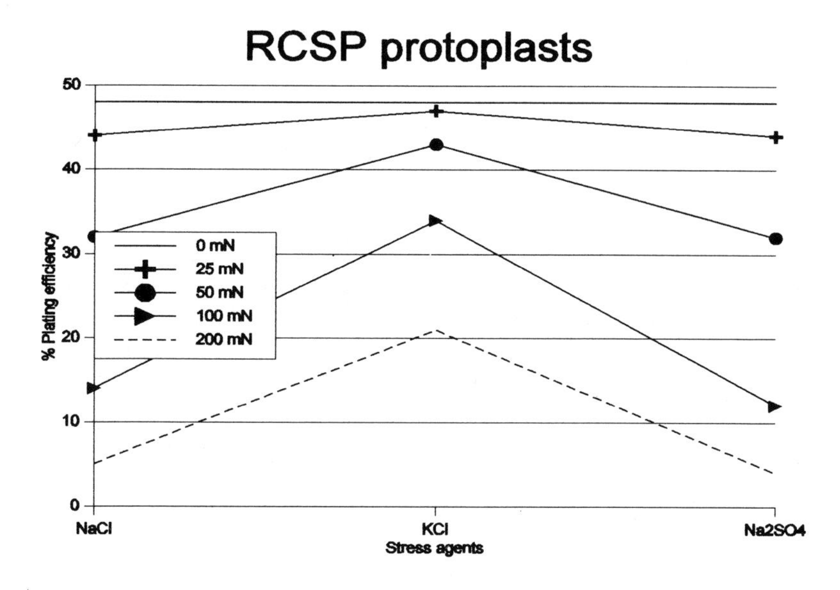


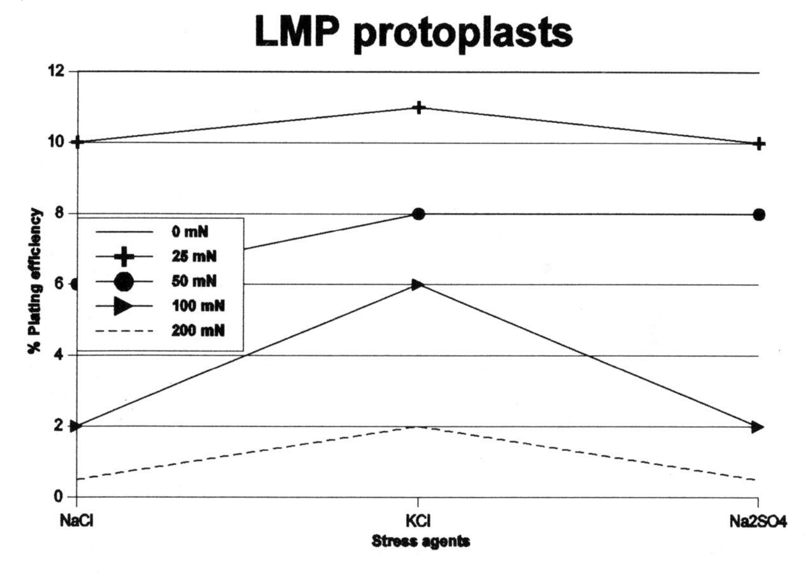


**Fig. 6.** Percentage plating efficiency (day 14) for Colt cherry root cell suspension (RCSP) and leaf mesophyll (LMP) protoplasts in media supplemented with salts but lacking DBN, following a 14-day culture period in similar media with DBN

### *2.5.2 Cell Wall Synthesis and the Link with Sensitivity to Salt Stress in Colt Cherry Protoplasts*

An assessment of plating efficiency, for nonstress (normal)-tolerant mesophyll and root cell suspension protoplasts of Colt cherry after 14 days of culture in media supplemented with salts confirmed a drop in cell viability (Fig. 5) as observed in previous experiments (cf. Fig. 4).

Conversely, when cell wall regeneration was inhibited in protoplasts, with 11.8 $\mu$M DBN, viability showed no statistically significant change to that of control protoplast populations cultured in salt-free BPM medium (Fig. 4). However, if such protoplasts were transferred to the appropriate media with salts but lacking DBN, after an initial 14-days culture period, as soon as cell wall regeneration proceeded, viability declined, with an associated reduction in plating efficiency, assessed after a further 14-days period (Fig. 6; Ochatt and Power 1989b).

In addition, the data in Fig. 6 show that the inhibition of cell wall synthesis, due to the addition of DBN to the media, disappears following the subsequent transfer of DBN-treated protoplasts to a medium which lacks DBN. It should be stated, however, that plant regeneration was not achieved from such salt-stressed, DBN-treated protoplasts, unlike protoplasts that were salt-stressed in media which lacked DBN.

The results for nonselected tissue protoplasts cultured in saline media with a cell wall inhibitor confirmed a reduced sensitivity to stress in protoplasts as compared to regenerating cells. The hypothesis, at least for Colt cherry, that the mechanism conferring an increased salt tolerance is associated with the onset of cell wall regeneration, would seem therefore to be sound (Ochatt and Power 1989b).

The reduced sensitivity to salt stress observed for protoplasts cultured in saline media, as compared to their cell counterparts, is concurrent with the proposed hypothesis that the mechanism conferring an increased salt tolerance is associated with the onset of cell wall regeneration. In this respect, and accepting that accumulation and vacuolar compartmentation was indeed the mechanism through which Colt cherry cells achieved salt tolerance, a change in permeability and ion transport properties associated with cell wall synthesis seems a likely explanation (Ochatt and Power 1989b). It is interesting to note that Hasegawa et al. (1986) also postulated this as a possible cause for the acquisition of salt tolerance in herbaceous genotypes. An alternative explanation for the clear difference in salt tolerance associated with cell wall regeneration is that the onset of cell wall synthesis depletes the metabolic energy charge which may be needed by the salt-tolerance mechanisms involved in the control of accumulation of ions and their subsequent exclusion through compartmentation in the vacuoles (Wink 1993). The presence of a newly formed cell wall, or membrane-associated processes (e.g., permeability and further transport of the permeant ions) operating at this time may therefore alter the selectivity to ions, leading to a relative sensitivity to salts and to ion toxicity in general. In this context, for a salt-tolerant trait to be stable, these changes in membrane properties would have to be permanent (Newton et al. 1991).

## 3 Conclusions

Although the general mechanisms which confer salt and water stress tolerance in cultured plant cells are, as yet, not fully understood (Newton et al. 1991; Funkhouser et al. 1993), the results with Colt cherry suggest that a recurrent selection strategy can be successful. This is likely to be of general interest in the context of fruit tree improvement and subsequent cultivation since, through tissue culture, sources of novel genetic diversity can be established. Moreover, due to the selection strategy adopted, a stable, genetic or epigenetic component would account for the acquired tolerance to salt and water stress. In this respect, little energy is needed to exclude salts (Wink 1993), as compared with the amount of metabolic energy needed to maintain a physiological adaptation (Chandler and Thorpe 1986). It is, therefore, most unlikely that those physiologically adapted cells that could survive the successive recurrent cycles might also have become irreversibly adapted. Confirmation of the stability of the acquired stress tolerance in Colt cherry is seen in the ability of callus taken from the first generation of regenerants to survive and proliferate under conditions of media-induced stress (Ochatt and Power 1989a).

The involvement of cell wall resynthesis on the response of Colt cherry protoplasts to salt stress clearly could provide information on the cellular mechanisms associated with the acquisition of salt tolerance (Funkhouser et al. 1993; Wink 1993). Moreover, it is apparent from these results that the isolation and culture of protoplasts from salt-stressed selected tissues and plants in stress media could represent a novel and useful tool to confirm the stability of such a selected trait, particularly for species with lengthy life cycles (e.g., trees) or where only asexual means of propagation are available (Ochatt and Power 1989b).

These results were the first examples of the regeneration of stable, salt/drought-tolerant plants from protoplasts, for both woody and herbaceous species (Ochatt and Power 1989a). The availability now of such stress-tolerant Colt cherry plants represents a new source of genetic variability. Protoplasts of such trees could be used in future somatic hybridization programs with other rosaceous fruit tree genotypes or, if micropropagated, could form the basis of a new generation of salt-tolerant cherry rootstocks.

## References


Ben-Hayyim G (1987) Relationship between salt tolerance and resistance to polyethylene glycol induced water stress in cultured *Citrus* cells. Plant Physiol 85: 430–433

Chandler SF, Thorpe TA (1986) Variation from plant tissue cultures: biotechnological application to improving salinity tolerance. Biotechnol Adv 4: 117–135

Fogle HW (1975) Cherries. In: Janick J, Moore JN (eds) Advances in fruit breeding. Purdue University Press, West Lafayette, pp 348–366

Funkhouser EA, Cairney J, Chang S, Dilip MA, Dias L, Newton RJ, Artlip TS (1993) Cellular and molecular responses to water deficit stress in woody plants. In: Pessarakli M (ed) Handbook of plant and crop stress. Marcel Dekker, New York, pp 347–362

Harms CT, Oertli JJ (1985) The use of osmotically adapted cell cultures to study salt tolerance in vitro. J Plant Physiol 120: 29–38

Hasegawa PM, Bressan RA, Handa S, Handa AK (1986) Cellular mechanisms of salinity tolerance. Hort Sci 21: 1317–1324

Maas EV (1985) Crop tolerance to saline sprinkling water. Plant Soil 89: 273–284

Murashige T, Skoog F (1962) A revised medium for rapid growth and bioassays with tobacco tissue cultures. Physiol Plant 15: 473–497

Newton RJ, Funkhouser EA, Fong F, Tauer CG (1991) Molecular and physiological genetics of drought tolerance in forest species. For Ecol Manage 43: 225–250
Ochatt SJ (1993) Regeneration of plants from protoplasts of some stone fruits (*Prunus* spp.). In: Bajaj YPS (ed) Biotechnology in agriculture and forestry, vol 23. Plant protoplasts and genetic engineering IV. Springer, Berlin, Heidelberg, New York, pp 78–96
Ochatt SJ, Power JB (1989a) Selection for salt/drought tolerance using protoplast- and explant-derived tissue cultures of Colt cherry (*Prunus avium* x *pseudocerasus*). Tree Physiol 5: 259–266
Ochatt SJ, Power JB (1989b) Cell wall synthesis and salt (saline) sensitivity of Colt cherry (*Prunus avium* x *pseudocerasus*) protoplasts. Plant Cell Rep 8: 365–367
Ochatt SJ, Cocking EC, Power JB (1987) Isolation, culture and plant regeneration of Colt cherry (*Prunus avium* x *pseudocerasus*) protoplasts. Plant Sci 50: 139–143
Ochatt SJ, Chand PK, Rech EL, Davey MR, Power JB (1988) Electroporation-mediated improvement of plant regeneration from Colt cherry (*Prunus avium* x *pseudocerasus*) protoplasts. Plant Sci 54: 165–169
Ochatt SJ, Patat-Ochatt EM, Power JB (1992) Protoplasts. In: Hammerschlag FA, Litz RE (eds) Biotechnology of perennial fruit crops. CAB International, Oxford, pp 77–103
Rosen A, Tal M (1981) Salt tolerance in the wild relatives of the cultivated tomato: responses of naked protoplasts isolated from leaves of *Lycopersicon esculentum* and *L. peruvianum* to NaCl and proline. Z Pflanzenphysiol 102: 91–94
Rossi J, Dorion N, Bigot C (1988) Some aspects of salinity tolerance of calli derived from tomato leaf protoplasts. In: Puite KJ, Dons JJM, Huizing HJ, Kool AJ, Koorneef M, Krens FA (eds) Progress in plant protoplast research. Kluwer Dordrecht, pp 401–402
Tal M (1990) Somaclonal variation for salt resistance. In: Bajaj YPS (ed) Biotechnology in agriculture and forestry, vol 11: Somaclonal variation in crop improvement I. Springer, Berlin, Heideleberg New York, pp 236–257
Tal M (1992) In vitro methodology for increasing salt tolerance in crop plants. Acta Hortic. 336: 69–79
Tyagi AK, Rashid A, Maheshwari SC (1981) Sodium chloride resistant cell line from haploid *Datura innoxia* Mill. A resistance trait carried from cell to plant and vice versa in vitro. Protoplasma 105: 327–332
Wink M (1993) The plant vacuole: a multifunctional compartment. J Exp Bot 44: 231–246

# Section II
# Somaclonal Variation in Medicinal and Aromatic Plants

# II.1 In Vitro Induction of Herbicide Resistance in *Atropa belladonna* L.

M. Yamazaki[1] and K. Saito[1]

## 1 Introduction

*Atropa belladonna* L. is a perennial solanaceous plant which is native to and cultivated in dry areas from Southern Europe to Western Asia. This plant is commonly known as nightshade because of its hallucinogenic activity, which is caused by some tropane alkaloids. The leaves and roots are used as sources of tropane alkaloids, mainly atropine (dl-hyoscyamine) with scopolamine as a minor component. These alkaloids inhibit the actions of acetylcholine competitively in synoptic transmission (Trease and Evans 1983), and they are used as parasympathetic agents. Atropine is used as a depressing drug for gastric secretion and gastrospasm by oral application or injection. It is also used as eye drops for mydriasis and local anesthetic. The activity of scopolamine as a mydriatic is shorter and stronger than that of atropine, and it has a sedative effect on the central nervous system which inhibits motion sickness. Because of its importance as a source of tropane alkaloids, numerous investigations regarding in vitro culture (reviewed by Bajaj and Simola 1991; Bajaj 1993) and genetic transformation (reviewed by Suzuki et al. 1993) have been carried out to improve the quality of this plant and to produce tropane alkaloids.

## 2 Somaclonal Variation and Genetic Engineering in *Atropa*

In vitro culture has been used for increasing genetic variability to obtain somaclonal variation in *A. belladonna*. Polyploid and aneuploid plants were obtained from culture of excised anthers and mesophyll protoplasts of anther-derived plants (Bajaj et al. 1978). Genetic variability has been induced also in intergenetic somatic hybrid plants between *A. belladonna* and *Nicotiana tabacum* (Kushnir et al. 1987; Babiychuk et al. 1992), *N. plumbaginifolia* (Gleba et al. 1988), *Datura innoxia* (Krumbiegel and Schieder 1979), and *Hyoscyamus muticus* (Ahuja et al. 1993). In most of these cases, regenerated plants were abnormal and sterile or partially fertile because of unexpected mutation in particular genes, rearrangement of chromosomes, and polyploidization.

[1] Faculty of Pharmaceutical Sciences, Laboratory of Molecular Biology and Biotechnology, Research Center of Medicinal Resources, Chiba University, Yoyoi-cho 1-33, Inage-ku, Chiba 263, Japan

Biotechnology in Agriculture and Forestry, Vol. 36
Somaclonal Variation in Crop Improvement II (ed. by Y.P.S. Bajaj)

Genetic transformation studies on *A. belladonna* were recently reviewed by Suzuki et al. (1993). This plant provides suitable material for transformation with *Agrobacterium*, being susceptible to infection by *Agrobacterium* and having a high morphogenetic potential (Bajaj 1993). Earlier reports were published on the transformation with non-disarmed Ti- or Ri-plasmids to obtain teratomas such as crown gall and hairy roots. Among these tissues, tropane alkaloids were increasingly produced in hairy roots (Kamada et al. 1986; Jung and Tepfer 1987). Subsequently, foreign genes for characters such as resistance to antibiotics and herbicides were introduced by using binary or integrated vectors on *Agrobacterium* (kanamycin resistance, Ondrei and Vlasák 1987; Mathews et al. 1990; Kurioka et al. 1992; herbicide resistance, Saito et al. 1992a).

The binary vector system based on *Agrobacterium*-Ri plasmids can be efficiently used to produce transgenic hairy roots containing the T-DNAs of a helper Ri plasmid and a second binary vector (Simpson et al. 1986). This technique depends on the fact that the T-DNA derived from the Ti plasmid can be mobilized in *trans* by *vir* gene products of the Ri plasmid. In some cases, the mature plants can be regenerated from hairy roots (Tepfer 1984; Tepfer et al. 1989). Some pharmaceutically important plants have been transformed by means of the *Agrobacterium*-Ri plasmid binary vector (Saito et al. 1990a, b, 1991, 1992a, b).

Recently, the biosynthetic pathway of tropane alkaloids in *A. belladonna* was modified by genetic engineering. The gene from *Hyoscyamus niger* encoding hyoscyamine 6$\beta$-hydroxylase (EC 1.14.11.11), which catalyzes the oxidative reaction in the biosynthetic pathway leading from hyoscyamine to scolopamine, was introduced into hyoscyamine-rich *A. belladonna* by means of the *Agrobacterium*-mediated transformation system (Yun et al. 1992). The transformants and their progeny possessing the transgene contained scopolamine as the main alkaloid. Such metabolic engineering, as well as herbicide resistance, is also useful to obtain improved medicinal plants.

## 3 Herbicide Resistance

Herbicide resistance is one of the agronomically useful traits that qualify plants for field cultivation and is a successful target in biotechnological research. Herbicide-tolerant plants have been obtained by selection of mutants and by genetic transformation with resistant genes.

Tobacco plants resistant to atrazine (Shigematsu et al. 1989) and sulfonylurea (Chaleff and Ray 1983) were obtained by in vitro culture of resistant cells. Cell lines of petunia and tomato tolerant to glyphosate (Steinruchken et al. 1986), phosphinothricin-tolerant alfalfa (Kishore and Shah 1988), and sethoxydim-tolerant maize (Parker et al. 1990), were selected by in vitro culture. These resistant traits were due to the overproduction or mutation of the target protein of herbicides. Generally, the acquired characters are unstable in regenerated plants and progeny. In some cases, selected lines lose the superior traits of the parent plants and gain undesirable traits during long-term culture.

To overcome the difficulties in mutant selection, genetic engineering is suitable for breeding. Plants resistant to herbicides, e.g., atrazine (Cheung et al.

1988), glyphosate (Comai et al. 1985; Shah et al. 1986), sulfonylurea (Lee et al. 1988), bromoxynil (Stalker and McBride 1988), and phosphinothricin (De Block et al. 1987) have been established by introduction of particular genes for modifying target proteins in plants or detoxifying herbicide molecules. In particular, phosphinothricin (PPT) is one of the most extensively studied herbicides. The *bar* gene from *Streptomyces hygroscopicus* encodes PPT acetyltransferase (PAT). PAT inactivates the synthetic herbicide PPT (Basta, Höchst), which inhibits glutamine synthase in plants and causes cell death. Another antibiotic herbicide, bialaphos (Herbiace, Meiji Seika Kaisha Ltd.), is a PPT derivative produced by *S. hygroscopicus* itself. Bialaphos releases PPT by endogenous peptidase in plant cells. The *bar* gene was transferred and expressed to confer the herbicide-resistant trait in transgenic plants of tobacco, potato, tomato (De Block et al. 1987), *Brassica* (De Block et al. 1989), maize (Spencer et al. 1990), *A. belladonna* (Saito et al. 1992a), and *Arabidopsis* (Sawasaki et al. 1994).

### 3.1 Genetic Transformation of *Atropa* with the Herbicide-Resistant Gene

Leaf disks of *A. belladonna* were infected with *Agrobacterium rhizogenes* harboring an Ri plasmid, pRi15834, and a binary vector, pARK5. The pARK5 contains the chimeric *bar* gene in its T-DNA under the control of the promoter for cauliflower mosaic virus 35S RNA. Within 2 weeks after infection, hairy roots appeared at the veins of the leaves (Fig. 1a). Sixteen out of 35 clones survived on the selection medium containing bialaphos. Adventitious shoots were regenerated spontaneously from hairy roots (Fig. 1c, d). Regenerated shoots could be isolated and rooted again to form plantlets (Fig. 1e) and transferred on to culture soil (Fig. 1f).

Finally, two clones of transgenic plants were regenerated (clones A1 and A8) from the hairy roots. The integration of the T-DNAs of both pRi15834 and pARK5 was confirmed by DNA-blot hybridization. The expression of the chimeric *bar* gene was analyzed by enzymatic assay of PAT. The regenerated plant of clone A8 was normal in shape, but that of A1 (Fig. 1f) showed the characteristic features of a regenerated plant from hairy roots caused by the expression of T-DNA genes of a Ri plasmid, such as wrinkled leaves and short internodes (Fig. 1).

The resistance of the regenerated transformants to commercial herbicide formulas was confirmed by application of a solution of Herbiace or Basta onto the leaves of transgenic plants. The control plant regenerated from hairy roots, transformed only with pRi15834, died 10 days after application of the herbicide solutions. In contrast, the transgenic plants possessing and expressing the *bar* gene were fully resistant to the herbicides (Fig. 2).

### 3.2 Inheritance and Segregation of a Transgenic Trait in Progeny

Self-fertilized progenies ($S_1$) were obtained from an A8 plant which showed normal phenotype shape. Seven out of 20 $S_1$ plants showed PAT activity (Table 1) and exhibited the herbicide-resistant trait. The A1 clone could not self-pollinate, because of the severe Ri syndrome. However, backcrossing of an A1 female or male with a nontransformed plant gave $F_1$ progeny. About 70% of the $F_1$ clones

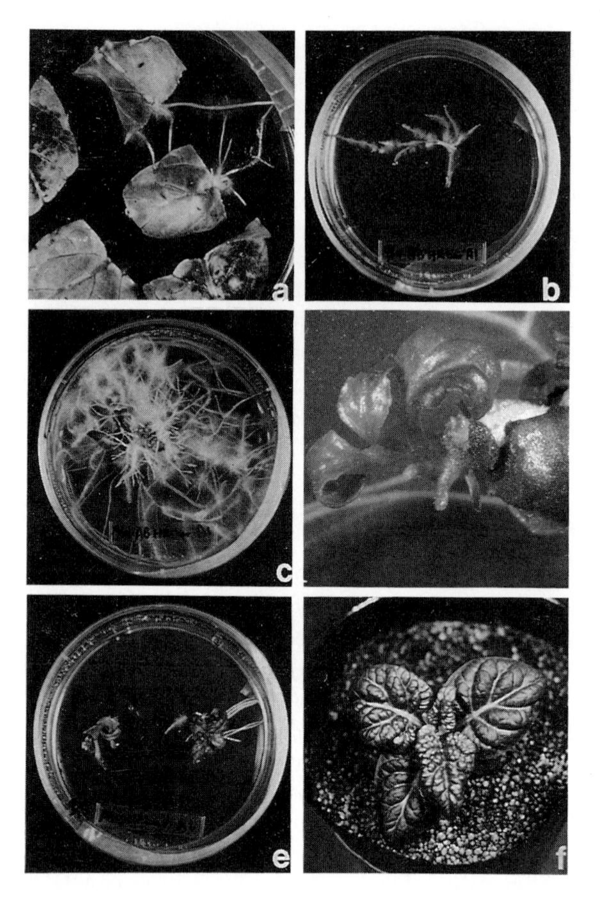

**Fig. 1a-f.** Regeneration of transgenic plant of *Atropa belladonna* (clone A1) from bialaphos-resistant hairy roots. (Saito et al. 1992a). **a** Induction of hairy roots on leaf disks. **b** Selection on B5 agar medium supplemented with 5 mg/l bialaphos. **c**, **d** Formation of adventitious shoots on B5 agar medium. **e** Rooting of regenerated shoots on B5 medium. **f** Regenerated plant A1 on culture soil showed characteristic phenotype due to the expression of Ri plasmid genes

in cases of both transgenic female and male parent exhibited positive PAT activity (Table 1) and herbicide resistance, indicating transmission of the dominant *bar* trait by inheritance. The trait of the Ri syndrome appeared only in the part of the $F_1$ progeny expressing the *bar* gene (Fig. 3). These results suggest that the integrated *bar* gene in clone A1 was segregated from the T-DNA of the Ri plasmid through meiosis and inherited by the progeny.

## 3.3 Production of Tropane Alkaloids in Regenerated Transgenic Plants

The production of tropane alkaloids in transgenic plants was determined (Table 2). All plants produced hyoscyamine as the major alkaloid, and scopolamine and 6$\beta$-hydroxyhyoscyamine as the minor bases. There were some differences in the

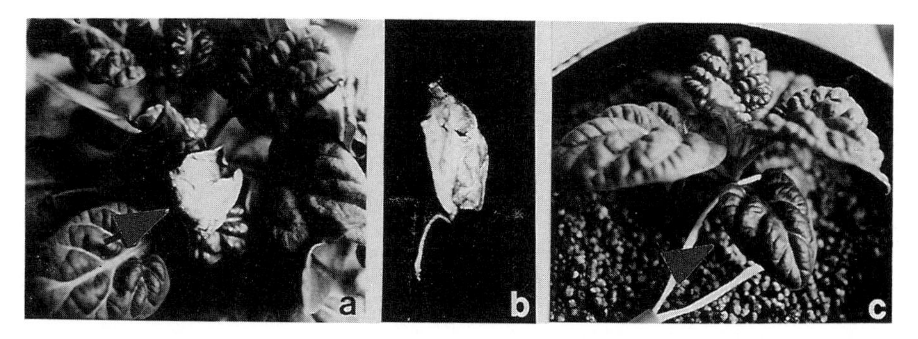


**Fig. 2 a–c.** Resistance of the regenerated transformants expressing *bar* gene towards bialaphos. (Saito et al. 1992a). The 0.2% aqueous solution of the commercially formulated Herbiace (bialaphos sodium salt content, 20%) was applied to one leaf of the plant as recommended by the supplier. **a, b** Clone HR1, control regenerated plant from hairy roots transformed only with pRi15834. After 10 days of application, the leaf with bialaphos applied (indicated by *arrow head*) died completely. **c** Clone A1 after 10 days of application. No change was observed in the leaf with bialaphos applied (indicated by *arrow head*)

**Table 1.** Segregation of PAT activity and Ri syndromes in progeny. (Saito et al. 1992a)

| Progeny | PAT(+)/total | Ri syn(+)/PAT(+) |
|---|---|---|
| A1 female x nontransformed male | 11/15 (73%) | 8/11 (72%) |
| A1 male x nontransformed female | 24/35 (69%) | 5/24 (21%) |
| A8, self-fertilized | 7/20 (35%) | 0/7 (0%) |

PAT activity was determined in the independent progenies by the method of De Block et al. (1987). The numbers of PAT positive clones [PAT (+)], total clones (total) and the clones showing Ri syndromes [Ri syn (+)] were indicated.

**Fig. 3.** Segregation of PAT activity and phenotype of Ri syndromes in progenies. *Left*, Progeny showing typical phenotype of Ri syndromes; *right* progeny exhibiting normal shape. PAT activity was positive in both clones and these clones showed herbicide resistance

**Table 2.** Accumulation of tropane alkaloids in leaves of transgenic regenerated plants of *Atropa belladonna.* (Saito et al. 1992a)

| Plant clone | Tropane alkaloid (%) | | |
|---|---|---|---|
| | Hyoscyamine | Scopolamine | 6β-Hydroxy-hyoscyamine |
| Untransformed control | 0.212[a] (0.017)[b] | 0.165 (0.014) | 0.066 (0.005) |
| HR-1[c] | 0.292 (0.023) | 0.028 (0.002) | 0.031 (0.002) |
| A1 | 0.278 (0.016) | 0.091 (0.005) | 0.015 (0.001) |
| A8 | 0.073 (0.009) | 0.030 (0.004) | 0.041 (0.005) |

[a] % Dry weight.
[b] % Fresh weight.
[c] Transformed with only pRi15834. The tropane alkaloids were extracted from dry leaves and determined by GC/MS by selected-ion-monitor mode using sparteine as an internal standard. Data are the means of triplicate determinations.

levels of alkaloid accumulation in the regenerated plants, in particular, low contents in clone A8. These differences may result either from the differences of age and physiological conditions of plants or the variations among clones derived from hairy roots.

## 4 Summary and Conclusions

The herbicide-resistant trait was obtained by transformation of *Atropa belladonna* with an Ri plasmid binary vector and plant regeneration from hairy roots. The *bar* gene encoding phosphinothricin acetyltransferase was transferred into genomic DNA of *A. belladonna.* Transformed hairy roots resistant to bialaphos were selected, and plantlets were regenerated. The integration of T-DNAs from pRi15834 and pARK5 were confirmed by DNA-blot hybridization. Expression of the *bar* gene in transformed $R_0$ tissues and in backcrossed $F_1$ progeny with a nontransformant and self-fertilized progeny was indicated by the enzymatic activity of the acetyltransferase. The transgenic plants showed resistance to bialaphos and phosphinothricin. Tropane alkaloids of normal amounts were produced in the regenerated transformants. These results present a successful application of transformation with an Ri plasmid binary vector for conferring an agronomically useful trait to medicinal plants.

The Ri plasmid vector has some characteristic features compared with a completely disarmed Ti vector: (1) Transgenic roots can be easily obtained integrated with any desirable foreign genes on a second binary vector in high frequency; in the present study 46% (16/35) gave double transformation; (2) this technique can be used for genetic manipulation of secondary metabolism of rapidly growing hairy roots that produce secondary products in high yield; (3) in some plant species, mature plants can be regenerated from hairy roots and offspring is also obtained (Tepfer 1984; Tepfer et al. 1989). These features could be of advantage in some cases of genetic engineering. However, the plants transformed with the Ri vector also show a "hairy root syndrome" unfavorable for application to transgenic plants.

In *A. belladonna*, the regeneration of plantlets occurred spontaneously from hairy roots on agar medium without addition of any phytohormones. A few reports are available on inheritance of T-DNAs and expression genes encoded on T-DNA in the progeny of plants regenerated from hairy roots (Costantino et al. 1984; Tepfer 1984; Sukhapinda et al. 1987). In some cases, the expression of integrated genes in offspring generation was suppressed in spite of the presence of full-length transgenes in the progeny. In the present study, about 70% of the backcrossed $F_1$ progeny and one third of $S_1$ progeny showed PAT activity (Table 1). The reason for these abnormal segregation ratios is not clear for the moment. However, the fact of transmission of the transgenic trait is a promising indication for success in molecular breeding of *A. belladonna* by the Ri vector, although detailed analysis is necessary to investigate inheritance and expression of transgenes at the molecular level.

## 5 Protocol

In binary expression pARK5, the *bar* gene was placed under the transcriptional control of the promoter for cauliflower mosaic virus 35S RNA (CaMV35S) and flanked with the terminator of the *nos* gene for nopaline synthase. The pRi15834 is a wild agropine-type Ri plasmid. *Agrobacterium* harboring both pRi15834 and pARK5 was cocultured with leaf disks of *A. belladonna* and used for transformation by the method described previously (Saito et al. 1992a). The hairy roots excised from leaf disks were selected on B5 agar medium (Gamborg et al. 1968) supplemented with 5 mg/l bialaphos for 2 weeks. The resistant hairy roots were cultured further in the same agar medium without bialaphos until adventitious shoots were regenerated. The shoots were isolated and transferred onto a new agar plate for rooting. The rooted plants were propagated as sterile shoot cultures and then transferred to culture soil. Progeny were obtained by backcrossing to a non-transformant by hand pollination or by self-fertilization.

## References

Ahuja PS, Laiq-ur-Rahman, Bhargava SC, Benerjee S (1993) Regeneration of intergeneric somatic hybrid plants between *Atropa belladonna* L. and *Hyoscyamus muticus* L. Plant Sci 92: 91–98

Babiychuk E, Kushnir S, Gleba YY (1992) Spontaneous extensive chromosome elimination in somatic hybrids between somatically congruent species *Nicotiana tabacum* L. and *Atropa belladonna* L. Theor Appl Genet 84: 87–91

Bajaj YPS (1993) Regeneration of plants from protoplasts of *Atropa belladonna* L. (deadly nightshade). In: Bajaj YPS (ed) Biotechnology in agriculture and forestry, vol 22. Plant protoplasts and genetic engineering III. Springer, Berlin Heidelberg New York, pp 13–21

Bajaj YPS, Simola LK (1991) *Atropa belladonna* L.: in vitro culture, regeneration of plants, cryopreservation, and the production of tropane alkaloids In: Bajaj YPS(ed) Biotechnology in agriculture and forestry, vol 15. Medicinal and aromatic plants III. Springer, Berlin Heidelberg New York, pp 1–23

Bajaj YPS, Gosch G, Ottma M, Weber A, Gröbler A (1978) Production of polyploid and aneuploid plants from anthers and mesophyll protoplasts of *Atropa belladonna and Nicotiana tabacum.* Indian J Exp Biol 16: 947–953

Chaleff RS, Ray TB (1984) Herbicide-resistant mutants from tobacco cell cultures. Science 223: 1148–1151

Cheung AY, Bogorad L, Van Montagu M, Shell J (1988) Relocating a gene for herbicide tolerance: a chloroplast gene is converted into a nuclear gene. Proc Natl Acad Sci USA 85: 391–395

Comai LD, Facciotti D, Hiatt WR, Thompson G, Rose RE, Stalker D (1985) Expression in plants of a mutant aroA gene from *Salmonella typhimurium* confers tolerance to glyphosphate. Nature (Lond) 317: 741–744

Costantino P, Spano L, Pomponi M, Benvenuto E, Ancora G (1984) The T-DNA of *Agrobacterium rhizogenes* is transmitted through meiosis to the progeny of hairy root plants. J Mol Appl Genet 2: 465–470

De Block M, Botterman J, Vandewiele M, Dockx J, Thoen C, Gossele V, Rao Movva N, Thompson C, Van Montagu M, Leemans J (1987) Engineering herbicide resistance in plants by expression of a detoxifying enzyme EMBO J 6: 2513–2518

De Block M, De Brouwer D, Tenning P (1989) Transformation of *Brassica napus* and *Brassica oleracea* using *Agrobacterium tumefaciens* and the expression of the *bar* and *neo* genes in the transgenic plants. Plant Physiol 91: 694–701

Gamborg OL, Miller RA, Ojima K (1968) Nutrient requirements of suspension cultures of soybean root cells. Exp Cell Res 50: 151–158

Gleba YY, Hinnisdaels S, Sidorov VA, Kaleda VA, Parokonny AS, Boryshuk NV, Cherep NN, Negrutiu I, Jacobs M (1988) Intergeneric asymmetric hybrids between *Nicotiana plumbaginifolia* and *Atropa belladonna* obtained by "gamma-fusion". Theor Appl Genet 76: 760–766

Jung G, Tepfer D (1987) Use of genetic transformation by the Ri T-DNA of *Agrobacterium rhizogenes* to stimulate biomass and tropane alkaloid production in *Atropa belladonna* and *Calystegia sepium* roots grown in vitro. Plant Sci 50: 145–151

Kamada H, Okamura N, Satake M, Harada H, Shimomura K (1986) Alkaloid production by hairy root cultures in *Atropa belladonna.* Plant Cell Rep 5: 239–242

Kishore GM, Shah DM (1988) Amino acid biosynthesis inhibitors as herbicides. Annu Rev Biochem 57: 627–663

Krumbiegel G, Schieder O (1979) Selection of somatic hybrids after fusion of protoplasts from *Datura innoxia* Mill. and *Atropa belladonna* L. Planta 145: 371–375

Kurioka Y, Suzuki Y, Kamada H, Harada H (1992) Promotion of flowering and morphological alternations in *Atropa belladonna* transformed with a CaMV 35S-*rolC* chimeric gene of the Ri plasmid. Plant Cell Rep 12: 1–6

Kushnir SG, Shlumukov LR, Pogrebnyak NJ, Berger S, Gleba YY (1987) Functional cybrid plants possessing a *Nicotiana* genome and an *Atropa* plastome. Mol Gen Genet 209: 159–163

Lee KY, Townsend J, Tepperman J, Black M, Chui C-F, Mazur B, Dunsmuir P, Bedbrock J (1988) The molecular bases of sulfonylurea herbicide resistance in tobacco. EMBO J 7: 1241–1248

Mathews H, Bharathan N, Litz RE, Narayanan KR, Rao PS, Bhatia CR (1990) The promotion of *Agrobacterium*-mediated transformation in *Atropa belladonna* L. by acetosyringone. J Plant Physiol 136: 404–409

Ondrej M, Vlasák J (1987) Expression of kanamycin resistance introduced by *Agrobacterium* binary vector into *Nicotiana tabacum* and *Atropa belladonna.* Biol Plant 29: 161–166

Parker WB, Somers DA, Wyse DL, Keith RA, Burton JD, Gronwald JW, Gengenbach BG (1990) Selection and characterization of sethoxydim-tolerant maize tissue culture. Plant Physiol 92: 1220–1225

Saito K, Kaneko H, Yamazaki M, Yoshida M, Murakoshi I (1990a) Stable transfer and expression of chimeric genes in licorice (*Glycyrrhiza uralensis*) using an Ri plasmid binary vector. Plant Cell Rep 8: 718–721

Saito K, Yamazaki M, Shimomura K, Yoshimatsu K, Murakoshi I (1990b) Genetic transformation of foxglove (*Digitalis purpurea*) by chimeric foreign genes and production of cardioactive glycosides. Plant Cell Rep 9: 121–124

Saito K, Yamazaki M, Kaneko H, Murakoshi I, Fukuda Y, Van Montagu M (1991) Tissue-specific and stress-enhancing expression of the TR promoter for mannopine synthase in transgenic medicinal plants. Planta 184: 40–46

Saito K, Yamazaki M, Anzai H, Yoneyama K, Murakoshi I (1992a) Transgenic herbicide-resistant *Atropa belladonna* using an Ri binary vector and inheritance of the transgenic trait. Plant Cell Rep 11: 219–224

Saito K, Yamazaki M, Murakoshi I (1992b) Transgenic medicinal plants: *Agrobacterium*-mediated foreign gene transfer and production of secondary metabolites. J Nat Prod 55: 149–161
Sawasaki T, Seki M, Anzai H, Irifune K, Morikawa H (1994) Stable transformation of *Arabidopsis* with the *bar* gene using particle bombardment. Transgenic Res 3: 279–286
Shah DM, Horsch RB, Klei HJ, Kishore GM, Winter JA, Tumer NE (1986) Engineering herbicide tolerance in transgenic plants. Science 233: 478–481
Shigematsu Y, Sato F, Yamada Y (1989) The mechanism of herbicide resistance in tobacco cells with a new mutation in the QB protein. Plant Physiol 89: 986–992
Simpson RB, Spielmann A, Margossian L, McKnight TD (1986) A disarmed binary vector from *Agrobacterium tumefaciens* functions in *Agrobacterium rhizogenes*. Plant Mol Biol 6: 403–415
Spencer TM, Gordon-Kamm WJ, Daines RJ, Start WG, Lemaux PG (1990) Bialaphos selection of stable transformants from maize cell culture. Theor Appl Genet 79: 625–631
Stalker DM, McBride KE (1988) Herbicide resistance in transgenic plants expressing a bacterial detoxification gene. Science 242: 419–422
Steinruchken HC, Schulz A, Amrhein N, Porter CA, Fraley RT (1986) Overproducing of 5-enolpyruvylshikimate-3-phosphate synthase in a glyphosate-tolerant *Petunia hybrida* cell line. Arch Biochem Biophys 244: 169–178
Sukhapinda K, Spivey R, Simpson RB, Shahin EA (1987) Transgenic tomato (*Lycopersicon esculentum* L.) transformed with a binary vector in *Agrobacterium rhizogenes*: non-chimeric origin of callus clone and low copy numbers of integrated vector T-DNA. Mol Gen Genet 206: 491–497
Suzuki Y, Kurioka Y, Ogasawara T, Kamada H (1993) Transformation in *Atropa belladonna*. In: Bajaj YPS (ed) Biotechnology in agriculture and forestry, vol 22. Plant protoplasts and genetic engineering III. Springer, Berlin Heidelberg New York, pp 136–143
Tepfer D (1984) Transformation of several species of higher plants by *Agrobacterium rhizogenes*: Sexual transmission of the transformed genotype and phenotype. Cell 37: 959–967
Tepfer D, Metzger L, Prost R (1989) Use of roots transformed by *Agrobacterium rhizogenes* in rhizosphere research: application in studies of cadmium assimilation from sewage sludges. Plant Mol Biol 13: 295–302
Trease GE, Evans WC (1983) Pharmcognosy, 12th edn. Bailliere Tindall, Eastbourne, UK
Yun D-J, Hashimoto H, Yamada Y (1992) Metabolic engineering of medicinal plants: transgenic *Atropa belladonna* with an improved alkaloid composition. Proc Natl Acad Sci USA 89: 11799–11803

# II.2 In Vitro Induction of Resistance to *Alternaria* Leaf Blight Disease in *Carthamus tinctorius* L. (Safflower)

R. E. KNEUSEL[1] and U. MATERN[1]

## 1 Introduction

### 1.1 Botany, Importance and Distribution of Safflower

Safflower (*Carthamus tinctorius* L.) is an annual or biennial plant of the family Asteraceae. The genus *Carthamus* consists of some 60 species which are native to the Canary Islands and the Mediterranean coasts from northern Africa to Egypt and from Spain along Greece to the Middle East. Among all the members of this genus, safflower is the only cultivated species.

The areas of origin for cultivated safflower were initially thought to include India, Afghanistan, and Ethiopia, based principally on the variability and location of wild species (Kupzow 1932; Vavilov 1951). This was modified later by Hanelt (1961a, b) and Knowles (1969a), who both postulated that safflower's origin is most likely in an area bounded by the eastern Mediterranean and the Persian Gulf because of its similarity to two closely related wild species: *C. flavescens* found in Turkey, Syria, and Lebanon, and *C. palaestinus*, present in western Iraq and southern Israel.

The plant domesticated long ago, initially for the orange/red pigments present in the florets, which have long been cherished by carpet weavers in Iran and Afghanistan, and led to its dissemination into southern Russian regions. In India, the plant was cultivated not only for the flower pigments; safflower oil had both medicinal and dietary value. In China, the pigments were employed as a dye for textiles and foods, and the plant was also valued as a medicine used in the treatment of a wide variety of disorders from broken bones to vascular diseases. Especially in the innermost territories of China, the tea prepared from safflower florets is still in widespread use, but its medicinal value is most likely due to constituents other than carthamins, because the petals are known to contain hydroxykaempferol glycosides that affect heart beat amplitude and possess anti-inflammatory properties (Hattori et al. 1992).

The original use of safflower as a dye or food color that was used, for example as an alternative to the expensive saffron, has regained interest in recent years, especially in the Japanese food industry. The metabolism of these pigments in the human intestine is being examined (Meselhy et al. 1993). Nevertheless, the most significant market for safflower has developed because of its oil. In the USA, the

[1] Biologisches Institut II, Lehrstuhl für Biochemie der Pflanzen, Albert-Ludwigs-Universität, Schänzlestrasse 1, D-79104 Freiburg, Germany

Biotechnology in Agriculture and Forestry, Vol. 36
Somaclonal Variation in Crop Improvement II (ed. by Y.P.S. Bajaj)

oil was produced initially as an industrial oil in nonyellowing paints and varnishes. Predominantly, however, safflower oil is being used in edible products such as salad oils and margarines. It is nutritionally very valuable because it contains a large proportion of polyunsaturated fatty acids. Other uses of safflower are as birdseed and livestock feed, either as the full-fat oilseed or as safflower meal, a high protein by-product remaining after oil extraction. The high-quality protein of the meal has, however, been underutilized due to the presence of phenolic glycosides with bitter taste and cathartic activity (Lyon et al. 1979), and conjugated serotonins that may cause neurological disorders (Sakamura et al. 1980; Sato et al. 1985). However, complete extraction of these contaminants can be achieved with aqueous alcohols.

The current status of safflower in Australia and the Americas was summarized by Mündel (1993). Within Europe, extensive genetic research on safflower has been conducted in Spain; however, the importance of safflower as an oil crop has nevertheless decreased steadily in recent years (Rojas et al. 1993)presumably due to EC guidelines in favor of sunflower. Instead, its drought-resistant qualities make safflower valuable as green fodder or silage in drier regions where other fodder crops do not have high yields (Weiss 1983) or for dryland farming (Tiwari and Namdeo 1993). Countries in the Far East, like India, boosted their safflower production and now hold the major share in the world's market. This development was also supported by the deep rooting capacity of safflower, which is increasingly exploited by planting late in the season as a second or paddy crop (Kumar 1993).

## 1.2 Breeding Objectives and Available Genetic Variability

The general aim of plant breeders is to produce safflower varieties with a high seed yield together with a number of other characteristics such as oil, protein, and hull percentage (Rojas et al. 1993), early maturation, flower color and pigment quality, oil content and quality, as well as pest and disease resistance (Kumar 1993).

Although many environmental factors and cultivation practices influence seed yield, there is great genetic diversity in the world collection of safflower regarding the head number per plant and head diameter, both significant determinants of seed yield (Abel and Driscoll 1976; Rao 1977; Kotecha 1981). There is also a positive correlation between yield and flower bloom days, plant height and length of the growing period (Ashri et al. 1975), as well as between branching and head number and, consequently, yield (Raghunatham et al. 1989). All these characteristics have been exhaustively investigated using the world collection of safflower germplasm, thereby aiding numerous breeding programs.

In Australia, northern China, and Canada, early maturing varieties with reasonable seed and oil yields would be helpful where the amount of rainfall during the ripening season is greater, resulting in increased disease outbreaks. In areas where the growing season tends to be shorter, quickly maturing varieties are also required (Knowles 1989). Saffire safflower, the first Canadian safflower variety, and its successor, AC Sterling, are two varieties with maturing periods of 120 days (Mündel et al. 1992). The oil content of Saffire is, however, too low (32%) for the oilseed market, whereas AC Sterling contains an average of 35% oil and has a 16% better yield than Saffire.

The regained interest in safflower as a source of food and cosmetic color additive (Saito and Fukushima 1986; Nakano et al. 1988; Saito et al. 1988, 1989; Fukushima et al. 1990; Saito 1991) has also stimulated breeding research. Flower color appears to be controlled by four genes, Y, C, O, and R, based on the segregation ratios from crosses of 25 genotypes (Narkhede and Deokar 1986). The pigments responsible for flower color are the orange-yellow, hydroquinone-based carthamin and yellow compounds (Fig. 1; Takahashi et al. 1982; Meselhy et al. 1993), although other unidentified compounds may also contribute to floret color (Hanagata et al. 1992). The enzymatic oxidation of precarthamin to carthamin appears to induce the shift in color from yellow to red pigmentation (Saito 1993).

The commercial development of safflower in the United States in the 1930s, then solely as an oil crop, was hindered at first by the lack of suitable genotypes

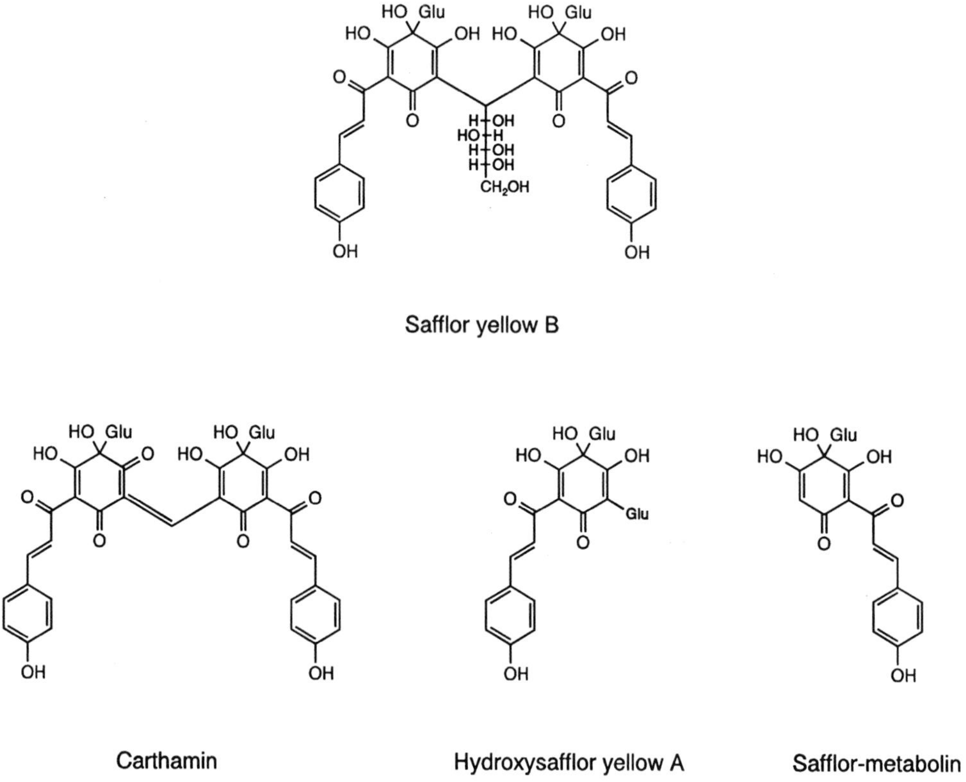

Safflor yellow B

Carthamin

Hydroxysafflor yellow A

Safflor-metabolin

**Fig. 1.** Pigments isolated from safflower florets; *Glu* glucose. (Takahashi et al. 1982; Meselhy et al. 1993)

containing the necessary levels of oil. Only after the introduction of germplasm from Egypt and the Sudan, which contained over 30% oil, was it possible to develop cultivars with 35% oil or better. Continuous breeding efforts have led to cultivars with higher oil contents, and, at present, cultivated varieties contain up to 44% oil. The breeding of partial hull genotypes may even result in the development of varieties containing more than 50% oil (Urie 1986).

The genetic variability for fatty acid composition of safflower oil is very large, and the genetic manipulation of oil quality in this plant is a model for other crops (Knowles 1969b, 1972, 1989). Safflower varieties exist with varying levels of linoleic, oleic, and stearic acid content. At present only the high linoleic type with ca. 80% linoleic acid and the high oleic type with ca. 80% oleic acid (Knowles et al. 1965; Fuller et al. 1967, 1968) are grown commercially.

A number of insect pests attack safflower, and, although some may cause appreciable losses, the majority do not justify control measures. Insects can cause losses in yield by reducing stands (e.g., wireworm, *Agriotes* spp. or cutworm, *Agrotis* spp), defoliating plants (leafhoppers, usually *Empoasca* spp.) or damaging developing buds or seeds (Mündel et al. 1992). Very little work has been done on the evaluation of world germplasm resources for resistance to insect pests, and eliminating damage to these pests has been accomplished by chemical control. Resistance breeding is necessary where safflower and related wild species are native, but in areas where the plant has only recently been introduced, breeding for resistance has not played an important role.

Safflower is susceptible to a variety of diseases caused by viruses, bacteria, and fungi, and a major breeding objective wherever the crop is grown has been greater resistance to disease. If cultivars with enhanced disease resistance were available, safflower would be much more competitive with other oil plants. Serious diseases that can be economically devastating are safflower rust caused by *Puccinia carthami* Cda. and *Phytophthora* root rot caused by *P. drechsleri* Tuck., both of which are more prevalent on irrigated fields. *Sclerotinia* head rot caused by *Sclerotinia sclerotiorium* is the most serious disease of safflower in western Canada (Mündel et al. 1992). *Verticillium* wilt (*Verticillium albo-atrum*), *Fusarium* wilt (*Fusarium oxysporum* f. sp. *carthami*), *Cercospora* leaf spot (*Cercospora carthami*), and head rot (*Botrytis cinerea*) are just a few other diseases which may cause severe decreases in crop yield and quality. Modern cultivars resistant to *Phytophthora*, *Verticillium*, and *Fusarium* have been identified (Urie and Knowles 1972; Urie et al. 1980).

## 1.3 Need for the Development of Resistance to *Alternaria carthami*

The leaf blight disease of safflower caused by the fungus *Alternaria carthami* Chowdhury is relatively common. It was first recorded in India (Chowdhury 1944) and was later identified in many other countries, including Australia (Irwin 1975) and the USA (Burns 1974). The disease spreads rapidly through a crop and can reduce seed yield and oil content. The fungus can be air-borne or seed-borne (Irwin 1976) and can also cause seed rot or damping-off of seedlings (Irwin 1976). Leaf blight is widespread on irrigated fields and in warmer climates where high temperatures and high relative humidities predominate. The losses resulting from the disease can be great, and in Australia during the 1978–1979 growing season it was responsible for an estimated crop loss of 20%, with some crops being completely destroyed (Jackson et al. 1982). Although some safflower cultivars have been reported to have increased resistance to *A. carthami* (Bergmann and Riveland 1983; Bergmann et al. 1985, 1987; Harrigan 1989), screening safflower for resistance to the fungus is difficult because disease outbreak is also dependent on environmental conditions (McRae et al. 1984).

A suitable resistance to the disease has not been clearly achieved and is of the utmost necessity if safflower is to be established as a economically valuable oilseed crop in areas such as the northern USA and in Australia (Jackson 1978, 1985; Jackson and Berthelsen 1986). A thorough understanding not only of the course of the disease is necessary but also of the mechanism by which the fungal parasite infects the plant in order to explicitly develop an adequate means of resistance. Plant tissue cultures are invaluable for the investigation of plant/fungal interactions, and they may also be utilized as a tool in safflower breeding.

## 2 Tissue Culture Studies on Safflower

Heterotrophic cell cultures of safflower have been employed in a number of investigations which deal with the suitability of explants (Zhanming and Biwen 1993), the formation and analysis of pigments (Saito and Fukushima 1986; Nakano et al. 1988; Saito et al. 1988, 1989; Hanagata et al. 1992), tocopherols (Furuya et al. 1987, the production of vitamin E in cell cultures (Furuya and Yoshikawa 1991) and phytoalexins (Allen and Thomas 1972; Tietjen and Matern 1984). In the latter study, the accumulation of linear polyacetylenes (Fig. 2), which are also present in plant roots (Stevens et al. 1990), could be induced by a cell wall preparation from *Alternaria carthami*, thereby stressing the suitability of these systems for the investigation of the resistance response. Furthermore, in vitro techniques have been used in the multiplication of safflower plants (George and Rao 1982; Tejovathi and Anwar 1987; Orlikowska and Dyer 1993), and elicitor-inducible hairy root cultures have been successfully employed in the investigation of polyacetylene accumulation (Flores et al. 1988). Recently, genetic transformation studies have been successfully conducted (Ying et al. 1992; Baker and Dyer 1996).

Somaclonal variation was detected in safflower, and variability could be increased by various mutagens (Anwar et al. 1993). Variations in plant type, flower color, seed morphology, and seed yield were observed. Oil enrichment in selected safflower tissue could also be achieved (Singh and Chatterji 1991), which may additionally assist breeding programs. However, in all of these examples, no screening of resistance to fungal parasites was carried out.

A further possibility for achieving resistance to pathogens is by adapting cell cultures to fungal toxic metabolites which may play a role in pathogenesis. Continuous efforts over a period of approximately 5 years were made in our own laboratory to adjust or select cells from tissue or cell suspension cultures of

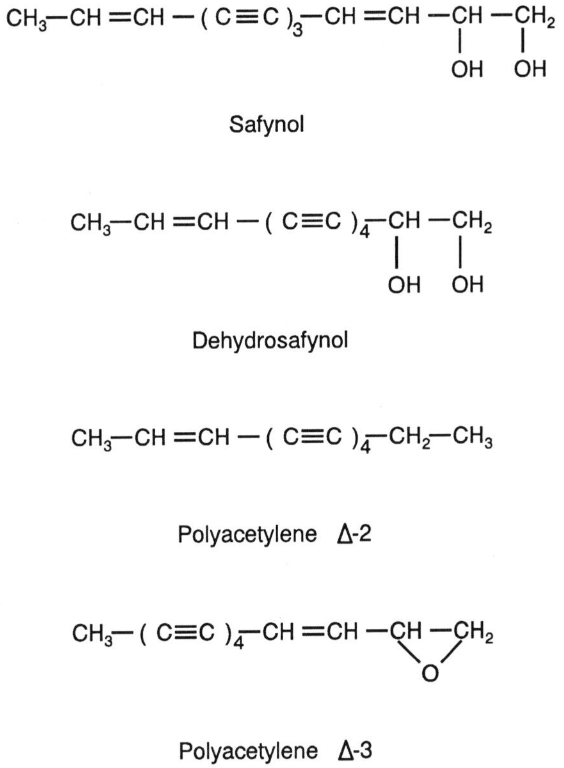

$$CH_3{-}CH{=}CH{-}(C{\equiv}C)_3{-}CH{=}CH{-}CH(OH){-}CH_2(OH)$$

Safynol

$$CH_3{-}CH{=}CH{-}(C{\equiv}C)_4{-}CH(OH){-}CH_2(OH)$$

Dehydrosafynol

$$CH_3{-}CH{=}CH{-}(C{\equiv}C)_4{-}CH_2{-}CH_3$$

Polyacetylene Δ-2

$$CH_3{-}(C{\equiv}C)_4{-}CH{=}CH{-}\underbrace{CH{-}CH_2}_{O}$$

Polyacetylene Δ-3

**Fig. 2.** Polyacetylenic phytoalexins isolated from infected safflower plants (Allen and Thomas 1972) and elicitor-induced cell cultures, respectively (Tietjen and Matern 1984)

safflower cv. US10 that might carry enhanced insensitivity to the fungal metabolite brefeldin A (see below; Matern and Tietjen 1989). Brefeldin A at nM concentrations had been shown before in these cultures (Tietjen and Matern 1984) to interfere with the elicitor-inducible accumulation of polyacetylenic phytoalexins (Fig. 2), and the capacity for brefeldin A synthesis in situ had been pinpointed as the pathogenicity factor of the safflower blight pathogen *Alternaria carthami* Chowdhury (Tietjen et al. 1985). Furthermore, the growth of cultured cells was severely inhibited by brefeldin A at elevated concentration (1 μM), making the cell mass increase the experimental parameter of choice for the determination of brefeldin A action. Supposing lack of growth inhibition in the presence of brefeldin A to be a satisfactory parameter for assessing disease resistance, resistance to *A. carthami* should increase parallel to tolerance to brefeldin A. A range of toxin concentrations (1.6–25.6 μM) was applied in MS medium (Murashige and Skoog 1962) in initial attempts to select insensitive safflower cells from suspension cultures. However, growth was observed only in control cultures, and the cells maintained on toxin failed to grow even when transferred into toxin-free medium after a one-day exposure. Long-term tests (over a period of 6 to 8 months) also failed to yield brefeldin A-tolerant cells.

The phenomenon of somaclonal variation had been demonstrated for the first time in potato cells employing toxins for screening (Matern et al. 1978). In those cases where a toxin was known to bind to a protein or to inhibit an enzyme, resistance was shown to be the result of altered enzyme activity, modification in the target enzyme or protein (Chaleff and Ray 1984), or increased levels of enzyme synthesis (Amrhein et al. 1983; Nafziger et al. 1984; Smart et al. 1985). Unlike all these toxicants, it is becoming clear from a number of recent studies in various plant and animal cells (see below) that even minute concentrations of brefeldin A cause severe impairments in the flux and processing of the secretory pathway. This most likely explains the inhibition of polyacetylene biosynthesis in safflower, which probably proceeds in the endoplasmic reticulum, and the absence of the desired brefeldin A-insensitive genotype cells. The complete failure of cells to grow in the presence of brefeldin A, moreover, suggests that processes of fundamental importance to the growth of plant cells were generally affected (Driouich et al. 1992). Although the target or receptor site for brefeldin A has not been unequivocally characterized, its modification might not be possible without sacrificing vital cellular functions. Similar reasons have also been discussed for the inability to adapt cell cultures to cercosporin, a nonselective phytotoxin from *Cercospora* spp. (Daub 1986). A different approach therefore appears to be required to achieve a durable resistance in safflower.

## 3 Introduction of Resistance to *Alternaria carthami*

The major goal of modern phytopathological research is the identification of single parameters as molecular triggers that might explain the compatibility which develops in highly complex plant-pathogen interactions. Geneticists therefore turn to cultivar and pathovar specificity to eventually pinpoint single genes as the cause of resistance and avirulence, respectively (Joosten et al. 1994). This approach, however, is very limiting and excludes a priori the entire group of nonhost-specific toxins as pathogenicity factors. If, however, a very close correlation can be drawn between the production of a phytotoxin and fungal phytopathogenicity, the plant/pathogen interaction is dissected to a single component which can be used directly for resistance screening of plant cells or indirectly for the generation of transformant plants that may overcome this hurdle. It is technically feasible to generate transgenic plants by transfer of individual genes or gene clusters that retain the otherwise beneficial properties of the variety, thus avoiding the genetic reassortment that occurs during normal plant breeding procedures. This strategy was followed in investigating the leaf blight disease of safflower.

### 3.1 Identification of Brefeldin A as Disease Determinant

*Alternaria carthami* causes the leaf and stem blight of safflower in which dark brown spots appear on older leaves and stems prior to flowering, and spread rapidly. A light green chlorotic region often surrounds the spots, which may join

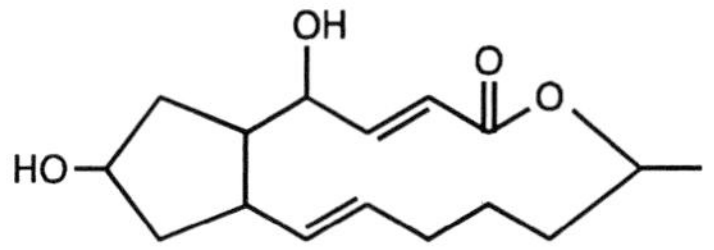

Brefeldin A

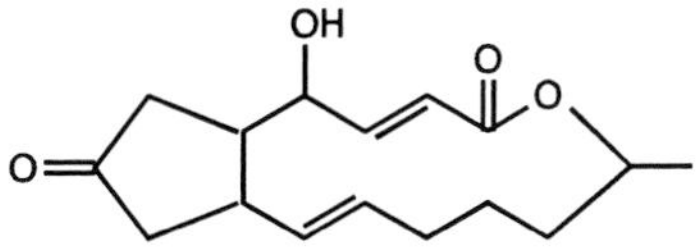

7-Oxobrefeldin A

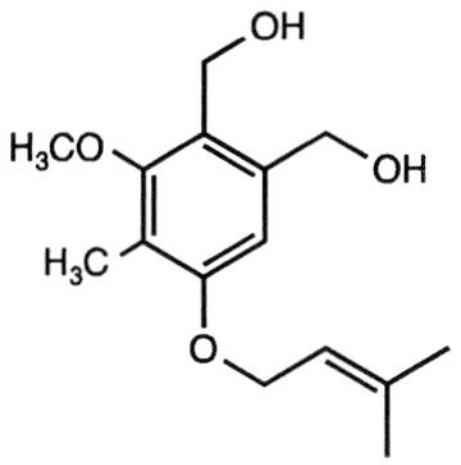

Zinniol

**Fig. 3.** Phytotoxins isolated from the culture broth of *Alternaria carthami* Chowdhury. (Tietjen et al. 1983)

to form large, irregular lesions. The fungus also infects safflower heads, resulting in the discoloration of infected seeds, and, in severe cases, the seeds are shriveled and empty. Infected, viable seeds can lead to high degrees of seedling blight and damping-off.

*A. carthami* produces three phytotoxins in culture—brefeldin A, 7-oxobrefeldin A, and zinniol (Fig. 3)—all of which produce symptoms when applied to safflower leaves (Tietjen et al. 1983). Of these toxins, only brefeldin A could be isolated from infected leaves, and much effort has since been dedicated to the correlation of toxin production and virulence. Toxin production in laboratory culture corresponded directly to virulence ratings for the various isolates of *A. carthami*. The fact that brefeldin A is a nonspecific phytotoxin tended to refute the hypothesis that brefeldin A is a pivotal factor in disease development, but toxin production by the fungus in situ is a further criterion. A highly sensitive radioimmunoassay, which enabled the determination of minute (nmol) quantities of brefeldin A in crude leaf homogenates (Tietjen et al. 1985), was used to determine the amounts and topological distribution of brefeldin A in safflower and nonhost plants after inoculation with *A. carthami* or other fungi known also to produce brefeldin A in culture. The toxin accumulated up to 3 mM in safflower tissue within 17 days following the inoculation with *A. carthami*. It was produced and diffused in advance of the growing hyphae, thereby providing a niche for successful growth of the fungus, and is most likely the cause of the chlorotic rings observed surrounding the infection sites. *Ascochyta imperfecta* and *Eupenicillium brefeldianum*, two fungi able to produce large amounts of brefeldin A in laboratory culture, failed to produce it in inoculated leaves, and neither could grow on safflower tissue. *A. carthami* did not produce the toxin on plants other than safflower, e.g., sunflower, and showed no extensive growth on nonhost plants.

These results support the hypothesis that the efficient production of brefeldin A in situ is the determining factor for infection of safflower by *A. carthami*, and that the plant qualifies as a host by particular metabolites that trigger fungal toxin biosynthesis. The specificity of the interaction thus resides within the plant, and this example provides a new conceptual basis for the role of nonhost-specific toxins in the etiology of plant diseases. One could surmise furthermore that, if safflower were resistant to brefeldin A, then *A. carthami* would not be able to establish itself in the plant and the respective blight disease would no longer occur.

## 3.2 Biosynthesis and Mechanism of Brefeldin A

Brefeldin A has attracted the attention of biochemists since it was first described from *Penicillium brefeldianum* (Härri et al. 1963). The reason lies in the structural similarity to the prostaglandins, and an analogous biosynthetic pattern starting from palmitic acid was assumed initially (Bu Lock and Clay 1969). However, precursor studies showed shortly thereafter that brefeldin A is a polyketide composed of multiple acetate units and formed by multienzyme complexes from malonyl-CoA (Coombe et al. 1969). Labeled palmitic or oleic acid was incorporated at a low rate only, and randomization of the label suggested an indirect incorporation via acetate (Cross and Hendley 1975; Tietjen 1982). This situation allows no detailed investigation or manipulation of the biosynthetic enzyme activities, which might be considered as one means to bridle toxin production and consequently pathogenicity.

Brefeldin A has always been classified as an antibiotic with antifungal, cytotoxic, and antiviral activities (Härri et al. 1963; Tamura et al. 1968). Early studies on the effects of brefeldin A on HeLa cells and *E. coli* indicated that glycolysis was stimulated, but that protein and nucleic acid biosyntheses were inhibited (Betina 1969; Betina and Montagier 1966). Brefeldin A was also shown to induce some morphological changes in pathogenic fungi (Betina et al. 1966). The mechanism of action was unknown at this time, but the growth inhibition of *Candida albicans* by brefeldin A was postulated to result from a disturbance of lipid metabolism (Hayashi et al. 1974). In wheat and onion, brefeldin A was shown to inhibit germination (Singleton et al. 1958) and root growth, respectively (Betina et al. 1963). Investigations on *Vicia faba* furthermore characterized brefeldin A as an inhibitor of mitosis (Betina and Murin 1964).

More detailed studies on the mode of brefeldin A action have been carried out only recently with animal cell cultures (for review see Betina 1992; Hurtley 1992 and Pelham 1991). At low concentrations (2–10 μM), brefeldin A impedes protein transport from the endoplasmic reticulum to the Golgi compartment (Misumi et al. 1986) and causes the *cis*-, medial- and *trans*-cisternae of the Golgi complex to redistribute to the ER, thereby eliminating the Golgi complex as a morphologically distinct organelle (Fujiware et al. 1988; Lippencott-Schwartz et al. 1989).

Considerably less insight is available into the mechanism of brefeldin A action in plants. Preliminary data from maize indicate effects analogous to those in animal cells, i.e., distortion and dissociation of the Golgi stacks concomitant with the appearance of numerous vesicles in the cytoplasm (Satiat-Jenuemaitre and

Hawes 1992, 1993). These effects occurred, however, at relatively high (100 μM) brefeldin A concentrations. In sycamore cells, brefeldin A (7.5 μM) blocks the transport from the ER to the Golgi, and although it does not induce the disassembly of the Golgi stacks, it causes an increase in the ratio of *trans* to *cis* and medial cisternae. Brefeldin A also leads to the accumulation of Golgi-derived, dense secretory vesicles within the ribosome-free Golgi matrix zone, and it induces the formation of large aggregates of Golgi stacks, possibly a result of the fusion of Golgi matrix zones (Driouich et al. 1992). It remains to be seen whether the same repertoire of effects applies also to safflower, but the generality of brefeldin A action observed so far makes this very likely. In addition to these subtle effects, however, brefeldin A exerts a very obvious devastating effect on safflower tissues that qualify the compound as a phytotoxin.

### 3.3 Detoxification of Brefeldin A

The failure to adapt safflower cells to brefeldin A prompted a search for toxin-degrading activities in other plant cells, as well as in plants that had withstood the toxin treatments in leaf bioassays (Tietjen 1982). This search was also spirited by the idea that plant genes encoding the presumed detoxifying enzymes would be better suited for the eventual transformation of safflower than a "heterologous" gene, i.e., from bacterial or fungal sources, which might give rise to a potentially toxic or unstable product. However, neither the cell cultures nor the plants showed a substantial metabolism of uniformly $^{14}C$-labeled brefeldin A fed at sublethal concentrations, even though the toxin was taken up readily from solutions upon incubation of cells or by immersion of the roots. In Chinese hamster ovary cells, for example, brefeldin A had been shown to become conjugated and inactivated by glutathione S-transferase action (GST; Brüning et al. 1992); such enzymatic conjugation was not measurable in safflower or in maize tissues, where GST activities have often been implicated in the detoxification of herbicides (Mozer et al. 1983). These results corroborated the previous findings that plant tissue is an inadequate source of enzymes for brefeldin A turnover, and the search for an active brefeldin A metabolism had to be extended to microorganisms.

*A. carthami* itself may be considered as a suitable producer of brefeldin A-degrading enzymes. In liquid stationary culture, the concentration of toxin usually progresses up to 35 days to reach values of about 0.5 mM and declines thereafter. This and the ability of the fungus to grow on such high concentrations of brefeldin A are indicative of the turnover capacity for toxic metabolite of its own, possibly similar to that described for *Cercospora* spp. (Daub et al. 1992). No activity towards brefeldin A, however, could be recovered in the crude fungal extracts at any stage of growth. Bacteria, on the other hand, are known to possess an outstanding repertoire for the degradation of toxic substances such as herbicides, toxic wastes, and antibiotics. Resistance of *E. coli* to the macrolide antibiotic erythromycin, for example, has been attributed to efficient enzymatic, one-step inactivations, e.g., hydrolysis (Barthélémy et al. 1984; Arthur et al. 1986, 1987) or phosphorylation (O'Hara et al. 1989). The enzymes involved, however, possess narrow substrate specificities and are inactive on macrolides of different ringsize. Identification of such a one-step enzymatic detoxification mechanism for the macrolide-substrate brefeldin A was expected to have an enormous impact on the control of the leaf blight disease. Bacteria isolated from local soil samples were indeed capable of growing in the presence of high concentrations of brefeldin A (360 μM) that would completely inhibit the growth of plant or animal cells. One particular isolate, later identified as *Bacillus subtilis* BG3, was especially proficient in degrading brefeldin A to a single hydrophilic substance (Kneusel et al. 1990). The metabolite, characterized as brefeldin A acid, resulted from hydrolysis of the lactone ring and release of the hydrophilic carboxylic acid. The catalytic activity was thus to be assigned to an esterase enzyme with substrate affinity to lactones and macrolides, respectively. The investigation of substrate specificity revealed later (Kneusel et al. 1994) an apparent $K_m$ for brefeldin A at 35 μM with a $V_{max}$ of 0.4 mkat/kg, and 7-epibrefeldin A or 7-oxobrefeldin A was hydrolyzed with similar efficiency. Ethyl valerate, an aliphatic ester that structurally resembles the ester portion of brefeldin A, was hydrolyzed much more effectively than brefeldin A with a $K_m$ of 48 μM and a $V_{max}$ of 51 mkat/kg. The esterase showed no activity toward the macrolides erythromycin and zearalenone or toward phenyl acetate (Kneusel et al. 1994).

The change in molecular conformation brought about by hydrolysis of the lactone was assumed to affect the toxicity of brefeldin A directly or, at least indirectly, by reducing its membrane permeability. Leaves treated with 1.4 or 14 μg brefeldin A in standard leaf bioassays (Tietjen 1982) developed large necrotic lesions, whereas those treated with equivalent amounts of brefeldin A acid showed no discoloration or necrosis. Since the compounds had been administered through puncture wounds in these assays, the data suggested that hydrolysis of brefeldin A abolishes the phytotoxic activity of the toxin rather than simply inhibiting its translocation in the plant tissue. Enzymatic hydrolysis thus represents a simple, one-step detoxification reaction for brefeldin A that is remarkable also from a chemical point of view: the double bonds in conjugation with the lactone carbonyl make brefeldin A a very inert molecule that is not selectively hydrolyzed even in strong, boiling acid or alkali (Tietjen 1982).

The hydrolysis of brefeldin A was also catalyzed by cell-free extracts of *Bacillus subtilis* BG3 without any cofactor requirements, and the esterase reaction was considered to give straightforward access to the generation of safflower plants insensitive to the toxin and resistant to *A. carthami*. Accordingly, the gene encoding the brefeldin A esterase had to be isolated. Direct cloning by simply selecting a BG3 DNA library in *E. coli* on brefeldin A-containing media (minimal media did not sustain growth and could not be used) failed, because nontransformed *E. coli* also grew to a certain extent without detoxifying brefeldin A. Therefore, the more laborious reverse genetic approach had to be followed which required the isolation of homogeneous esterase from *B. subtilis* BG3 followed by peptide sequencing and the generation of antisera. Antisera or degenerate oligonucleotides complementary to the partial enzyme sequence could then be used to screen a *B. subtilis* BG3 genomic library. The esterase was purified by a six-step procedure (Kneusel et al. 1994) and shown to consist of a monomeric polypeptide of 40 kD[a] apparent molecular mass that cross-reacted specifically in Western blots with a rabbit polyclonal antiserum generated to the pure native esterase (Kneusel et al. 1994). Maximal activity of the homogeneous esterase in vitro was measured at 37 °C with substantial activity (75%) at 25 °C, which is an essential prerequisite for the enzyme to be useful in transformed safflower plants to be grown in the field.

Microsequencing of the homogeneous esterase protein revealed the sequences of an N-terminal stretch of 26 amino acids and of two tryptic peptides of 9 and 11 residues, respectively. This sequence information was sufficient to design two specific oligonucleotide probes, and two rounds of screening of a genomic *B. subtilis* BG3 library resulted in the isolation of seven clones. The translated sequence of one of these clones with an insert size of roughly 2 kb revealed the N-terminal amino acids of the esterase protein preceded by 40 noncoding nucleotides. The insert containing the full esterase coding sequence was furthermore subcloned into *E. coli* pT7-7, which resulted in an enzymatically active fusion protein and yielded crude cellular extracts with a specific activity three-fold greater than that of the purified *B. subtilis* esterase (Kneusel et al. 1994). The isolation and cloning of the *B. subtilis* brefeldin A esterase gene has set the stage for the resistance transformation of safflower. Investigations regarding the stable expression of the gene in transgenic tobacco are currently under way, as well as the transformation and regeneration of safflower. Using a modified version of a safflower regeneration protocol (Orlikowska and Dyer 1993), we have already successfully transformed and regenerated shoots from two different safflower varieties. The regeneration of complete transgenic plants will be achieved in the near future.

## 4 Summary and Conclusions

Since standard techniques of somaclonal variation failed to yield safflower plants resistant to brefeldin A and *Alternaria carthami*, we turned to the reverse genetic approach to achieve the goal. Brefeldin A had been identified as a pathogenicity factor from *A. carthami* that suppresses the plant's defense response. Thus, the detoxification of brefeldin A in planta would be an effective means to protect safflower from *Alternaria* leaf spot in the field. As a first step toward this goal, a strain of *B. subtilis* was isolated that is capable of detoxifying brefeldin A by hydrolysis of the macrolide ring. The enzyme responsible was purified and the encoding gene was cloned. The esterase clone provides the basis for the desired resistance transformation of safflower, and the protocol for regeneration of transformed plants is being tested.

The stable expression of the esterase at a satisfactory level in transgenic safflower may represent yet another hurdle that needs to be overcome. Constitutive expression from the commonly employed CaMV 35S promotor may exert deleterious effects on the basic cellular metabolism of safflower. The rationale, therefore, is the expression of the esterase in the plant under the control of an elicitor-responsive promoter, which functions as a switch only upon challenge by a fungal pathogen. The eventual deleterious side effects of esterase expression in transgenic safflower may also be diminished by the integration of a cellular export leader peptide. Such a strategy has been successfully employed in the excretion of functional monoclonal antibodies and T4 lysozyme from transgenic tobacco plants (Düring et al. 1990). Targeting of the esterase to the intercellular space in transgenic safflower appears most useful for the in situ detoxification of brefeldin A. The resistance transformation of safflower with the brefeldin A esterase from *B. subtilis* BG3 differs considerably from previous reports on the adaptation to and toleration of obnoxious chemicals and metabolites, since the plants will be capable of permanently degrading the toxin. Such plants will not only be of benefit to a classical breeding scheme, but they will prove for the first time that a nonspecific phytotoxin can indeed play the pivotal role in the development of a specific plant-fungal interaction.

*Acknowledgments.* Financial support for the work quoted from our laboratory was received from Deutsche Forschungsgemeinschaft and Fonds der Chemischen Industrie, which is gratefully acknowledged.

## References

Abel GH, Driscoll MF (1976) Sequential trait development and breeding for high yields in safflower. Crop Sci 16: 213–216

Allen EH, Thomas CA (1972) Relationship of safynol and dehydrosafynol accumulation to *Phytophthora* resistance in safflower. Phytopathology 62: 471–474

Amrhein N, Johänning D, Schab J, Schulz A (1983) Biochemical basis for glyphosate-tolerance in a bacterium and a plant tissue culture. FEBS Lett 157: 191–196

Anwar SY, Tejovathi G, Khadeer MA, Seeta P, Rajendra Prasad B (1993) Tissue culture and mutational studies in safflower (*Carthamus tinctorius* L.). Proc 3rd Int Safflower Conf, Institute of Botany, Chinese Academy of Sciences, Beijing, China, pp 124–136

Arthur M, Autissier D, Courvalin P (1986) Analysis of the nucleotide sequence of the ereB gene encoding the erythromycin esterase type II. Nucleic Acids Res 14: 4987–4999

Arthur M, Brisson-Noël A, Courvalin P (1987) Origin and evolution of genes specifying resistance to macrolide, lincosamide and streptogramin antibiotics: data and hypotheses. J Antimicrob Chemother 20: 783–802

Ashri A, Zimmer DE, Urie AL, Knowles PF (1975) Evaluation of the germplasm collection of safflower *Carthamus tinctorius* L. VI. Length of planting to flowering period and plant height in Israel, Utah and Washington. Theor Appl Genet 46: 395–396

Baker CM, Dyer WE (1996) Genetic transformation of safflower (*Carthamus tinctorius* L.). In: Bajaj YPS (ed) Biotechnology in agriculture and forestry, vol 38. Plant protoplasts and genetic engineering VII. Springer, Berlin Heidelberg New York (in press)

Barthélémy P, Autissier D, Gerbaud G, Courvalin P (1984) Enzymic hydrolysis of erythromycin by a strain of *Escherichia coli*. A new mechanism of resistance. J Antibiot 37: 1692–1696

Bergman JW, Riveland NR (1983) Sidwell safflower. Crop Sci 23: 1012–1013

Bergman JW, Carlson G, Kushnak G, Riveland NR, Stallknecht G (1985) Registration of Oker safflower. Crop Sci 25: 1127–1128

Bergman JW, Baldridge DE, Brown PL, Dubbs AL, Kushnak GD, Riveland NR (1987) Registration of Hartman safflower. Crop Sci 27: 1090–1091

Betina V (1969) Effect of the macrolide antibiotic, cyanein on HeLa cell growth and metabolism, Neoplasma 16: 23–32

Betina V (1992) Biological effects of the antibiotic brefeldin A (decumbin, cyanin, ascotoxin, synergisidin)—a retrospective. Folia Microbiol 37: 3–11

Betina V, Montagier L (1966) Action of cyanein on the synthesis of nucleic acid and protein in animal cell and bacterial protoplasts. Bull Soc Chim Biol 48: 194–198

Betina V, Murin A (1964) Inhibition of mitotic activity in root tips of *Vicia faba* by the antibiotic cyanein. Cytologia (Tokyo) 29: 370–374

Betina V, Nemec P, Baráth Z (1963) Growth inhibition of *Allium cepa* roots by the antibiotic cyanein. Naturwissenschaften 50: 696

Betina V, Betinová M, Kutková M (1966) Effects of cyanein on growth and morphology of pathogenic fungi. Arch Mikrobiol 55: 1–16

Brüning A, Ishikawa T, Kneusel RE, Matern U, Lottspeich F, Wieland F (1992) Brefeldin A binds to glutathione S-transferase and is secreted as glutathione and cysteine conjugates by Chinese hamster ovary cells. J Biol Chem 267: 7726–7732

Bu Lock JD, Clay PT (1969) Fatty acid cyclization in the biosynthesis of brefeldin A; a new route to some fungal metabolites. Chem Commun 1969: 237–238

Burns EE (1974) Identification and etiology of *Alternaria carthami* on safflower in Montana. Proc Am Phytopathol Soc USA 1: 41–44

Chaleff RS, Ray TB (1984) Herbicide-resistant mutants from tobacco cell cultures. Science 223: 1148–1151

Chowdhury S (1944) An *Alternaria* disease of safflower. J Ind Bot Soc 23: 59–65

Coombe RG, Foss PS, Jacobs JJ, Watson TR (1969) The biosynthesis of brefeldin A. Aust J Biochem 22: 1943–1950

Cross BE, Hendley P (1975) The biosynthesis of brefeldin A. J Chem Soc Chem Commun 1975: 124–125

Daub ME (1986) Tissue culture and the selection of resistance to pathogens. Annu Rev Phytopathol 24: 159–186

Daub ME, Leisman GB, Clark RA, Bowden EF (1992) Reductive detoxification as a mechanism of fungal resistance to singlet oxygen-generating photosensitizers. Proc Natl Acad Sci USA 89: 9588–9592

Driouich A, Zhang GF, Staehelin LA (1992) Effect of brefeldin A on the structure of the Golgi apparatus and on the synthesis and the secretion of proteins and polysaccharides in sycamore suspension cultured cells. Plant Physiol 101: 1363–1373

Düring K, Hippe S, Kreuzaler F, Schell J (1990) Synthesis and self-assembly of a functional monoclonal antibody in transgenic *Nicotiana tabacum*. Plant Mol Biol 15: 281–293

Flores HE, Pickard JJ, Hoy MW (1988) Production of polyacetylenes and thiophenes in heterotrophic and photosynthetic root cultures of Asteraceae. In: Lam J, Breteler H, Arnason T (eds) Naturally occurring acetylenes and related compounds. Elsevier, Amsterdam, pp 233–254

Fujiwara T, Oda K, Yokota S, Takatsuki A, Ikehara Y (1988) Brefeldin A causes dissambly of the Golgi complex and accumulation of secretory proteins in the endoplasmic reticulum. J Biol Chem 263: 18545–18552

Fukushima A, Takahashi, Y, Ashihara H, Saito K (1990) Relationship between floret elongation and pigment synthesis in the flowers of *Carthamus tinctorius*. Ann Bot 65: 361–363

Fuller G, Kohler O, Appelwhite TH (1967) High-oleic acid safflower oil: a new stable edible oil. J Am Oil Chem Soc 43: 477–478

Fuller G, Diamond MJ, Appelwhite TH (1968) High-oleic safflower oil. Stability and chemical modification. J Am Oil Chem Soc 44: 264–266

Furuya T, Yoshikawa T (1991) *Carthamus tinctorius* L. (safflower): Production of vitamin E in cell cultures. In: Bajaj YPS (ed) Biotechnology in agriculture and forestry, vol 15. Medicinal and aromatic plants III. Springer, Berlin Heidelberg New York, pp 142–155

Furuya T, Yoshikawa T, Kimura T, Kaneko H (1987) Production of tocopherols by cell culture of safflower. Phytochemistry 26: 2741–2747

George L, Rao PS (1982) In vitro multiplication of safflower (*Carthamus tinctorius* L.) through tissue culture. Proc Ind Natl Sci Acad B (Biol Sci) 48: 791–794

Hanagata N, Ito A, Fukuju Y, Murata K (1992) Red pigment formation in cultured cells of *Carthamus tinctorius* L. Biosci Biotech Biochem 56: 44–47

Hanelt P (1961a) Systemic study of the genus *Carthamus* L.—a monographic review. PhD Thesis, Martin-Luther Universität, Halle

Hanelt P (1961b) Contributions to our knowledge of *Carthamus tinctorius* L. Kulturpflanze 9: 114–145

Härri E, Loeffler HP, Sigg HP, Staehelin H, Tamm C (1963) Über die Isolierung neuer Stoffwechselprodukte aus *Penicillium brefeldianum* Dodge. Helv Chim Acta 46: 1235–1243

Harrigan EFS (1989) Review of research of safflower in Australia. Abstr 2nd Int Safflower Conf, Hyderabad, India, 31 pp

Hattori M, Huang X-L, Che Q-M, Kawata Y, Tezuka Y, Kikuchi T, Namba T (1992) 6-Hydroxykaempferol and its glycosides from *Carthamus tinctorius* petals. Phytochemistry 31: 4001–4004
Hayashi T, Takatsuki A, Tamura G (1974) The action mechanism of brefeldin A I. Growth recovery of *Candida albicans* by lipids from the action of brefeldin A. J Antibiot 27: 65–72
Hurtley SM (1992) Now you see it, now you don't: the Golgi disappearing act. TIBS 17: 325–327
Irwin JAG (1975) *Alternaria carthami* on safflower. Aust Plant Pathol Soc News 3: 24–28
Irwin JAG (1976) *Alternaria carthami*, a seed-borne pathogen of safflower. Aust J Exp Agric Anim Husb 16: 921–925
Jackson KJ (1978) Safflower variety testing—what's happening? Queensl Agric J (Aust) 104: 257–263
Jackson KJ (1985) Safflower production in Australia. Sesame and safflower status and potentials. FAO Plant Prod. Prot Pap 66: 23–29
Jackson KJ, Berthelsen JE (1986) Production of safflower *Carthamus tinctorius* L. in Queensland. J Aust Inst Agric Sci 52: 63–72
Jackson KJ, Irwin JAG, Berthelsen JE (1982) Effect of *Alternaria carthami* on the yield components and seed quality of safflower. Aust J Exp Agric Anim Husb 22: 221–225
Joosten MHAJ, Cozijnsen TJ, De Wit PJGM (1994) Host resistance to a fungal tomato pathogen lost by a single base-pair change in an avirulence gene. Nature 367: 884–886
Kneusel RE, Matern U, Wray V, Klöppel K-D (1990) Detoxification of the macrolide toxin brefeldin A by *Bacillus subtilis* FEBS Lett 275: 107–110
Kneusel RE, Schiltz E, Matern U (1994) Molecular characterization and cloning of an esterase which inactivates the macrolide phytotoxin brefeldin A. J Biol Chem 269: 3449–3456
Knowles PF (1969a) Centers of plant diversity and conservation of crop germplasm: safflower. Econ Bot 23: 324–329
Knowles PF (1969b) Modification of quantity and quality of safflower oil through plant breeding. J Am Oil Chem Soc 46: 130–132
Knowles PF (1972) The plant geneticist's contribution toward changing lipid and amino acid composition of safflower. J Am Oil Chem Soc 49: 27–29
Knowles PF (1989) Safflower. In: Downey RK, Röbbelen G, Ashri A (eds) Oil crops of the world. McGraw-Hill, New York, pp 363–374
Knowles PF, Hill AB, Ruckman JE (1965) High oleic acid content in new safflower UC-1. Calif Agric 19: 15
Kotecha A (1981) Inheritance of seed yield and its components in safflower. Can J Gen Cytol 23: 111–117
Kumar H (1993) Current trends in breeding research for enhancing productivity of safflower in India. In: Fernández Martinez J (ed) Sesame and safflower newsletter, vol 8. Institute of Sustainable Agriculture, CSIC, Córdoba, Spain, pp 70–73
Kupzow AJ (1932) The geographical variability of the species *Carthamus tinctorius* L. Bull Appl Bot Genet Plant Breed 9th Ser 1: 99–181
Lippencott-Schwartz J, Donaldson JG, Schweizer A, Berger EG, Hauri H-P, Yuan LC, Klausner RD (1989) Rapid distribution of golgi proteins into the ER in cells treated with brefeldin A: evidence for membrane cycling from Golgi to ER. Cell 56: 801–813
Lyon CK, Gumbmann MR, Betschart AA, Robbins DJ, Sauders RM (1979) Removal of deleterous glucosides from safflower meal. J Am Oil Chem Soc 56: 560–562
Matern U, Tietjen KG (1989) Metabolism of the phytotoxin brefeldin A in safflower (*Carthamus tinctorius* L.) plants. In: Graniti A, Durbin RD, Ballio A (eds) Phytotoxins and plant pathogenesis. Nato ASI Series, vol H 27. Springer, Berlin Heidelberg New York pp 419–421
Matern U, Strobel G, Shepard J (1978) Reaction to phytotoxins in a potato population derived from mesophyll protoplasts. Proc Natl Acad Sci USA 75: 4935–4939
McRae CF, Harrigan EKS, Brown JF (1984) Effect of temperature, dew period and inoculation density on blight of safflower caused by *Alternaria carthami*. Plant Dis 68: 408–410
Meselhy MR, Kadota S, Hattori M, Namba T (1993) Metabolism of safflor yellow B by human intestinal bacteria. J Nat Prod 56: 39–45
Misumi Y, Misumi Y, Miki K, Takatsuki A, Tamura G, Ikehara Y (1986) Novel blockade by brefeldin A of intracellular transport of secretory proteins in cultured rat hepatocytes. J Biol Chem 261: 11398–11403

Mozer TJ, Tiemeier DC, Jaworski EG (1983) Purification and characterization of glutathione S-transferase. Biochemistry 22: 1068–1080
Mündel H-H (1993) Keynote address. Proc 3rd Int Safflower Conf, Institute of Botany, Chinese Academy of Sciences, Beijing, China, pp 8–13
Mündel H-H, Morrison RJ, Blackshaw RE, Roth B (eds) (1992) Safflower production on the Canadian prairies. Graphcom Printers, Lethbridge, Alberta
Murashige T, Skoog F (1962) A revised medium for rapid growth and bioassays with tobacco tissue cultures. Physiol Plant 15: 473–497
Nafziger ED, Widholm JM, Steinrücken HC, Kilmer JC (1984) Selection and characterization of a carrot cell line tolerant to glyphosate. Plant Physiol 76: 571–579
Nakano K, Sekino Y, Yomo N, Wakayama S, Miyano S, Kusaka K, Daimon E, Imaizumi K, Totsuka Y, Oda S (1988) High-performance liquid chromatography of carthamin, safflor yellow A and a precursor of carthamin. Application to the investigation of an unknown red pigment produced in cultured cells of safflower. J Chromatogr 438: 61–72
Narkhede BN, Deokar AB (1986) Inheritance of corolla color in safflower. J Maharashtra Agric Univ 11: 278–281
O'Hara K, Kanda T, Ohmiya K, Ebisu T, Kono M (1989) Purification and characterization of macrolide 2'-phosphotransferase from a strain of *Escherichia coli* that is highly resistant to erythromycin. Antimicrob Agents Chemother 33: 1354–1357
Orlikowska TK, Dyer WE (1993) In vitro regeneration and multiplication of safflower (*Carthamus tinctorius* L.). Plant Sci 93: 151–157
Pelham HRB (1991) Multiple targets for brefeldin A. Cell 67: 449–451
Raghunatham G, Jagdish Chandra A, Satyanarayana A (1989) Studies on relationship of some genetic parameters in safflower. Abstr 2nd Int Safflower Conf, Hyderabad, India, 37 pp
Rao VR (1977) An analysis of association of components of yield and oil in safflower (*Carthamus tinctorius*). Theor Appl Genet 50: 185–191
Rojas P, Ruso J, Osorio J, de Haro A, Fernández Martinez J (1993) Variability in protein and hull content of the seed of a world collection of safflower. In: Fernández Martinez J (ed) Sesame and safflower newsletter, vol 8. Institute of Sustainable Agriculture, CSIC, Córdoba, Spain, pp 122–126
Saito K (1991) A new method for reddening dyer's saffron florets: evaluation of carthamin productivity. Z Lebensm Unters Forsch 192: 343–347
Saito K (1993) The catalytic aspects of glucose oxidase in the red colour shift of *Carthamus tinctorius* capitula. Plant Sci 90: 1–9
Saito K, Fukushima A (1986) Effect of external conditions on the stability of enzymatically synthesized carthamin. Acta Soc Bot Pol 55: 639–652
Saito K, Fukushima A (1988) On the mechanism of the stable red color expression of cellulose-bound carthamin. Food Chem 29: 161–176
Saito K, Daimon E, Kusaka K, Wakayama S, Sekino Y (1988) Accumulation of a novel red pigment in cell suspension cultures of floral meristem tissues from *Carthamus tinctorius* L. Z Naturforsch (Biosci) 43c: 862–870
Saito K, Fukushima A, Sasamoto H, Ashihara H (1989) Variation in activities of phenol oxidizing enzymes in tissue cultures of *Carthamus tinctorius* L. Acta Physiol Plant 11: 233–240
Sakamura S, Terayama Y, Kawakatsu S, Ichihara A, Saito H (1980) Conjugated serotonins and phenolic consituents in safflower seeds (*Carthamus tinctorius* L.). Agric Biol Chem 44: 2951–2954
Satiat-Jeunemaitre B, Hawes C (1992) Reversible dissociation of the plant Golgi apparatus by brefeldin A. Biol Cell 74: 325–328
Satiat-Jeunemaitre B, Hawes C (1993) The distribution of secretory products in plant cells is affected by brefeldin A. Cell Biol Int 17: 183–193
Sato H, Kawagishi H, Nishimura T, Yoneyama S, Yoshimoto Y, Sakamura S, Furusaki A, Katsuragi S, Matsumoto T (1985) Serotobenine, a novel phenolic amide from safflower seeds (*Carthamus tinctorius* L.) Agric Biol Chem 49: 2969–2974
Schmitt D, Pakusch AE, Matern U (1991) Molecular cloning, induction and taxonomic distribution of caffeoyl-CoA 3-O-methyltransferase, an enzyme involved in disease resistance. J Biol Chem 266: 17416–17423
Singh HP, Chatterji AK (1991) Oil enrichment in leaf callus culture of safflower. N D J Agric Res 6: 171–175

Singleton VL, Bohonos N, Ullstrup AJ (1958) Decumbin, a new compound from a species of *Penicillium*. Nature 181: 1072–1073

Smart CC, Johänning D, Müller G, Amrhein N (1985) Selective overproduction of 5-enolpyruvylshikimic acid 3-phosphate synthase in a plant cell culture which tolerates high doses of the herbicide glyphosate. J Biol Chem 260: 16338–16346

Stevens KL, Witt SC, Turner CE (1990) Polyacetylenes in related thistles of the subtribes Centaureinae and Carduinae. Biochem Syst Ecol 18: 229–232

Takahashi Y, Miyasaki N, Tasaka S, Miura I, Urano S, Ikura M, Hikichi K, Matsumoto T, Wada M (1982) Constitution of two coloring matters in the flower petals of *Carthamus tinctorius* L. Tetrahedron Lett 23: 5163–5166

Tamura G, Ando K, Suzuki S, Takatsuki A, Arima K (1968) Antiviral activity of brefeldin A and verrucarin A. J Antibiot 21: 160–161

Tejovathi G, Anwar SY (1987) Plantlet regeneration from cotyledonary cultures of safflower (*Carthamus tinctorius* L.) In: Reddy GM (ed) Plant cell and tissue culture of economically important plants, Proc Symp, Hyderabad, India, pp 347–353

Tietjen KG (1982) Zur Rolle von Phytotoxinen und Elicitoren in Pflanzenkrankheiten. Untersuchungen zu Struktur und physiologischer Bedeutung von Metaboliten des Fäberdistelpathogens *Alternaria carthami*. PhD Thesis, Universität Freiburg

Tietjen KG, Matern U (1984) Induction and suppression of phytoalexin biosynthesis in cultured cells of safflower, *Carthamus tinctorius* L., by metabolites of *Alternaria carthami* Chowdhury. Arch Biochem Biophys 229: 136–144

Tietjen KG, Schaller E, Matern U (1983) Phytotoxins from *Alternaria carthami* Chowdhury: structural indentification and physiological significance. Physiol Plant Pathol 23: 387–400

Tietjen KG, Hammer D, Matern U (1985) Determination of toxin distribution in *Alternaria* leaf spot diseased tissue by radioimmunoassay. Physiol Plant Pathol 26: 241–257

Tiwari KP, Namdeo KN (1993) Study on special arrangement and fertility levels on the spiny and spineless genotypes of safflower (*Carthamus tinctorius* L.). In: Fernández Martinez J (ed), Sesame and safflower newsletter, vol 8, Institute of Sustainable Agriculture, CSIC, Córdoba, Spain, pp 97–100

Urie AL (1986) Inheritance of partial hull in safflower. Crop Sci 26: 493–498

Urie AL, Knowles PF (1972) Safflower introductions resistant to *Verticillium* wilt. Crop Sci 12: 545–546

Urie AL, DaVia DJ, Knowles PF, Zimmerman LH (1980) Registration of 14–5 safflower germplasm. Crop Sci 20: 115–116

Vavilov NI (1951) The origin, variation, immunity and breeding of cultivated plants. Ronald Press, New York, pp 364

Weiss EA (ed) (1983) Oilseed crops. Longman, London

Ying M, Dyer WE, Bergman JW (1992) *Agrobacterium tumefaciens*-mediated transformation of safflower (*Carthamus tinctorius* L.) cv. Centernnial. Plant Cell Rep 11: 581–585

Zhanming H, Biwen H (1993) The tissue culture of safflower and its histological and cytological study. Proc 3rd Int Safflower Conf, Beijing, China, Beijing Botanical Garden, Institute of Botany, Chinese Academy of Sciences, pp 184–195

# II. 3 Somaclonal Variation in *Hypericum perforatum* (St. John's Wort)

E. ČELLÁROVÁ[1] and K. BRUŇÁKOVÁ[1]

## 1 General Account

### 1.1 Botany, Importance, Distribution

*Hypericum perforatum* L. (syn. *H. officinarum* Crantz, *H. vulgare* Lam., *H. veronense* Schrank incl., *H. stenophyllum* Opiz incl.), a member of the family Clusiaceae in an erect perennial herb with branched stems near the top, and simple, opposite, sessile, or subsessile leaves. Flowers are bright yellow, 13–30 mm in diameter, and are arranged in broad corymbs. *Hypericum perforatum* has five petals, irregular, with marginal black glandular dots, sometimes also with superficial black dots or streaks. The fruit consists of a three-celled capsule, hard seed, black or brown, reticulate testa, without endosperm.

*Hypericum perforatum* L. is considered to be an important source of pharmaceuticals which occur in the aerial parts of the plant (Herba hyperici). The most important constituents are the naphtodianthrones hypericin and pseudohypericin. These compounds are found only in certain members of the genus *Hypericum*, and are particularly prevalent in *Hypericum perforatum*. Herba hyperici also contains tannins, essential oil, flavonoids, and other secondary metabolites.

The importance of this plant is due in particular to the content of anthraquinones, namely hypericin and its derivatives. Hypericin (1, 3, 4, 6, 8, 13-hexahydroxy-10, 11-dimethylphenanthro-[1, 10, 9, 8-opgra]perylene-7, 14-dione; Thomas et al. 1992; Fig. 1) is a photodynamic pigment found in certain members of the genus *Hypericum* (Giese 1980; Knox and Dodge 1985). This pigment occurs in no other higher plant species. Pseudohypericin is a 2-methoxy derivative of hypericin. Hypericin is known to produce singlet oxygen (and possibly radicals) on exposure to visible light (Duran and Song 1986), and is responsible for the phototoxic symptoms called hypericism in grazing animals ingesting large quantities of *Hypericum* plants. However, this plant can be administered safely to humans and other animals (Giese 1980; Duran and Song 1986).

It has recently been found that hypericin and pseudohypericin are effective against certain viruses and retroviruses, for example Friend Leukemia Virus (FV) and Radiation Leukemia Virus (RadLV) in vivo, Human Immunodeficiency Virus type 1 (HIV-1) in vitro (Meruelo et al. 1988), a DNA murine

[1] Department of Experimental Botany and Genetics, Faculty of Science, P.J. Šafárik University, Mánesova 23, 04154 Košice, Slovakia

Biotechnology in Agriculture and Forestry, Vol. 36
Somaclonal Variation in Crop Improvement II (ed. by Y.P.S. Bajaj)

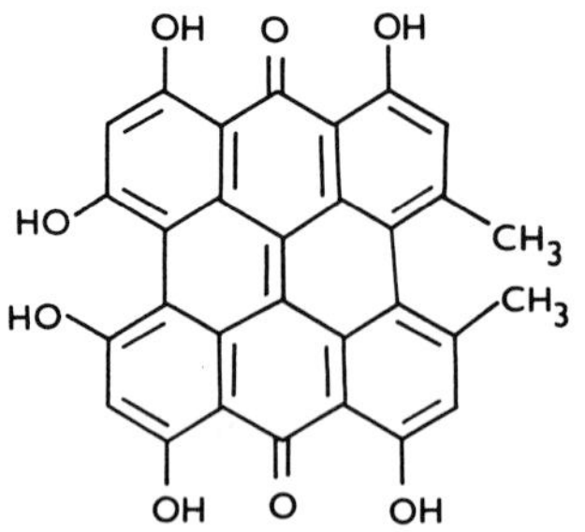

Fig. 1. Structural formula of hypericin

cytomegalovirus (MCMV), and a RNA Sindbis virus (SV; Hudson et al. 1991, 1993; Lopez-Bazzocchi et al. 1991). Antineoplastic activity of hypericin has been reported by Thomas and Pardini (1992) and Thomas et al. (1992).

Lavie et al. (1990) studied the structural features of hypericin essential for antiretroviral activity, using molecular analogues and precursors of hypericin. They reported on the role of carbonyl groups in the antiretroviral activity of these molecules which are involved in the photodynamic action.

Hypericin is also reported to have antidepressant effects, which was determined by inhibition of monoamine oxidase, type A and B in vitro (Suzuki et al. 1984).

Apart from the isolation of hypericin and its derivatives from natural plant sources, successful attempts have also been made in the synthesis of emodin anthrone, the immediate precursor of hypericin synthesis (Fig. 2). Hypericin can be obtained by dimerization of emodin derivatives (Falk and Schoppel 1991). Kraus et al. (1990) found an antiretroviral activity of synthetic hypericin against equine infectious anemia virus (EIAV).

Another group of important constituents of the plant with antibiotic effects is acylphloroglucinols hyperforin (Bystrov et al. 1978), adhyperforin, and their polar derivatives (Maisenbacher and Kovar 1992).

Systematic investigations of flavonoids in *Hypericum perforatum* have been made (Michaluk et al. 1956; Michaluk 1961a, b). The flavonoid glycosides (rutin, hyperosid, isoquercitrine, quercitrine) and aglycons (quercetin, kaempferol, and luteolin) are also considered to be potentially therapeutic compounds due to their antiinflammatory and spasmolytic effects (Berghöfer and Hölzl 1987; Hölzl and Ostrowski 1987). The biflavonoids I3, II8- and I3', and II8'-Biapigenin are effective in the treatment of stomach diseases (Berghöfer and Hölzl 1987, 1989) because of their sedative activity.

*H. perforatum* occurs naturally in Europe, apart from the arctic regions, as far east as western Siberia and northwest China, as well as in Minor Asia, northern Iran, and northern Africa.

*H. perforatum* is a hemicryptophyte. It grows on sunny hillsides, drier meadows, old fields, pastures, rocks, in light forests and forest clearings, river

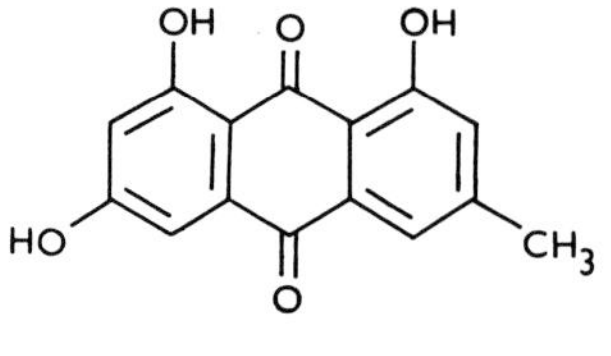

Fig. 2. Emodin – an immediate precursor of hypericin

banks, roadsides, railway land, and on other similar types of habitat. It is a species with wide ecological and coenological amplitude.

### 1.2 Breeding Objectives and Available Genetic Variability

The important breeding objective is high content of secondary metabolites with significant pharmaceutical activity, namely naphtodianthrones; hypericin and pseudohypericin, acylphloroglucinols; hyperforin and adhyperforin, and flavonoids.

*Hypericum perforatum* plants growing in natural populations are very variable. In addition, the variability of the species can be broadened by several in vitro approaches with respect to spectrum and specificity of product synthesis. From this point of view, another important objective is genetic stability of high-producing genotypes and their offspring, which may depend, with the exception of other factors, on the proportion of apomictic and sexual reproduction.

### 1.3 Need to Induce Somaclonal Variation

Due to the specific manner of reproduction in *H. perforatum*, the use of conventional breeding methods is rather limited and restricted to the selection. Therefore, new approaches to broaden the variability, such as somaclonal variation, somatic hybridization, and genetic transformation, have to be considered. The induction of somaclonal variation in *H. perforatum* is convenient due to its very effective regeneration capacity in vitro, which leads to hundreds of somaclones originating from one particular genotype. Tissue culture of Saint John's Wort followed by regeneration can lead to plants varying in different characters: growth habit, biomass production, foliar morphology, ploidy, and hypericin content.

## 2 In Vitro Culture Studies

Tissue culture studies on the genus *Hypericum* have been recently reviewed (Yazaki and Okuda 1994; Čellárová et al. 1995). A very effective regeneration pattern was reported from seedlings of *H. perforatum* L., and studies on the variability of several morphological characters and histological structures in regenerants have been carried out (Čellárová et al. 1992). *H. perforatum* can be micropropagated easily from seedling explants on basal Linsmaier-Skoog medium (RM) supplemented with 0.1 to 2 mg/l BAP. Green shoot primordia appear within 10 days on the cut sites of leaf, stem, and root explants. After four weeks all explants produce multiple shoots with minimal callus formation (Figs. 3, 4). Regenerated shoots are easily rooted on RM medium without growth regulators (Fig. 5). This regeneration system provides hundreds of shoots per explant within several months. A similar response of leaf and node explants to BAP and thidiazuron (TDZ) has recently been reported by Berger-Büter et al. (1994).

Fig. 3. Multiple shoot differentiation from seedling explant after 1 month in culture. (Photo Ms. Markušová)

Fig. 4. Eight-week-old shoots before transfer into hormone-free medium for rooting. (Photo Ms. Markušová)

Fig. 5. Plantlets with developed root system on hormone-free medium. (Photo Ms. Markušová)

Brutovská et al. (1994) studied the effects of the nonionic surfactant, Pluronic F-68, on the growth of shoots regenerated from seedlings of *H. perforatum*, and found that supplementation of agar-solidified medium with 0.001% (w/v) of Pluronic increased the mean fresh weight of regenerants by 40% and the mean number of shoots per seedling by 34%. The growth of seedling-derived callus was unaffected; however, there was a tendency for callus cells grown in the presence of Pluronic to be more highly pigmented with anthocyanins.

Variability in hypericin content in somaclones of *H. perforatum*, along with the quantitative evaluation of hypericin-containing glands and their ultrastructure, was studied by Čellárová et al. (1994). Studies have also been performed on tissue culture of a related *Hypericum erectum* species in order to determine the formation of procyanidins and to analyze quantitatively individual polyphenols belonging to the procyanidins (Yazaki and Okuda 1990).

# 3 Somaclonal Variation

The potential of somaclonal variation to contribute genetic variation in the improvement of plants has been widely studied (see Bajaj 1990). The tissue culture-derived regenerants exhibit a range of altered characteristics which can be inherited. The maintainance of desirable properties in variants created from tissue cultures in *H. perforatum* may be expected if apomictic pseudogamy is predicted as the prevalent method of reproduction. Therefore, an evaluation of morphological, cytogenetic, and biochemical characters in tissue culture-derived regenerants is supplemented with a correlation between ploidy and the characters studied.

## 3.1 Morphological Alterations

In the first year of cultivation, the regenerants had an unusual trailer-plant habit, and rarely reached the stage of flowering. However, in the second year of cultivation the habit of regenerants resembled that of control plants. As shown in Table 1, higher concentrations of BAP resulted in more extensive branching and an increased biomass production.

The leaf shape in regenerants was very variable. On the base of the index determined by the ratio width/length of a leaf, broad-leaved, intermediate and narrow-leaved plants were found even amongst those originating from one particular genotype. In contrast with majority of control plants, which were mostly narrow-leaved, in regenerants the intermediate leaf shape was prevalent (Table 1; Čellárová et al. 1994).

## 3.2 Cytogenetic Changes

Numerous chromosome changes were observed amongst plants regenerated from seedling explants. In comparison with the control, which was always tetraploid ($2n = 4x = 32$), almost half of the total 81 evaluated somaclones exhibited an altered ploidy with occurrence of diploid (11.1%), triploid (27.2%), and mixoploid metaphases with 40 or 48 chromosomes (9.8%). Aneuploid metaphases were very rare.

*Hypericum perforatum* is a tetraploid (Robson 1968, 1981; Löve and Löve 1974). According to earlier data (Noack 1941), in the reproduction of *H. perforatum*, a normal reduced embryo sac occurred in only 3% of the ovules observed. In the majority of plants (97%), aposporous embryo sacs with 32 chromosomes are present. These unreduced egg cells are occasionally fertilized. The pollen is produced by normal meiosis. *H. perforatum* is considered to be a facultative apomict. However, the variability observed in natural populations suggests the necessity for a more detailed study of the reproduction. Chromosome alterations detected in $R_0$ somaclones were transmitted to the seed offspring. The offspring of all diploid somaclones remained stable with the diploid

**Table 1.** Variability of biomass production and morphological characters of regenerants in *Hypericum perforatum* L. (Čellárová et al. 1994)

| Group | C | 1 | 2 | 3 | 4 |
|---|---|---|---|---|---|
| | | | Mean values | | |
| Fresh weight (g) | 9.7[b] ±6.8 | 9.4[b] ±5.6 | 14.8[b] ±3.8 | 29.6[b] ±17.2 | 16.2[b] ±6.8 |
| Dry weight (g) | 5.3[b] ±3.4 | 4.7[b] ±2.3 | 6.6[b] ±2.2 | 9.4[b] ± 5.2 | 4.7[b] ±1.9 |
| Height of regenerants (cm) | 26.4[b] ±5.9 | 23.7[b] ±2.4 | 23.6[b] ±2.3 | 23.8[b] ± 3.4 | 20.7[b] ±4.1 |
| Leaf shape[a] | n, i | n, i | n, b, i | n, b, i | n, b, i |
| No. of branches per plant | 5.9[b] ±3.4 | 5.4[b] ±3.4 | 5.9[b] ±1.8 | 8.1[b] ± 3.8 | 9.0 ±4.0 |

Approximately 100 regenerants of each experimental group and 100 control plants were investigated. C, control.
Regenerants were obtained on the basal RM medium supplemented with 0.1 mg/l of (1), 0.5 mg/l (2), 1 mg/l (3) and 2 mg/l (4) BAP.
[a] Leaf shape was determined on the basis of the ratio length/width of a leaf and grouped as follows:
narrow-leaved (n): 0.20–0.35
intermediate (i): 0.36–0.50
broad-leaved (b): 0.51–0.65.
The prevalent leaf shape is underlined.
[b] Significant difference within a group at $\alpha = 0.05$.

chromosome sets, and the chromosome number was retained also in 61% of tetraploid plants. As expected, triploids and mixoploids produced mixoploid progeny. These alterations may affect the mechanism of reproduction in terms of the proportion of apomictic and sexual processes and the ploidy level in following generations.

## 3.3 Hypericin Content

An extensive polymorphism among somaclones was determined in hypericin content. Great variability was observed among somaclones originating from a particular genotype, except for several clones with extremely high and balanced hypericin content during the 2-year cultivation of $R_0$ somaclones. These plants originated in vitro from tetraploid plants but exhibited a diploid chromosome number. These clones were selected for the study of following generations in order to determine the stability of the hypericin content in their offspring.

Hypericin is accumulated, and probably synthetized, in special morphological structures consisting of a core of large cells surrounded by an irregular uni- to biseriate sheath of flattened cells. The cells proliferate to some maximum number, which is probably reached when the two sheathing layers of flat cells are produced (Curtis and Lersten 1991). Hypericin is accumulated within these cells.

**Table 2.** Variability in hypericin content and morphological structures containing hypericin. (Čellárová et al. 1994)

| Group | C | 1 | 2 | 3 | 4 |
|---|---|---|---|---|---|
| | | | Mean values | | |
| Density of hypericin glands per $mm^2$ of leaf area | 0.29[a] ± 0.07 | 0.27[a] ± 0.17 | 0.22[a] ± 0.10 | 0.20[a] ± 0.13 | 0.28[a] ± 0.23 |
| No. of glands per leaf | 10.50[a] ± 6.7 | 11.39[a] ± 6.4 | 9.39[a] ± 5.6 | 8.51[a] ± 5.5 | 11.05[a] ± 4.2 |
| Hypericin content (μg/g dry matter) | 253.9[a] ± 137.2 | 287.7[a] ± 113.5 | 257.5[a] ± 93.1 | 264.5[a] ± 72.8 | 260.5[a] ± 58.3 |

[a] Significant difference within a group at $\alpha = 0.05$.

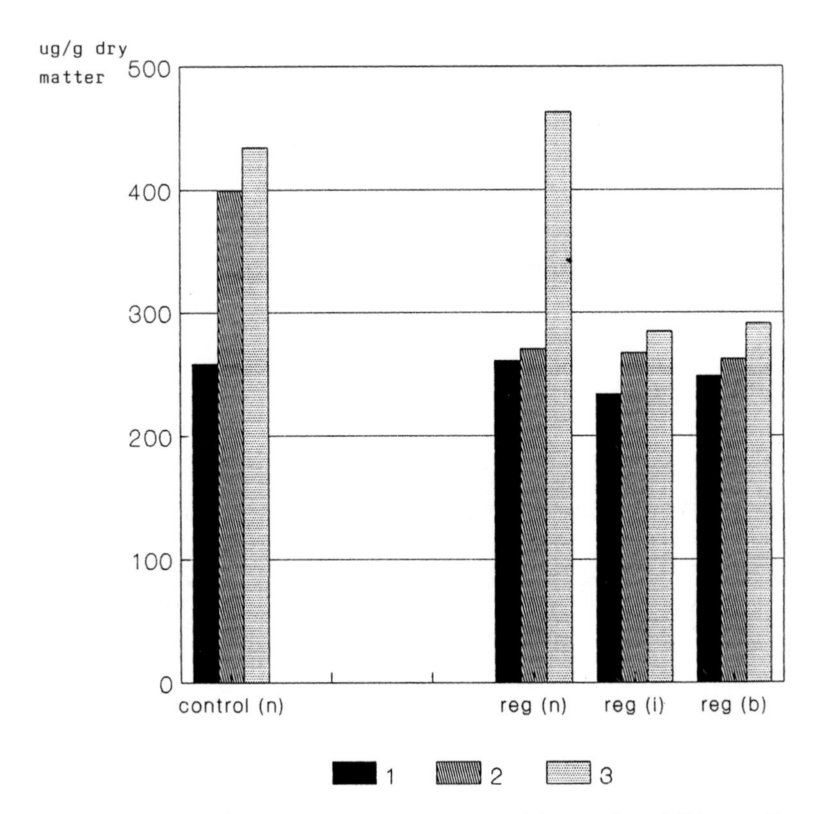


**Fig. 6.** Hypericin content in control and regenerants; reg regenerants; (n) narrowleaved; (i) intermediate; (b) broadleaved; 1 number of hypericin containing glands per leaf blade $\leqslant 6$; 2 7–15; 3 $\geqslant 16$

The nodules do not conform to any internal secretory structures known from any other group of plants. Ultrastructural observation shows that cells surrounding the core of a gland contain a large central vacuole with either dark homogenous or transparent content, and are surrounded with a thickened cell wall. The core is composed of cells of irregular shape filled with lamellar structures and vesicles containing hypericin (Čellárová et al. 1994). The variability of the number of hypericin containing multicellular nodules is shown in Table 2.

From this point of view, a comparison of hypericin content and gland density or number of glands per leaf showed no correlation between these variables, even with respect to the leaf shape of somaclones. However, an increased dependence has been observed using the partition method (Fig. 6).

### 3.4 Correlations Between Ploidy and Morphological and Biochemical Characters

Changes in ploidy level have many phenotypic effects, including changes in plant form, flowering time, biomass production, secondary metabolite formation, etc. The relationship between response variables and independent variables (ploidy level) can be quantified by regression analysis. For most characters, the simplest linear model was sufficient. $R^2$ values estimate the amount of variability described by a linear model. High $R^2$ values for biological data may range from 0.50 to 0.90 (Compton 1994).

**Table 3.** Correlation between ploidy and fresh weights of regenerants. (Bruňáková 1994)

| Fresh weight | s(16) | s(24) | s(32) |
|---|---|---|---|
| Average (g) | 10.911 | 28.47 | 25.307 |
| Median | 7.78 | 26.58 | 15.0 |
| Standard deviation | 9.312 | 16.436 | 22.380 |
| Variation range | 2.0–32.2 | 4.0–58.4 | 4.8–86.0 |
| Correlation coefficient (r) | | 0.77 | |
| ($R^2$) | | 0.5915 | |

s, Somaclones; 2n chromosome number in parentheses.

**Table 4.** Correlation between ploidy and dry matter of regenerants. (Bruňáková 1994)

| Dry matter | s(16) | s(24) | s(32) |
|---|---|---|---|
| Average (g) | 7.523 | 4.591 | 10.344 |
| Median | 5.575 | 3.04 | 11.0 |
| Standard deviation | 3.776 | 3.619 | 5.029 |
| Variation range | 4.4–14.0 | 1.2–12.3 | 1.2–17.8 |
| Correlation coefficient (r) | | 0.51 | |
| ($R^2$) | | 0.2597 | |

### 3.4.1 Fresh Weight

The lowest mean values were found in control and diploid plants. Triploid and tetraploid somaclones exhibited higher mean values. The differences between individual groups are significant. Individual groups were characterized by an occurrence of extreme values and a great variation range. As expected, the fresh weight of somaclones increases with increased ploidy and almost 60% of the variability can be described by a linear model $y = a + bx$ (Table 3).

### 3.4.2 Dry Weight

Different values of arithmetic means and medians indicate that individual groups are not symmetric and the occurrence of extreme values is predicted (Table 4). The presented data do not show a clear correlation between ploidy and dry matter.

### 3.4.3 Height of Plants

The means and medians were not influenced by the occurrence of extreme values. The regenerants with a diploid chromosome number exhibited the lowest mean values for height. The higher ploidy levels resulted in a slight increase in height. There is a clear correlation between ploidy and height (Table 5).

**Table 5.** Correlation between ploidy and height of regenerants. (Bruňáková 1994)

| Height of plants | s(16) | s(24) | s(32) |
|---|---|---|---|
| Average (g) | 20.44 | 25.33 | 26.04 |
| Median | 20.00 | 25.04 | 26.00 |
| Standard deviation | 6.366 | 6.044 | 3.255 |
| Variation range | 10–30 | 12–40 | 19–34 |
| Correlation coefficient (r) | | 0.92 | |
| ($R^2$) | | 0.8434 | |

**Table 6.** Correlation between ploidy and number of branches of regenerants. (Bruňáková 1994)

| Number of branches | s(16) | s(24) | s(32) |
|---|---|---|---|
| Average (g) | 10.555 | 7.285 | 8.219 |
| Median | 8.00 | 7.00 | 8.00 |
| Standard deviation | 5.981 | 3.132 | 4.992 |
| Variation range | 4–22 | 3–17 | 2–24 |
| Correlation coefficient (r) | | −0.69 | |
| ($R^2$) | | 0.4809 | |

**Table 7.** Correlation between ploidy and number of glands per $mm^2$ of leaf area of regenerants. (Bruňáková 1994)

| Gland density | s(16) | s(24) | s(32) |
|---|---|---|---|
| Average (g) | 0.373 | 0.158 | 0.188 |
| Median | 0.366 | 0.162 | 0.192 |
| Standard deviation | 0.049 | 0.072 | 0.089 |
| Variation range | 0.2–0.72 | 0.05–0.3 | 0.05–0.4 |
| Correlation coefficient (r) | | −0.79 | |
| ($R^2$) | | 0.6286 | |

**Table 8.** Correlation between ploidy and hypericin content of control and regenerants. (Bruňáková 1994)

| Hypericin content | s(16) | s(24) | s(32) |
|---|---|---|---|
| Average (g) | 375.732 | 280.216 | 265.253 |
| Median | 332.71 | 245.96 | 261.66 |
| Standard deviation | 143.541 | 149.445 | 54.446 |
| Variation range | 223–597 | 187–814 | 181–387 |
| Correlation coefficient (r) | | −0.92 | |
| ($R^2$) | | 0.8464 | |

### *3.4.4 Number of Branches per Plant*

As shown in Table 6, increased ploidy resulted in decreased branching, although the differences between individual groups were not significant.

### *3.4.5 Gland Density and Hypericin Content*

A clear negative correlation was found between gland density per leaf blade and chromosome number (Table 7) as well as between hypericin content and ploidy (Table 8). The high $R^2$ values indicate that most of the variability can be described by the linear model.

The presented results show that in regenerants of *Hypericum perforatum*, arranged according to the ploidy level, a correlation exists between the number of chromosomes and most of the characters studied. While the fresh and dry weight and the height of regenerants increase with increased ploidy, the branching, hypericin gland density per leaf blade, and hypericin content show an opposite effect.

# 4 Summary and Conclusions

The tissue culture system was used to broaden the variability in *Hypericum perforatum*. Somaclones originating from individual genotypes were evaluated with respect to their morphological alterations, ploidy, and hypericin content.

The concentration of BAP, which had promoted shoot differentiation, affected some morphological characters such as biomass production and branching. Hypericin content among individual regenerants varied significantly. Nevertheless, a correlation between hypericin content and BAP concentration has not been found. Similarly, no relation has been determined between hypericin content and gland density.

Due to the occurrence of a high number of regenerants with altered ploidy, the correlation between the chromosome number and all characters studied has also been determined. Increased ploidy resulted in higher biomass production and height of regenerants but lowered gland density and hypericin content. The highest values of hypericin content in diploid somaclones and their cytogenetic and biochemical stability in $R_1$ generation suggest their potential use in breeding programs.

In conclusion, the proportion of apomictic and sexual reproduction in the species studied seems to be a crucial point in the evaluation of somaclonal variation in further generations. The use of haploids and the determination of DNA fingerprints may bring more understanding to this problem.

# 5 Protocol

*1. Source of Explant and Culture Media.* Shoots of *Hypericum perforatum*, cv. Topas were obtained in vitro from seedlings (14 days postgermination) on basal RM medium (Linsmaier and Skoog 1965) supplemented with B5 vitamin solution (Gamborg et al. 1968), 30 g/l sucrose, 2 mg/l glycine, 100 mg/l meso-inositol, 0.1, 0.5, 1 and 2 mg/l 6-benzylaminopurine (BAP), and 0.6% agar. pH was adjusted to 5.6 before autoclaving. These organically connected shoots were separated and transferred to the medium lacking phytohormones for rooting. One hundred regenerants that differentiated on each concentration of BAP arising from 10 different genotypes and 100 controls originating from seeds on the basal medium were used for evaluation of somaclonal variation.

*2. Acclimatization and Field Cultivation of Somaclones.* Regenerants were transferred to Jiffy pots and put into the growing chamber with high relative humidity for 1 month. After this acclimatization period, they were transplanted to field conditions.

*3. Evaluation of Morphological Characters and Ploidy.* $R_0$ plants were evaluated in respect of the following characters: biomass production, height, branching, and density of glands containing hypericin per leaf blade.

Chromosome counts were determined in squash preparations of meristems followed by Giemsa staining.

*4. Hypericin Assay.* Dried and homogenized aerial parts of the plants were extracted with chloroform in a sonicator and filtered. The dried powder was then extracted with methanol and the filtrate allowed to evaporate in a water bath. This was followed by addition of chloroform, and the mixture was shaken. The supernatant was then discarded and the solid phase containing hypericin and its derivatives was dissolved in methanol and filtered (Genius 1971, modified). Methanol extracts were further analyzed by spectrophotometry (Shimadzu UV 3000) at 592 nm wavelength.

# References


Bajaj YPS (1990) Biotechnology in agriculture and forestry, vol 11. Somaclonal variation in crop improvement I. Springer, Berlin Heidelberg New York

Berger-Büter K, Büter B, Schaffner W (1994) In vitro propagation of *Hypericum perforatum* L.: impact of thidiazuron and 6-benzylaminopurine. Abstr VIIIth Int Congr Plant Tissue and Cell Culture, Firenze, June 12–17, 1994, 62 pp

Berghöfer R, Hölzl J (1987) Biflavonoids in *Hypericum perforatum*; Part 1. Isolation of I3, II8-biapigenin. Planta Med 53: 216–217

Berghöfer R, Hölzl J (1989) Isolation of I3′, II8-Biapigenin (Amentoflavone) from *Hypericum perforatum*. Planta Med 55: 91

Bruňáková K (1994) Hodnotenie somaklonálnej variability vybraných znakov $R_0$ generácie regenerantov *Hypericum perforatum* L. (Evaluation of somaclonal variation of selected $R_0$ generation characters in regenerants of *Hypericum perforatum* L.) Diploma Thesis, P. J. Šafárik University, Košice, 73 pp

Brutovská R, Čellárová E, Davey MR, Power JB, Lowe KC (1994) Stimulation of multiple shoot regeneration from seedling leaves of *Hypericum perforatum* L. by Pluronic F-68. Acta Biotechnol 14: 347–353

Bystrov NS, Dobrynin VN, Kolosov MN, Popravko SA, Chernov BK (1978) Chemistry of hyperforin. VI. General chemical characterization. Bioorg Khim 4: 791–797 (in Russian)

Compton ME (1994) Statistical methods suitable for the analysis of plant tissue culture data. Plant Cell Tissue Organ Cult 37: 217–242

Curtis JD, Lersten NR (1991) Internal secretory structures in *Hypericum* (Clusiaceae): *H. perforatum* L. and *H. balearicum* L. New Phytol 114: 571–580

Čellárová E, Kimáková K, Brutovská R (1992) Multiple shoot formation and phenotypic changes of $R_0$ regenerants in *Hypericum perforatum* L. Acta Biotechnol 12: 445–452

Čellárová E, Daxnerová Z, Kimáková K, Halušková J (1994) The variability of hypericin content in the regenerants of *Hypericum perforatum*. Acta Biotechnol 14: 265–271

Čellárová E, Kimáková K, Daxnerová, Mártonfi P (1995) *Hypericum perforatum* (St. John's Wort): In vitro culture and the production of hypericin and other secondary metabolites. In: Bajaj YPS (ed) Biotechnology in agriculture and forestry, vol 33. Medicinal and aromatic plants VIII. Springer, Berlin Heidelberg New York, pp 261–275

Duran N, Song PS (1986) Hypericin and its photodynamic action. Photochem Photobiol 43: 677–680

Falk H, Schoppel G (1991) A synthesis of emodin anthrone. Monatsh Chem 122: 739–744

Gamborg OL, Miller RA, Ojima K (1968) Nutrient requirements of suspension cultures of soybean root cells. Exp Cell Res 50: 151–158

Genius OB (1971) Bundesrepublik Deutschland Deutsches Patentamt, Offenlegungsschrift 1569849

Giese AC (1980) Hypericism. Photochem Photobiol Rev 5: 229–255

Hölzl J, Ostrowski H (1987) Johanniskraut (*Hypericum perforatum* L.). HPLC-Analyse der wichtigen Inhaltsstoffe und deren Variabilität in einer Population. Dtsch Apoth Ztg 127: 1227–1230

Hudson JB, Lopez-Bazzocchi I, Towers GHN (1991) Antiviral activities of hypericin. Antiviral Res 15: 101–112

Hudson JB, Harris L, Towers GHN (1993) The importance of light in the anti-HIV effect of hypericin. Antiviral Res 20: 173–178

Knox JP, Dodge AD (1985) Isolation and activity of the photodynamic pigment hypericin. Plant Cell Environ 8: 19–25

Kraus GA, Pratt D, Tossberg J, Carpenter S (1990) Antiretroviral activity of synthetic hypericin and related analogs. Biochem Biophys Res Commun 172: 149–153

Lavie G, Mazur Y, Lavie D, Levin B, Ittah Y, Meruelo D (1990) Hypericin as an antiretroviral agent. In: Aids: Anti-HIV agents, therapies, and vaccines. Ann NY Acad Sci 616: 556–562

Linsmaier EM, Skoog F (1965) Organic factor requirement of tobacco tissue cultures. Physiol Plant 18: 100–127
Lopez-Bazzocchi I, Hudson JB, Towers GHN (1991) Antiviral activity of the photoactivated plant pigment hypericin. Photochem Photobiol 54: 1–5
Löve A, Löve D (1974) Atlas of the Slovenian flora. J Cramer, Leutershausen
Maisenbacher P, Kovar KA (1992) Adhyperforin: a homologue of hyperforin from *Hypericum perforatum*. Planta Med 58: 291–292
Meruelo D, Lavie G, Lavie D (1988) Therapeutic agents with dramatic antiretroviral activity and little toxicity at effective doses: aromatic polycyclic diones: hypericin and pseudohypericin. Proc Natl Acad Sci USA 85: 5230–5234
Michaluk A (1961a) Flavonoids in species of the genus *Hypericum* II. The flavonols. Diss Pharm 13: 73–79
Michaluk A (1961b) Leucoanthocyanidins in *Hypericum perforatum*. Diss Pharm 13: 81–88
Michaluk A, Brunarska Z, Beonarska D (1956) Tannins and flavones in different species of *Hypericum*. Diss Pharm 8: 47–62
Noack KL (1941) Geschlechtsverlust und Bastardierung beim Johanniskraut. Forsch Fortschr 17: 13–15
Robson NKB (1968) Hypericum. In: Tutin TG et al. (eds) Flora Europea 2. Cambridge University Press, Cambridge, pp 261–269
Robson NKB (1981) Studies in the genus *Hypericum* L. (Guttiferae). 2. Characters of the genus. Bull Br Mus Nat Hist (Bot) 8: 55–226
Suzuki O, Katsumata Y, Oya M, Bladt S, Wagner H (1984) Inhibition of monoamine oxidase by hypericin. Planta Med 50: 272–274
Thomas C, Pardini RS (1992) Oxygen dependence of hypericin-induced phototoxicity to EMT6 mouse mammary carcinoma cells. Photochem Photobiol 55: 831–837
Thomas C, Macgill RS, Miller GC, Pardini RS (1992) Photoactivation of hypericin generates singlet oxygen in mitochondria and inhibits succinoxidase. Photochem Photobiol 55: 47–53
Yazaki K, Okuda T (1990) Procyanidins in callus and multiple shoot cultures of *Hypericum erectum*. Planta Med 56: 490–491
Yazaki K, Okuda T (1994) *Hypericum erectum* Thunb. (St. John's Wort): In vitro culture and the production of procyanidins. In: Bajaj YPS (ed) Biotechnology in agriculture and forestry, vol 26, Medicinal and aromatic plants VI. Springer, Berlin Heidelberg New York pp 167–178

# II.4 Somaclonal Variation in *Lavatera* Species

J.M. Iriondo[1] and C. Pérez[1]

## 1 Introduction

### 1.1 Botany, Importance and Distribution

The genus *Lavatera* (Malvaceae) consists of about 45 species, mainly distributed in the Mediterranean region but also extending to the Canary Islands, N.W. Himalayas, Central Asia, E. Siberia, Australia, and the USA (California)/(Fernandes 1968b).

*Lavatera* species are taxonomically described as herbs or soft-wooded shrubs, usually stellate-pubescent, with solitary or clustered flowers in the leaf axils. The epicalyx has three segments, more or less united at the base, at least in buds. Petals are emarginate and stigmas are lateral and filiform. The fruit, a schizocarp, is composed of numerous one-seeded mericarps, usually indehiscent, arranged in a single whorl (Fernandes 1968a).

Several of these species have an important ornamental value. The shrub known in gardens as *L. olbia* has become widely cultivated because of its profuse production of large pink flowers all through the summer, its fast growth, and easy vegetative multiplication. Cultivars of this species have been well known and some of them have the Royal Horticultural Society Award of Merit (Cheek 1989). However, it must be observed that there has been an important problem of classification and nomenclature with several so-called *L. olbia* cultivars, such as the cvs. Barnsley and Rosea, since they actually belong to the species *L. thuringiaca* (Cheek 1989). In most modern plant catalogs, these errors have been amended, but in certain places the old names are still used.

*L. trimestris,* an example of an annual resistant plant, is also cultivated in gardens. Several cultivars of *L. trimestris* are presently on the market, such as cv. Mont Blanc with compact white flowers, cv. Silver Cup with slightly fluting pink flowers and cv. Ruby Regis with variegated pink and red flowers.

*L. kashimiriana* Cool Ice is a well-shaped shrub that produces beautiful white flowers for many months. Other cultivars of this species have pink flowers and, in general, they are all frost-resistant. *L. assurgentiflora* is another frost-tolerant shrub, with dark cherry flowers, that is also well appreciated in gardening.

There are other noncommercialized species, which, due to the beauty of their flowers, are of potential ornamental interest and could be cultivated or used in breeding programs. This is the case, for instance, with *L. oblongifolia* (Fig. 1) an

[1] Departamento de Biología Vegetal. ETSI Agrónomos Universidad Politécnica, 28040 Madrid, Spain

Biotechnology in Agriculture and Forestry, Vol. 36
Somaclonal Variation in Crop Improvement II (ed. by Y.P.S. Bajaj)

**Fig. 1.** *L oblongifolia*, an endangered species of great ornamental value

endangered species, endemic of S. E. Spain, that has locally been used as an ornamental plant and has been proposed as the symbol of La Alpujarra, the region where it is most abundant.

Some *Lavatera* species such as *L. kashimiriana* and *L. trimestris* are also of medicinal importance (Kapur and Sarin 1987; Pyasyatskene et al. 1988). Other possible uses of *Lavatera* species are as uncommon fiber plants (Rumyantseva 1987) and as late-flowering nectar plants, i.e., *L. thuringiaca*, which secretes up to 4.8 mg nectar per flower a day (Kucherov and Siraeva 1981).

Moreover, *Lavatera* species have been especially valuable for basic and applied research in different fields of plant biology. *Lavatera cretica* (lesser tree mallow) has frequently been used to study the mechanism by which the leaf of certain plants reorients its lamina to face the sun throughout the day (i.e., Werker and Koller 1987; Koller and Levitan 1989; Koller et al. 1990). *Lavatera arborea* (tree mallow), a maritime cliff species, has been used as a model in the study of the ecology and physiology of plants adapted to these particular environmental conditions, especially in the investigation of the mechanism to salinity tolerance (Malloch et al. 1985; Okusanya and Fawole 1985). *Lavatera* species are also very interesting in studies of cytokinesis, where they represent the model of a simultaneous type (Longly and Waterkeyn 1978, 1979; Kudlicka and Rodkiewicz 1991).

In applied research, *Lavatera trimestris, L. arborea*, and *L. cretica* (as other weed members of the Malvaceae) have been frequently used as test plants in studies regarding identification, characterization, and epidemiology of viruses that attack important crops (especially economically important cultivars of the Cucurbitaceae family; Marco 1975; Lecoq et al. 1981; Al- Musa 1989).

### 1.2 Breeding Objectives and Available Genetic Variability

The main breeding objectives for *Lavatera* as an ornamental plant should be focused on obtaining new cultivars with good esthetic qualities mainly based upon flower form, size, and color. However, obtaining disease resistance as well as tolerance to hardy conditions, such as frost, would also be highly desirable and of great economic importance. *Lavatera* plants are damaged by *Tyrophagus longior* (Acaridae), a pest of ornamental plants grown under protection. At present, this pest is controlled with organophosphorus pesticides (Buxton 1989). Anthracnose symptoms have also been observed on *L. trimestris* cultivars Mont Blanc and Silver Cup (Mortensen 1991).

The propagation of *Lavatera* plants or cell lines highly productive in specific secondary metabolites of medicinal interest, in fiber, or in nectar may also become relevant breeding objectives in the near future.

Available genetic variability within the genus should be explored, evaluated, and screened in search of new and novel traits. A mere 10% of the species of this genus is presently being utilized for commercial purposes. Many of the remaining 90% have, at least, great potential ornamental value and could be incorporated into the gene pool of ornamental breeding. Nevertheless, the genetic variability in *Lavatera* is presently being threatened, as a relevant number of species belonging to this genus is on the verge of extinction. In the near future, this situation could instigate an irreparable loss of germplasm of great potential use in breeding programs.

## 2 In Vitro Culture Studies

Plantlets of *Lavatera oblongifolia* were regenerated from calli obtained from cotyledons excised from young seedlings (Iriondo and Pérez 1992). Abundant callus and adventitious buds developed in cultures on modified MS (Murashige and Skoog 1962) medium plus 4.44 μM BAP and 0.54 μM NAA (Fig. 2). Isolated shoots proliferated well on MS medium plus 4.92 μM IBA or 0.44 μM BAP and 0.05 μM NAA. Rooting of shoots hardly ever took place in any of the media tested (MS without growth regulators and MS supplemented with different concentrations and combinations of IBA, NAA, and BAP). However, when transferred to pots containing 3:1 mixture (v/v) of sterilized peat moss and vermiculite, 82% rooting occurred. Acclimatized plantlets looked normal and showed vigorous growth. In the second growing season after acclimatization, the plants produced standard and functional flowers. Continuous callus formation and dedifferentiation of organized tissue was the main problem throughout the process of regeneration.

Image analysis techniques were used to help determine the developmental patterns found in in vitro regenerated plantlets (Iriondo and Pérez 1991a). The parameters measured indicate that acclimatized vitroplants of *L. oblongifolia* behave similarly to plantlets from cuttings in a greenhouse when the first new

**Fig. 2.** Callus growth and shoot organogenesis of *L oblongifolia* on MS + 4.44 μM BAP + 0.54 μM NAA

leaves of acclimatized vitroplants and the first new leaves of cuttings just rooted are considered as the starting point. The juvenile features observed in the first stages developed rapidly into adult forms. Explants of this species were successfully maintained without subculture for 6 months at 5 °C using slow growth in vitro storage techniques (Iriondo and Pérez 1991b).

*Lavatera acerifolia* has been micropropagated at the Royal Botanic Gardens at Kew (M. F. Fay, pers. comm.) using soft stem sections (the new season's growth) as starting plant material. Successful multiplication was obtained in MS plus 0.44 or 4.44 μM BAP, although there were some problems with leaf drop. Rooting was achieved in MS without growth regulators or MS plus 5.37 μM NAA. Rooted plantlets were potted in standard loam-based compost with added vermiculite. They were maintained in a mist bench for 2 weeks, after which the humidity was gradually lowered until the normal requirements for this species were reached.

## 3 Somaclonal Variation

Clearcut conclusions as to the extent and nature of somaclonal variation in *Lavatera* are difficult to reach because so little work has been done so far. All the information available up to now refers to *Lavatera oblongifolia*. The plantlets subject to analysis were regenerated according to the procedure given above using MS plus 0.44 μM BAP and 0.05 μM NAA for shoot proliferation. No particular agents were used to purposely induce somaclonal variation. The studies were oriented to evaluate several somaclonal-variation detection techniques in plantlets of different endangered species obtained through in vitro culture techniques for conservation purposes. In *Lavatera oblongifolia*, somaclonal variation was studied at morphological, enzymatic, and chromosome levels.

## 3.1 Somaclonal Variation in *Lavatera oblongifolia* Cultures

### *3.1.1 Morphological Variation*

Plantlets of *Lavatera* regenerated from calli showed relevant morphological differences from adult plants observed in their natural habitat. In shoots, a progressive rounding of leaf limbs, a decrease in leaf hairiness, and an apparent swelling of petioles (almost dedifferentiated) was observed. Moreover, all leaves had disproportionately long petioles and small limbs.

A gross development of the shoot on one plane of its growth axis was often observed. In some cases, this gross development occurred on all possible planes of the axis, making the shoot look extremely thick and surrounded by a large number of leaves arranged in an abnormal phylotaxia (Fig. 3).

Most of the plantlets that lived through the acclimatization process did not present markedly abnormal morphological features. One plantlet with large development of the shoot on one plane of its growth axis was successfully transplanted to the greenhouse. Once there, it experienced an intense branching at the top of the abnormal shoot, with no shoot being dominant over the rest. After the spring growing season, it died with the increase of temperatures in the summer.

**Fig. 3.** Abnormal shoots of *L oblongifolia* with a gross development in one plane of the axis (*left*) and all around the axis (*right*)

The morphological variability detected in leaves of regenerated *Lavatera* plantlets was further examined through computer-assisted image analysis techniques. Image analysis techniques are widely used in morphometry and have often been used for morphological studies and evaluation of micropropagated plants (Smith et al. 1989, 1990; McClelland et al. 1990). In general terms, a computer-assisted image analysis system comprises a computer, a high-resolution monitor, a video-camera, and a printer (Fig. 4). It is operated by specific software. In spite of their potential to accurately measure and process a vast amount of morphological data, they have seldom been used to assess somaclonal variation.

Acclimatized vitroplants were taken to a greenhouse and, from this point, the formation and development of new leaves was carefully monitored. Leaves were numbered in order of appearance and measured when they became fully developed. A similar procedure was followed with plantlets obtained from cuttings.

The features measured in leaves were petiole length, leaf hairiness and limb length, area, perimeter, apical angle, and form factor (which varies from 0 for a straight line to 1 for a perfect circle).

Nonacclimatized vitroplants had a significantly higher petiole length/limb length ratio, limb width/limb length ratio, form factor, and apical angle than plants obtained from cuttings and adult plants living in their natural habitat. However, no significant differences were found in these ratios between acclimatized vitroplants and plants obtained from cuttings. Thus, these morphological variations must be epigenetic in nature. Anyhow, it must be observed that these ratios are highly dependent on the developmental stage of the plant, in the sense that the proportionally long petioles and the wide and round limbs are a sign of juvenility (Iriondo and Pérez 1991a; Iriondo 1991). These features were quite uniform among vitroplants (at least as uniform as in other groups). No single variants were found in leaf shape, a situation commonly described in plants of

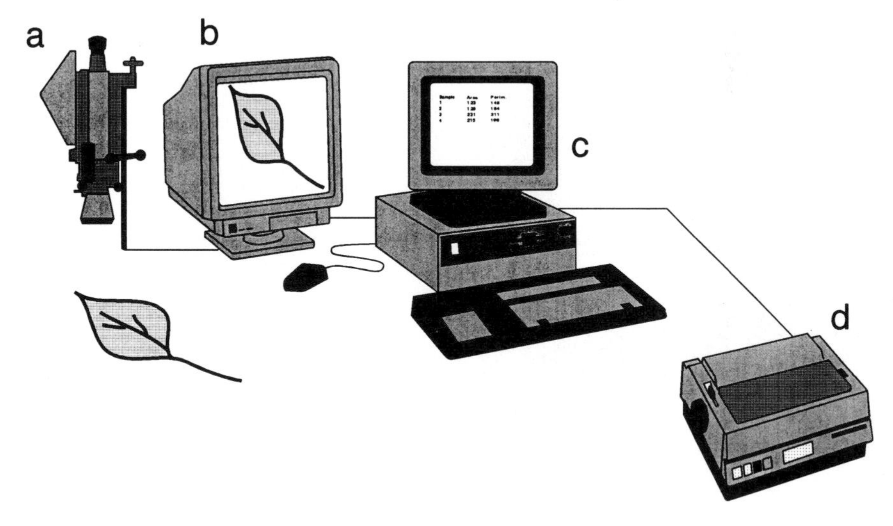


**Fig. 4a–d.** Basic hardware configuration of a computer-assisted image analysis system. **a** Video-camera. **b** High-resolution color monitor to observe and treat captured images. **c** Computer system with graphics card, monitor, and mouse. **d** Printer

other species regenerated from calli, i.e., *Saintpaulia ionantha* (Cassells and Plunkett 1986), *Chrysanthemum* spp. (Miyazaki and Tashiro 1978; Sutter and Langhans 1981).

Similarly, in other species, differences in petiole length between acclimatized vitroplants and plants obtained from cuttings (Cassells and Plunkett 1986) or between regenerated plants and the donors of the original explants (Reisch and Bingham 1981; Latunde-Dada and Lucas 1983) have been reported. However, in

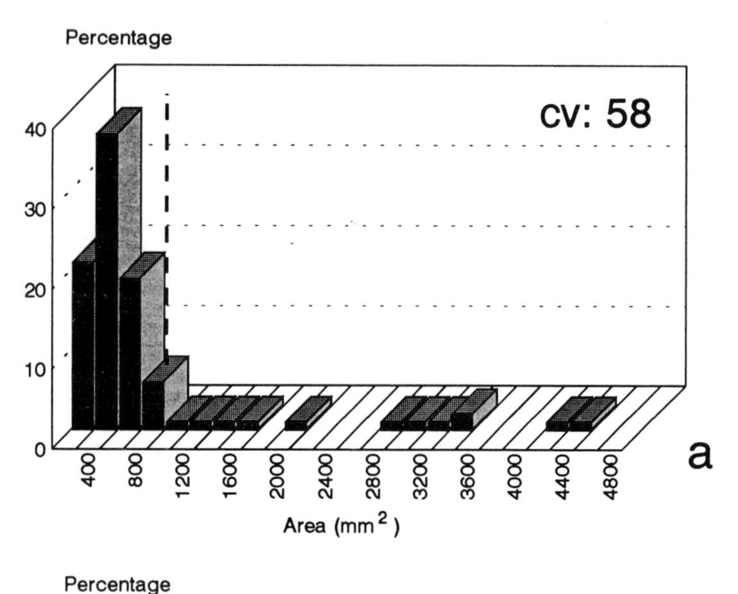


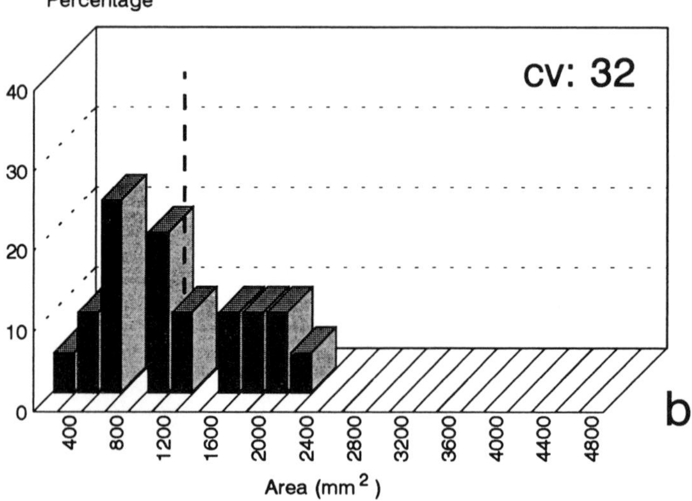


**Fig. 5a, b.** Leaf-area frequency distribution in *L oblongifolia* plants of two different origins. **a** Acclimatized vitroplants. **b** Plants from cuttings. *cv* Coefficient of variation. Mean values are expressed by a *discontinuous line*

*L. oblongifolia*, no significant differences were found in petiole length among nonacclimatized vitroplants, acclimatized vitroplants, and plants obtained from cuttings.

Acclimatized vitroplants showed greater variability in leaf size than plants obtained from cuttings, of the same age and grown under the same conditions. The study of leaf-area frequency distributions (Figs. 5, 6), showed that the existence of a high inner variation in leaf area among acclimatized vitroplants was due to the presence of a group of plants with larger leaves. On the contrary, in a similar type of study, Cassells and Plunket (1986), working with *Saintpaulia ionantha*, observed that the coefficient of variation of leaf area in micropropagated plants was smaller than in plants obtained from cuttings. The analysis of limb perimeter and limb length showed results similar to those obtained for leaf area. All these parameters are dependent, to a greater or lesser degree, on a common fundamental biological variable.

The hairiness on the adaxial side of leaves of nonacclimatized vitroplants (3.3 hairs/$mm^2$) was significantly lower than that of 9-month-old plantlets obtained from cuttings (4.4 hairs/$mm^2$) and that of adult plants in their natural habitats (5.6 hairs/$mm^2$). However, adaxial leaf hairiness in acclimatized vitroplants was not significantly different (4.8 hairs/$mm^2$) from that of plantlets obtained from cuttings. On the other hand, the hairiness on the abaxial side of leaves of both acclimatized and nonacclimatized vitroplants was significantly lower than that of plantlets obtained from cuttings and adult plants in their natural habitats. These differences in hairiness observed in the population of vitroplants as a whole with respect to the other groups seemed to be epigenetic in nature and dependent on the type of environment the plants lived in. Differences in leaf hairiness have been observed by several authors in different species (Liu and Chen 1976; Skirvin and Janick 1976; Sutter and Langhans 1981; Thomas 1981; Cancellier and Cossio 1988).

Variation in leaf-hairiness in nonacclimatized vitroplants was significantly greater than in the rest of the groups (Fig. 7). Accounting for this variation, the presence of plants with no hairs or almost no hairs was detected in this group. The decrease in inner variation in leaf-hairiness from nonacclimatized vitroplants to acclimatized vitroplants may be interpreted in two ways. (1) Variation in leaf hairiness is epigenetic in nature and ex vitro conditions induce the presence of a shorter range in leaf hairiness. (2) Variation in leaf hairiness is at least partially genetic in nature and acclimatization acts as a process of selection of variants eliminating those with low hairiness.

Moreover, qualitative differences in leaf hairiness between nonacclimatized plants and the rest of the groups were also found. Stellate hairs of nonacclimatized plants were shorter in height and in diameter and were more transparent (Fig. 8).

### *3.1.2 Enzymatic Variation*

Esterase zymograms obtained from leaves of acclimatized regenerants derived from a single clone are shown in Fig. 9. Four types of zymograms were found. The

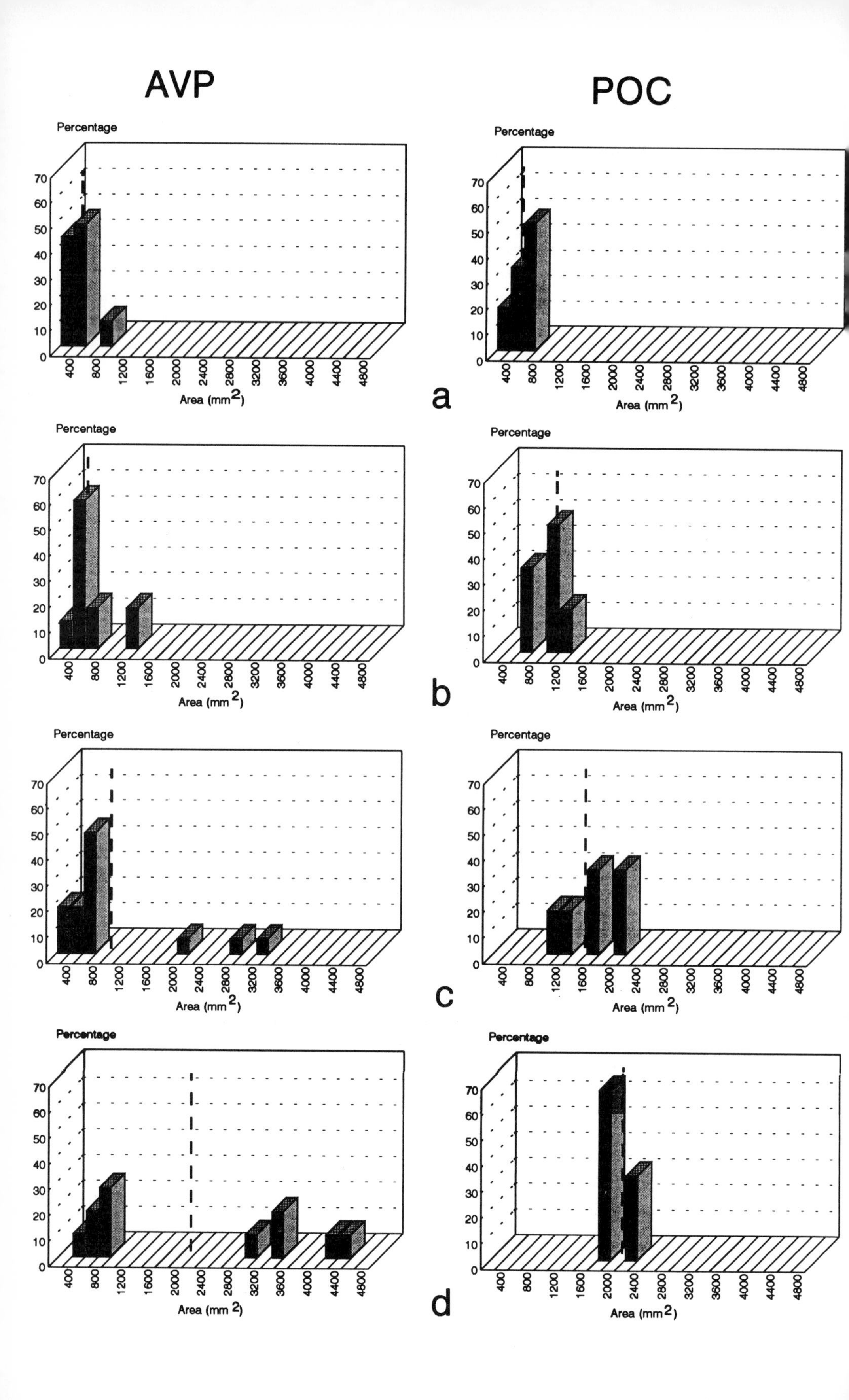

AVP
POC
Percentage
70
60
50
40
30
20
10
0
400
800
1200
1600
2000
2400
2800
3200
3600
4000
4400
4800
Area (mm2)
a
b
c
d

type III zymogram had one highly active band (5), five minor bands (1–2, 6–8, 11), and three faint bands (3–4, 9). It was the most abundant among regenerants. Type I and II lacked bands 9, 11 but type II had two additional bands (7, 12). Type IV had just one additional band (10). Differences in esterase banding patterns among regenerants have also been found in other species (Ogihara 1981; Taliaferro et al. 1989; Nagl 1990). No further differences among regenerants were detected when MDH, GOT, Cy-O, Ac-P, Co, and PER systems were assayed.

#### *3.1.3 Chromosome Counts*

*L. oblongifolia* has a chromosome number of $2n = 42$ ($x = 7$, hexaploid) in natural plants (Luque and Devesa 1986). No changes in the ploidy level were found among regenerants, although the chromosome counts found ($2n = 36$–$38$) suggest possible hypo-aneuploids. Aneuploidy is generally induced in vitro through either nuclear fragmentation followed by mitosis or defective chromosome behavior during mitosis (Tonelli 1990). The most frequent numeric variation in polyploid species is aneuploidy (Karp and Maddock 1984). Since these species have multiple chromosome sets, they are able to bear this type of numeric change better than diploids (Edallo et al. 1981).

#### *3.1.4 Somaclonal Variation During In Vitro Storage*

In vitro storage of *L. oblongifolia* shoots at 5 °C in MS medium for 1 year resulted in different survival rates, varying from 13 to 28%, among different subclonal lines derived from a single clone. Keeping in mind that minimum growth conditions impose definite selection pressures that may lead to directional genetic change (Scowcroft 1984), these results suggest the presence of in vitro-cultured-induced variability in terms of vigor or resistance to stress conditions among regenerants. Although the experiment was designed to search for additional ways of preserving *Lavatera* germplasm, similar experiments assaying a wide range of low temperatures could be used to obtain cold-tolerant somaclonal variants.

#### *3.1.5 Discussion*

It must be noted that all the studies in *Lavatera oblongifolia* with regard to somaclonal variation were carried out on regenerated plants ($R_0$).

As in other species (Breiman and Rotem-Abarbanell 1990), it is expected that a greater variation than that observed at the $R_0$ level exists at the cellular level in calli. On the other hand, the observed variation would probably be

**Fig. 6a–d.** Leaf-area frequency distribution in *L oblongifolia* plants of two different origins classified by leaf categories. *AVP* Acclimatized vitroplants; *POC* plants from cuttings. Leaves are grouped in classes according to their order of appearance. **a** Leaves 1–3. **b** Leaves 4–6. **c** Leaves 7–9. **d** Leaves 10–12. Mean values are expressed by a *discontinuous line*.

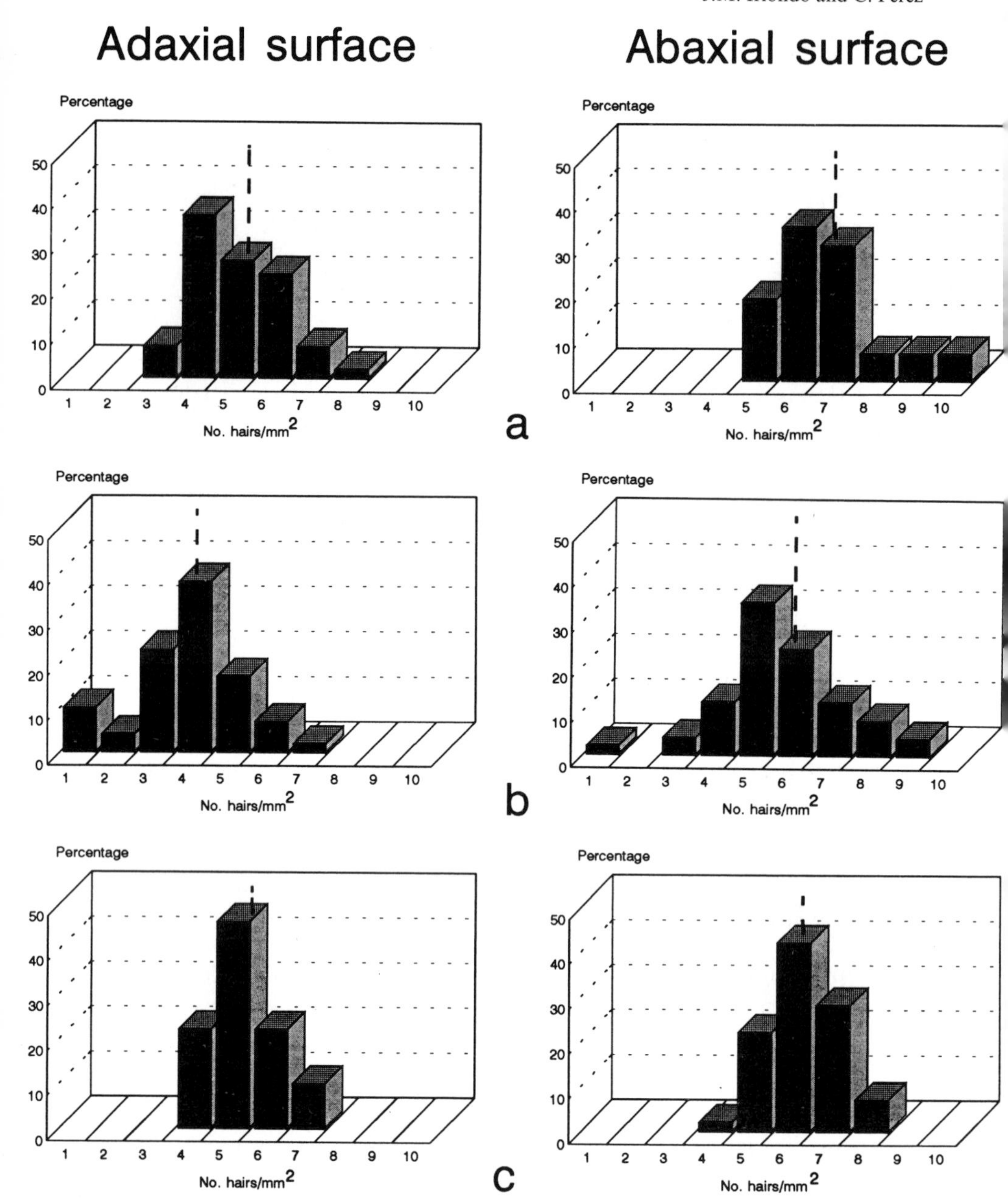


**Fig. 7a–c.** Leaf-hairiness frequency distribution in *L oblongifolia* plants of different origins. **a** Plants from cuttings. **b** Nonacclimatized vitroplants. **c** Acclimatized vitroplants

reduced by sexual reproduction, resulting in a lower rate of heritable somaclonal variation.

An important issue is whether alterations seen in $R_0$ plants are heritable or simple ephimeral responses to culture conditions. As previously observed,

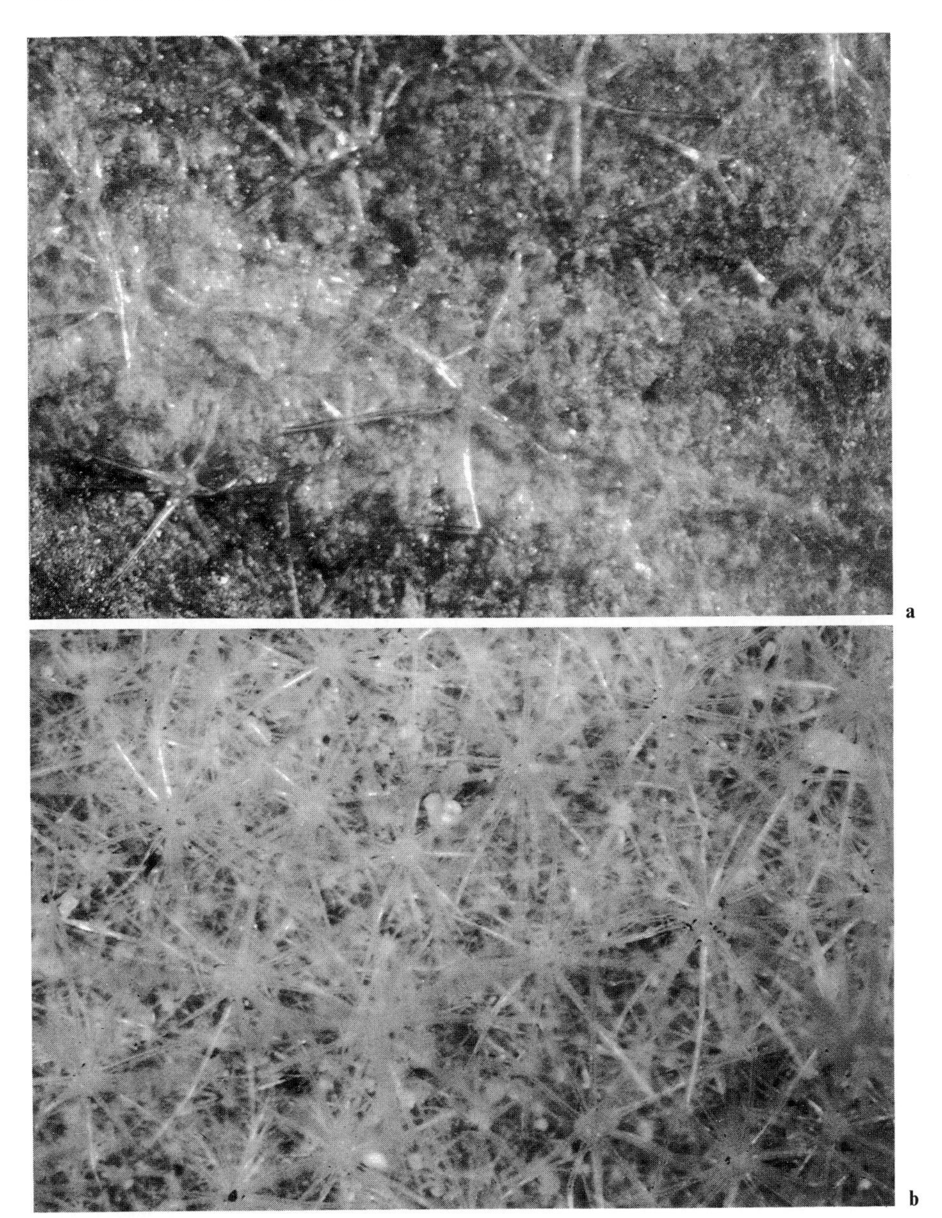

**Fig. 8a, b.** Hair density and morphology on the adaxial side of *L oblongifolia* leaves. **a** Leaf of nonacclimatized vitroplant. **b** Leaf of acclimatized vitroplant

differences observed in petiole length/limb length ratio, limb width/limb length ratio, form factor, and apical angle are epigenetic. However, for other parameters, no clearcut conclusions can be derived at the moment. Further studies, including the examination of progeny from regenerated plants, are essential

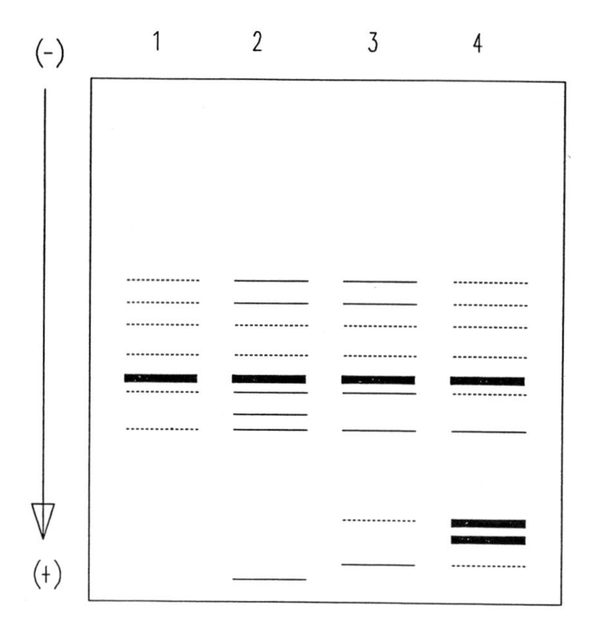


**Fig. 9.** Esterase zymograms of acclimatized vitroplants of *L. oblongifolia*

to evaluate the heritability of observed variation. In this respect, it must be observed that attention to $R_0$ plants alone may give misleading or incomplete results.

All the cultures were done in the absence of artificial selection. Hence, a random array of genetic variation characteristic of that produced in the undifferentiated culture must be expected, the only exception being a bias attributable to natural selection for cellular vigor and for the ability to differentiate into a plant.

## 3.2 Potential Uses of Somaclonal Variation in *Lavatera*

The use of tissue culture techniques in ornamental plants as a method for in vitro induction of variants for flower color, shape, and other esthetic traits is well documented (see Bajaj 1992). Thus, the potential uses of somaclonal variation for creating "new types" of *Lavatera* is evident. Several *Lavatera* cultivars available on the market have arisen as chimeras (Cheek 1989). In this context, there is a strong possibility that a plant regenerated from tissue cultures will not be genetically homogeneous. The spontaneous appearance or deliberate creation of genetically heterogeneous plants may result in novel plants with desirable characteristics (Marcotrigiano 1990).

The production of *Lavatera* secondary metabolites from selected plants or cell lines and of variants that would be used as controls in previously mentioned physiological research are other lines of investigation where somaclonal variation could be exploited.

The induction of somaclonal variation also offers the opportunity to enhance genetic variability in plants when the genetic base of one species is so narrow that its mid-term viability in the natural habitat is discarded (Jacobsen and Dohmen 1990), as may soon be the case of some *Lavatera* species. Accordingly, it may help in improving the chances of success in the reintroduction of almost extinct species in their natural habitats. In vitro technology must therefore be regarded as a powerful tool in the induction of much-needed variability (Bajaj 1990).

## 4 Summary and Conclusions

In vitro studies in *Lavatera* are very scarce, and to the best of our knowledge only two species have been assayed. Plantlets of *L. oblongifolia* have been regenerated from calli obtained from cotyledons excised from young seedlings, whereas *L. acerifolia* has been micropropagated using soft stem sections as starting material.

In *L. oblongifolia* different methods have been used to detect somaclonal variation among regenerants. Several variants were found at the morphological level, most of them epigenetic in nature. Additionally, differences in the esterase banding pattern were found among acclimatized regenerants. Hypoaneuploid variants were found at the chromosome level.

The success of any long-term breeding program in *Lavatera* depends on the genetic variability present in the gene pool. In order to increase the genetic variability available to the breeder, the natural diversity in *Lavatera* must be preserved. Based on this, in vitro technology can profitably be used for the induction and selection of additional variants that may meet the needs expressed above for *Lavatera* plants.

*Acknowledgments.* In vitro studies on *L. oblongifolia* were partially supported by CICYT Projects NAT 89-0865 and AMB 93-0092. The information provided by Dr. M. F. Fay from Royal Botanic Gardens at Kew on *L. acerifolia* is also acknowledged.

## References

Al-Musa MA (1989) Oversummering hosts for some cucurbit viruses in the Jordan Valley. J Phytopathol 127(1): 49–54

Bajaj YPS (1990) Somaclonal variation—origin, induction, cryopreservation, and implications in plant breeding. In: Bajaj YPS (ed) Biotechnology in agriculture and forestry, vol 11. Somaclonal variation in crop improvement I. Springer, Berlin Heidelberg New York, pp 3–48

Bajaj YPR (ed) (1992) Biotechnology in agriculture and forestry, vol 20. High-Tech and micropropagation IV. Springer, Berlin Heidelberg New York

Breiman A, Rotem-Abarbanell D (1990) Somaclonal variation in barley (*Hordeum vulgare* L.). In: Bajaj YPS (ed) Biotechnology in agriculture and forestry, vol 11. Somaclonal variation in crop improvement I. Springer, Berlin Heidelberg New York, pp 352–375

Buxton JH (1989) *Tyrophagus longior* (Gervais) (Acarina, Acaridae) as a pest of ornamentals grown under protection. Plant Pathol 38(3): 447–448

Cancellier S, Cossio F (1988) Field observations on a clone of Corvina veronese (*Vitis vinifera* L.) multiplied by in vitro culture. Acta Hortic 227: 508–513

Cassells AC, Plunkett A (1986) Habit differences in African Violets produced from leaf cuttings and in vitro from leaf discs and recycled axenic leaves. In: Withers LA, Alderson PG (eds) Plant tissue culture and its agricultural applications. Butterworths, London, pp 105–111
Cheek M (1989) The *Lavatera* imbroglio. Garden 114 (1): 23–27
Edallo S, Zucchinalli C, Perenzin M, Salamini F (1981) Chromosomal variation and frequency of spontaneous mutation association with in vitro culture and plant regeneration in maize. Maydica 26: 39–56
Fernandes R (1968a) *Lavatera* L. In: Tutin TG, Heywood VH, Burges NA, Moore DM, Valentine DH, Walters SM, Webb DA (eds) Flora Europaea, vol 2. Cambridge University Press, Cambridge, pp 251–253
Fernandes R (1968b) Contribuições para o conhecimento do género *Lavatera* L. I Notas sobre algumas espécies. Collect Bot (Barcinone) 7(1): 393–447
Iriondo JM (1991) Aplicación de técnicas de cultivo in vitro a la conservación de *Coronopus navasii* Pau, *Lavatera oblongifolia* Boiss. y *Centaurium rigualii* Esteve, tres especies endémicas de la Península Ibérica en peligro de extinción. Doctoral Thesis, Universidad Politécnica de Madrid, Madrid, 395 pp
Iriondo JM, Pérez C (1991a) Use of image analysis techniques for developmental studies in in vitro-regenerated plants of *Lavatera oblongifolia.* Acta Hortic 289: 335–336
Iriondo JM, Pérez C (1991b) In vitro storage of three endangered species from S.E. Spain. Bot Gard Microp News 1(4): 46–48
Iriondo JM, Pérez C (1992) In vitro plant regeneration of *Lavatera oblongifolia*, an endangered species. Bot Gard Microp News 1(5): 54–57
Jacobsen HJ, Dohmen G (1990) Modern plant biotechnology as a tool for reestablishment of genetic variability in *Sophora toromiro.* Cour Forschungsinst Senckenb 125: 233–237
Kapur SK, Sarin YK (1987) Vegetable raw materials of Jai-Barthal tract of Bhadararwah Hills Jammu and Kashmir State India. J Econ Taxon Bot 11(1): 25–40
Karp A, Maddock SE (1984) Chromosome variation in wheat plants regenerated from cultured immature embryos. Theor Appl Genet 67: 249–256
Koller D, Levitan I (1989) Diurnal phototropism in leaves of *Lavatera cretica* L. under conditions of simulated solar-tracking. J Exp Bot 40(218): 1059–1064
Koller D, Shak T, Briggs WR (1990) Enhanced diaphotrophic responses to vectorial excitation in solar-tracking leaves of *Lavatera cretica* by an immediately preceding opposite vectorial excitation. J Plant Physiol 135(5): 601–607
Kucherov EV, Siraeva SM (1981) Nectar plants of late summer. Pchelovodstvo 8: 18 (in Russian)
Kudlicka K, Rodkiewicz B (1991) Organelle coating of meiotic nuclei during microsporogenesis in Malvaceae. Phytomorphology 40(1–2): 33–42
Latunde-Dada AO, Lucas JA (1983) Somaclonal variation and reaction to *Verticillium* in *Medicago sativa* L. plants regenerated from protoplasts. Plant Sci Lett 32: 205–211
Lecoq H, Pitrat M, Clement M (1981) Identification and characterization of a potyvirus that induces muskmelon yellow stunt. Agronomie 1(10): 827–834
Liu MC, Chen WH (1976) Tissue and cell culture as aids to sugar cane breeding. I. Creation of genetic variation through callus culture. Euphytica 25: 393–403
Longly B, Waterkeyn L (1978) Study of cytokinesis 2. Structure and isolation of micro sporocyte cellular plates. Cellule 72(3): 225–242
Longly B, Waterkeyn L (1979) Study of cytokinesis 3. Simultaneous and successive partitions in microsporocites. Cellule 73(1): 65–80
Luque T, Devesa JA (1986) Contribución al estudio citotaxonómico del género *Lavatera* (Malvaceae) en España. Lagascalia 14(2): 227–239
Malloch AJC, Bamidele JF, Scott AM (1985) The phytosociology of British UK sea cliff vegetation with special reference to the ecophysiology of some maritime cliff plants. Vegetation 62(1–3): 309–318
Marco S (1975) Occurrence of alfalfa mosaic virus and malva vein clearing virus on weed members of Malvaceae in Israel. Plant Dis Rep 59(1): 34–36
Marcotrigiano M (1990) Genetic mosaics and chimeras: implication in biotechnology. In: Bajaj YPS (ed) Biotechnology in agriculture and forestry, vol 11. Somaclonal variation in crop improvement I. Springer Berlin Heidelberg New York, pp 85–111
McClelland MT, Smith MA, Carothers ZB (1990) The effects of in vitro and ex vitro root initiation on subsequent microcutting root quality in three woody plants. Plant Cell Tiss Organ Cult 23: 21–26

Miyazaki S, Tashiro Y (1978) Tissue culture of *Chrysanthemum morifolium* Ramat. III Variation in chromosome number and flower color of plants regenerated from different parts of shoots in vitro. Agric. Bull Saga Univ 44: 13–31

Mortensen K (1991) *Colletotrichum gloeosporioides* causing anthracnose of *Lavatera* sp. Can Plant Dis Surv 71(2): 155–159

Murashige T, Skoog F (1962) A revised medium for rapid growth and bioassays with tobacco tissue cultures. Physiol Plant 15: 473–497

Nagl W (1990) Gene amplification and related events. In: Bajaj YPS (ed) Biotechnology in agriculture and forestry, vol 11. Somaclonal variation in crop improvement I. Springer, Berlin Heidelberg New York, pp 153–201

Ogihara Y (1981) Tissue culture in *Haworthia*. 4. Genetic characterization of plants regenerated from callus. Theor Appl Genet 60: 353–363

Okusanya OT, Fawole T (1985) The possible role of phosphate in the salinity tolerance of *Lavatera arborea*. J Ecol 73(1): 317–322

Pyasyatskene AA, Vaichyunene YA, Biveinis YU (1988) Localization of mucous cells in plants introduced in the Lithuanian SSR and in local plant species 3. Formation of mucus in annual Malvaceae species in ontogeny. Liet Tsr Mokslu Akad Darb Ser C Biol Mokslai 0(4): 33–48 (in Russian)

Reisch B, Bingham ET (1981) Plants from ethionine-resistant alfalfa tissue cultures: Variations in growth and morphological characteristics. Crop Sci 21: 781–788

Rumyantseva LT (1987) Characteristics of uncommon fibre plants. Sb Nauchn Tr Prik Bot Gen Selek 113: 108–111 (in Russian)

Scowcroft WR (1984) Genetic variability in tissue culture: impact on germplasm conservation and utilisation. IBPGR Report, Rome

Skirvin RM, Janick J (1976) Tissue culture-induced variation in scented *Pelargonium* spp. J Am Soc Hortic Sci 101: 281–290

Smith MA, Spomer LA, Meyer MJ, McClelland MT (1989) Non-invasive image analysis evaluation of growth during plant micropropagation. Plant Cell Tissue Organ Cult 19: 91–102

Smith MA, Spomer LA, McClelland MT (1990) Direct analysis of root zone data in a microculture system. Plant Cell Tissue Organ Cult 23: 115–123

Sutter E, Langhans RW (1981) Abnormalities in *Chrysanthemum* regenerated from long term cultures. Ann Bot 48: 559–568

Taliaferro CM, Dabo SM, Mitchell ED, Johnson BB, Metzinger BD (1989) Morphologic, cytogenetic, and enzymatic variation in tissue culture regenerated plants of apomictic old-world bluestem grasses (*Bothriochloa* sp.). Plant Cell Tissue Organ Cult 19: 257–266

Thomas E (1981) Plant regeneration from shoot culture-derived protoplasts of tetraploid potato (*Solanum tuberosum* cv. Maris Bard). Plant Sci Lett 23: 81–88

Tonelli C (1990) Somaclonal variation in cereals. In: Bajaj YPS (ed) Biotechnology in agriculture and forestry, vol 11. Somaclonal variation in crop improvement I. Springer, Berlin Heidelberg, New York, pp 271–287

Werker E, Koller D (1987) Structural specialization of the site of response to vectorial photoexcitation in the solar-tracking leaf of *Lavatera cretica*. Am J Bot 74(9): 1339–1349

# II. 5 Tobacco Somaclones Resistant to Tomato Spotted Wilt Virus

I.S. Scherbatenko[1] and L.T. Oleschenko[1]

## 1 Introduction

Tomato Spotted Wilt Virus (TSWV) occurs in a number of strains, and causes economic losses in many important crops (Best 1968; Reddy and Wightman 1988; Francki et al. 1991). It is able to produce genetic recombinants, is transmitted by thrips in a persistent manner, and has an extremely broad host range.

This virus is one of the most dangerous infecting tobacco (Schmidt and Kleinhempel 1986). Although tobacco genotypes differ somewhat in susceptibility, as well as in incidence and severity of the disease, generally most cultivars show similar systemic symptoms: chlorotic or necrotic spots and rings on infected leaves followed by chlorosis and necrosis along veins, mosaic patterning and distortion of the leaves, tip necrosis and bending of the apical shoots, and stunting and premature defoliation of the plants (Kovalenko et al. 1987). Severe strains of TSWV usually kill most of the young infected plants and reduce the yield by as much as 70% under field conditions.

The control of epidemics caused by TSWV is so far a serious problem. Some progress was achieved in vector control (Reddy and Wightman 1988), diagnostics of viral infection (Rice et al. 1990; Wang and Gonsalves 1990), breeding for resistance by interspecific hybridization (Ternowsky 1971; Gajos 1981; Schmidt and Kleinhempel 1986), and genetic engineering (Gielen et al. 1991; MacKenzie and Ellis 1992; Pang et al. 1992).

A promising source of virus-resistant plants is somaclonal variation (Evans 1989; Bajaj 1990). It was successfully used for selection of Fijivirus-resistant sugarcane (Krishnamurthi and Thaskal 1974), PVX- and TMV-resistant tobacco (Shepard 1975; Murakishi and Carlson 1976, 1982; Saha and Gupta 1989; Toyoda et al. 1989), TMV-resistant tomato (Barden et al. 1986), PVX-, PVY-, and PLRV-resistant potato (Jellis et al. 1984; Wenzel and Uhrig 1981), BaYMV-resistant barley (Foroughi-Wehr and Friedt 1984), and BaYDV-resistant wheat (Comeau and Plourde 1987). Our investigations on TSWV resistance in tobacco somaclones (Kovalenko et al. 1987, 1989; Scherbatenko et al. 1989, 1991a; Scherbatenko and Oleschenko 1993, 1995) are summarized here.

Experiments were performed on plant material and a severe strain of TSWV obtained from the Crimean Experimental Station on tobacco from the village Tabachnoe, Ukraine.

[1] Zabolotny Institute of Microbiology and Virology, Academy of Sciences of Ukraine, Zabolotny Street 154, Kiev 143, 252623, Ukraine

Biotechnology in Agriculture and Forestry, Vol. 36
Somaclonal Variation in Crop Improvement II (ed. by Y.P.S. Bajaj)

## 2 Somaclone Regeneration and Testing for Resistance

The regenerants were obtained from leaf explant-derived callus tissues (somaclones SCO), mesophyll protoplasts (protoclones PO), and anthers (androclones AO). For in vitro culture were used: protoplast culture medium $K_3$ NM of Nagy and Maliga (1976); anther culture medium LSU (Scherbatenko and Oleschenko 1993); callus culture medium MSK; shoot induction medium MSP; shoot elongation medium MS42; and root induction medium MSR (Scherbatenko et al. 1991b). All MS media were based on Murashige and Skoog (1962).

The regenerated somaclones were inoculated mechanically with TSWV using crude extracts of infected tobacco plants (1 g of leaf tissues per 10 ml of GMS buffer) as inoculum. The response of plants was recorded over a 10-day period up to 60 days post first inoculation. Infected plants were removed, and noninfected ones were inoculated once more. Resistant somaclones (noninfected after four subsequent inoculations) were tested for TSWV resistance in successive selfed generations: SC1, SC2, SC3...; P1, P2...; A1, A2..., and so on.

## 3 Tomato Spotted Wilt Virus Resistance in Tobacco Somaclones

### 3.1 Callus-Derived Somaclones

Callus cultures were produced from leaf explants of healthy as well as TSWV-infected tobacco plants. The tobacco varieties tested showed some differences in response to tissue culture, but all of them were able to initiate callus growth on MSK medium. No differences in callus appearance, growth rate, or regeneration capacity between tissues from healthy and TSWV-infected tobacco varieties were found.

The infected callus samples of Immunny 580 tobacco (immune to peronosporosis) showed an irregular distribution of TSWV in tissues (Table 1). After 9 weeks of culture (three transfers to fresh MSK medium) infection was from 1.3

**Table 1.** Assay of TSWV infectivity in five samples of callus tissue of Immunny 580 tobacco cultivar[a]. (Scherbatenko et al. 1989)

| Weeks of culture | No. of transfers | Sample No. | | | | |
|---|---|---|---|---|---|---|
| | | 1 | 2 | 3 | 4 | 5 |
| 9 | 3 | 2.4 | 1.3 | 4.7 | 2.3 | 7.0 |
| 12 | 4 | 5.0 | 8.0 | 8.0 | 62.0 | 9.6 |
| 15 | 5 | 0 | 0 | 0 | 5.0 | 0.5 |
| 18 | 6 | 0 | 0 | 0 | 0 | 0 |

[a] Callus samples (1 g) were homogenized in 10 ml of GSM buffer and tested on *Petunia hybrida* leaves. Data are given as the number of local lesions per leaf.

to 7 local lesions per leaf of *Petunia hybrida* L. During subsequent subculturing, the amounts of infective TSWV increased to 5–62 lesions per leaf, and then decreased, followed by elimination of the virus after five to six transfers. Similar results were obtained with infected callus tissues of all tobacco varieties tested.

Most callus explants (70–80%) produced shoots on MSP medium. More than 90% of shoots formed a root system when transplanted onto MSR medium. The regenerants (somaclones SCO) were similar to source tobacco varieties, but some of them (20–50%) showed particular alterations in morphological traits. These somaclones were discarded. The plants regenerated from infected callus tissues showed no disease symptoms as detectable TSWV. The results indicate that this virus is spontaneously eliminated from callus tissues and does not translocate to plantlets. The differences between our results and those previously reported (Shepard 1975; Toyoda et al. 1985) could be due to the different viruses and/or in vitro methods used. It is clear that the in vitro technique used in our investigations is suitable for regeneration of virus-free somaclones from TSWV-infected tobacco plants.

Testing for resistance of infected plant-derived somaclones revealed differences from source plants in their response to TSWV. Plants of all varieties tested showed severe disease symptoms after 8–10 days postinoculation. In contrast to this, the somaclones displayed a broad range of resistance to the virus (Table 2). The majority of the clones were infected after two successive inoculations till 20 days post first inoculation (susceptible). Some plants, which had relative resistance, become infected after two to four inoculations, and others remained noninfected after four inoculations (resistant). The resistant somaclones producing a viable seed were tested for resistance in subsequent selfed generations.

The results of progeny testing (Table 3) revealed some differences in the inheritance of resistance between both the selected clones and tobacco varieties.

**Table 2.** The response of tobacco somaclones to artificial inoculation with TSWV. (Scherbatenko et al. 1989 with additions)

| Source of plant | No. of somaclones | | | | |
|---|---|---|---|---|---|
| | Tested | Susceptible[a] | Relatively resistant[b] | Resistant[c] | Resistant and fertile |
| American 19 × American 5 | 8 | 7 | 1 | 0 | 0 |
| American 307 | 2 | 2 | 0 | 0 | 0 |
| American 3 | 39 | 30 | 6 | 3 | 3 |
| American 3j | 29 | 27 | 2 | 0 | 0 |
| American 5j | 75 | 61 | 10 | 4 | 2 |
| American 361 | 18 | 16 | 1 | 1 | 1 |
| Immunny 580 | 99 | 79 | 15 | 5 | 3 |
| Krymsky stepovy | 25 | 16 | 2 | 7 | 4 |
| Krupnolistny B3 | 14 | 9 | 3 | 2 | 0 |
| Ostrolist | 42 | 37 | 5 | 0 | 0 |
| Peremozhets 83 | 23 | 17 | 2 | 4 | 4 |

[a] Infected by 20 days post first inoculation.
[b] Infected after two to four successive inoculations.
[c] Not infected after four successive inoculations.

**Table 3.** Inheritance of TSWV resistance in progenies of tobacco somaclones derived from virus-infected plants. (Scherbatenko and Oleschenko 1995)

| Tobacco cultivar | Soma-clone | No. of | | Resistant plants (%) |
|---|---|---|---|---|
| | | Plants tested | Clones tested | |
| Immunny 580 | SC0 | 99 | 99 | 6.1 |
| | SC1 | 391 | 4 | 0.0–8.1 |
| | SC2 | 179 | 1 | 0 |
| Peremozhets 83 | SC0 | 23 | 23 | 17.4 |
| | SC1–SC3 | 566 | 3 | 0.3–5.9 |
| | SC4 | 90 | 2 | 0 |
| Krymsky stepovy | SC0 | 25 | 25 | 28.1 |
| | SC1–SC4 | 1248 | 19 | 0.0–11.8 |
| | SC5 | 51 | 1 | 9.8 |
| | SC6 | 76 | 3 | 11.1–15.4 |
| American 361 | SC0 | 18 | 18 | 5.6 |
| | SC1 | 372 | 1 | 0.8 |
| | SC2–SC3 | 544 | 6 | 1.3–4.3 |
| | SC4 | 368 | 7 | 0 |
| American 3 | SC0 | 39 | 39 | 7.7 |
| | SC1 | 60 | 1 | 3.3 |
| | SC2 | 51 | 1 | 0 |
| American 5j | SC0 | 75 | 75 | 5.3 |
| | SC1 | 111 | 2 | 20.7–51.7 |
| | SC2 | 293 | 7 | 2.4–50.0 |

The percentage of resistant plants in SC1 was, as a rule, lower than in SCO. The preselected resistant somaclones of some varieties lost resistance in SC1 (Immunny 580 and Krimsky stepovy) or in SC2–SC4 (American 3, American 361, Immunny 580, and Peremozhets 83). In contrast to this, some somaclones of Krimsky stepovy produced from 11.1 to 15.4% resistant plants in SC6. An increase in resistant plants in subsequent selfed generations was observed in somaclones of American 5j. For further comparative study, investigations on protoclones were performed.

## 3.2 Protoplast-Derived Somaclones

As mesophyll protoplasts isolated from TSWV-infected tobacco were not viable, protoclones were regenerated from noninfected plants (Scherbatenko et al. 1991b). The regeneration procedure was similar to that for somaclones with the exception of producing protoplast-derived mini calli on $K_3MN$ medium.

Protoclones of healthy tobacco varieties responded to TSWV similarly to somaclones of infected plants. The frequency of resistant variants in PO was from 1.3 to 13% (Table 4). Some resistant protoclones selected in PO were sterile (American 5j), and others lost resistance in the first (Immunny 580) or second generation (Immunny 580 × Krimsky stepovy). However, protoclones PO of American 3j, as well as their somaclones SCOAO, displayed an increased proportion of resistant plants in subsequent generations.

**Table 4.** Resistance to TSWV in tobacco protoclones regenerated from mesophyll protoplasts of virus-free plants. (Scherbatenko and Oleschenko 1995)

| Source of plant | Proto-clone | No. of | | Resistant plants (%) |
|---|---|---|---|---|
| | | Plants tested | Clones tested | |
| Immunny 580 | P0 | 53 | 53 | 13.0 |
| | P1 | 349 | 7 | 0 |
| Immunny 580 × Krymsky stepovy | P0 | 77 | 77 | 1.3 |
| | P1 | 82 | 2 | 12.5–81 |
| | P2 | 0 | 0 | 0 |
| American 5j | P0 | 56 | 56 | 3.6 |
| | P1 | 0 | 0 | 0 |
| American 3j | P0 | 61 | 61 | 3.3 |
| | P1 | 50 | 1 | 18.0 |
| | P2 | 85 | 2 | 12.8–50 |
| | SC0P0 | 24 | 24 | 4.2 |
| | SC1P0 | 38 | 1 | 5.3 |
| | SC2P0 | 35 | 1 | 54.3 |

The results demonstrate the presence of from 1.3 to 28.1% TSWV-resistant plants in both infected callus-derived somaclones, SCO, and healthy protoplast-derived protoclones, PO, regenerated from susceptible tobacco varieties. Most of the resistant clones selected in SCO or PO display high segregation, and lose resistance in later selfed generations. Nevertheless, in the progeny of some somaclones (American 5j SCO, American 3j PO, and SCOPO) the percentage of resistant plants increases.

The results suggest that resistance to TSWV in tobacco somaclones is due to spontaneous genetic and epigenetic variability, with many genes involved. This variability depends on plant genotype, and leads to differences between tobacco varieties in the regeneration capacity of explants, morphological traits, and fertility of regenerants, as well as in the yield of resistant somaclones and inheritance of resistance.

Taking into account the frequent somaclonal alterations in fertility, we attempted to regenerate a fertile somaclone from sterile plants of TSWV-resistant interspecific tobacco hybrids.

## 3.3 Somaclones of Interspecific Hybrids

Since numerous sexual crosses between tobacco varieties and TSWV-resistant *Nicotiana glauca* or *N. sanderae* resulted in only five self-sterile plants (Kovalenko et al. 1987), all five were used for somaclone regeneration with the aim of overcoming the sterility.

The regenerated somaclones showed significant variability in both resistance to TSWV and fertility (Table 5). The percentage of resistant plants was from 7.1 to 86%. Hybrids of *N. tabacum* × *N. glauca* produced from 11.1 to 16.7% fertile somaclones. By contrast, no self-fertile plants were found in somaclones of *N.*

**Table 5.** Resistance to TSWV and plant fertility in somaclones of interspecific tobacco hybrids. (Kovalenko et al. 1989; Scherbatenko and Oleschenko 1995)

| Hybrid | No. of Soma-clones tested | Somaclones (%) | | |
|---|---|---|---|---|
| | | Susceptible | Resistant | Fertile |
| American 361 × *N. sanderae* | 141 | 14.0 | 86.0 | 0 |
| Khurchavy 73 × *N. sanderae* | 3 | 33.4 | 66.6 | 0 |
| Khurchavy 140 × *N. glauca* | 79 | 38.4 | 61.6 | 11.1 |
| Harmanly × *N. glauca* | 44 | 87.5 | 12.5 | 16.7 |
| Trapesond 19 × *N. glauca* | 14 | 92.9 | 7.1 | 13.3 |
| *N. tabacum* N 10 × *N. glauca* | 68 | 80.9 | 19.1 | 12.5 |

**Table 6.** Inheritance of TSWV resistance in progenies of somaclone-derived interspecific tobacco hybrids. (Kovalenko et al. 1989; Scherbatenko and Oleschenko 1995)

| Hybrid | Genera-tion | No. of | | Resistant plants (%) |
|---|---|---|---|---|
| | | Plants tested | Clones tested | |
| American 361 × Asan[a] | $F_1$ | 205 | 4 | 0.0–3.4 |
| | $F_2$ | 134 | 3 | 0 |
| Asansan[b] | $F_1$ | 174 | 3 | 75.6–100 |
| | $F_2$ | 0 | 0 | 0 |
| American 361 × Asansan[b] | $F_1$ | 2 | 1 | 50.0 |
| | $F_2$ | 123 | 1 | 1.6 |
| | $F_3$ | 76 | 2 | 0 |
| (American 19 × American 5) × *N. alata* | $F_1$ | 4 | 1 | 0.25 |
| | $F_2$ | 0 | 0 | 0 |
| American 3 × *N. alata* | $F_1$ | 189 | 1 | 4.2 |
| | $F_2$ | 349 | 7 | 0.0–0.25 |
| | $F_3$ | 169 | 6 | 0 |
| (American 3 × *N. alata*) × *N. alata* | $F_1$ | 4 | 1 | 50.0 |
| | $F_2$ | 43 | 1 | 48.8 |
| | $F_3$ | 118 | 3 | 73.7–83.5 |

[a] Asan = American 361 × *N. sanderae.*
[b] Asansan = Asan × *N. sanderae.*

*tabacum* × *N. sanderae*, but some of them were able to set viable seeds when backcrossed with somaclones of tobacco or *N. sanderae*.

All somaclones of American 361 × *N. sanderae* (Asan) produced seeds when pollinated with pollen of *N. sanderae* (Table 6). Most of the hybrid plants obtained (Asansan) were resistant to TSWV. They reached maturity and flowered, but no progeny were recovered.

Surprisingly, the crosses between two resistant somaclones (American 361 × Asan or Asansan) resulted in losing resistance in the first, second, or third generation. This phenomenon was also observed in crosses between resistant tobacco somaclones and resistant species *N. alata*. However, pollination of resistant hybrid American 3 × *N. alata* with pollen of *N. alata* resulted in a high percentage of resistant plants in all progeny tested.

The results of this study indicate that it is possible to produce TSWV-resistant tobacco plants by regeneration of somaclones from sterile interspecific hybrids. The self-fertile somaclones regenerated are already being used in the selection of tobacco for resistance to ISWV. The self-infertile resistant somaclones can be used as a male or female parent in further crosses.

Sterility in intergeneric virus-resistant hybrids of wheat was also overcome (Comeau and Plourde 1987) by regeneration of diploid somaclones from anthers.

## 3.4 Androgenic Clones

Androgenic (anther-derived) plants: androclones AO, and somaclones SCOAO were obtained from tobacco somaclones, protoclones, or hybrids preselected for TSWV resistance. Androclones AO were regenerated from anthers by production of embryoids, shoots, and rooted plantlets on LSU, MS42, and MSR media, respectively (Scherbatenko and Oleschenko 1993). Somaclones SCOAO were obtained from leaf explants of haploid androclones AO selected for resistance to TSWV.

In five out of seven haploid androclones tested, the percentage of resistant plants was somewhat higher or lower than that in the donors of the anthers (Table 7). Two androclones AO gave 6.2 and 19.1% relatively resistant plants, which showed a mild mosaic but subsequently, recovered. The diploid somaclones SCOAO (doubled haploids) regenerated from both resistant and recovered haploid androclones showed from 71.9 to 91.7% resistant plants. Thus, the selection of TSWV-resistant haploid androclones followed by production of a

**Table 7.** Yield of resistant plants in selfed progenies and androgenic specimens of resistant tobacco somaclones or hybrids. (Scherbatenko and Oleschenko 1995)

| Source of plants | Selfed progenies | | Androgenic specimens | |
|---|---|---|---|---|
| | Clones | Resistant plants (%) | Clones[a] | Resistant plants (%) |
| Krymsky stepovy | SC3 | 1.7 | A0 | 2.6 |
| | SC4 | 3.8 | SC0A0 | 78.3 |
| | SC3 | 8.5 | A0 | 3.2 |
| | SC4 | 3.3 | SC0A0 | 71.9 |
| (Immunny 580 × Krymsky stepovy) $F_1$ | P1 | 12.5 | A0 | 33.3 |
| American 5j | SC1 | 39.7 | A0 | 7.7 |
| American 3 | SC1 | 1.7 | A0 | 19.1[b] |
| | SC2 | 0 | SC0A0 | 83.8 |
| American 3 × *N. alata* | $F_1$ | 4.2 | A0 | 8.4 |
| | $F_2$ | 3.1 | A0 | 6.2[b] |
| | $F_3$ | 0 | SC0A0 | 91.7 |

[a] A0 = Haploid androclones; SC0A0 = doubled haploids.
[b] Plants showed a mild mosaic followed by recovery.

doubled haploid gives a yield of resistant plants many times higher than successive selection in generations of diploid somaclones. Our results strongly confirm the high efficiency of androgenic technology in plant breeding for virus resistance (Wenzel and Uhrig 1981; Comeau and Foroughi-Wehr and Friedt 1984; Plourde 1987).

## 4 Summary and Conclusions

Investigations demonstrate that somaclones regenerated from TSWV-susceptible tobacco varieties show significant variability in response to this virus, from delay in symptom development to complete resistance. The yield of resistant variants depends on both tobacco variety and regeneration procedure (type of somaclones).

Somaclonal variation can be used to overcome interspecific incompatibility in some crosses between tobacco varieties and TSWV-resistant *Nicotiana* species. Fertile virus-resistant plants can be produced by regeneration of somaclones from self-sterile interspecific hybrids, as well as by crossing tobacco somaclones with resistant species or somaclones of interspecific hybrids.

TSWV-resistant somaclones and somaclone derived interspecific hybrids display a high segregation in selfed progenies and lose resistance in the first to fifth generation. However some clones have heritable resistance, which leads to an increase in the proportion of resistant plants in successive generations.

A high yield of resistant clones can be obtained after selection for TSWV resistance in haploid androclones followed by regeneration of doubled haploids.

The yield of virus-resistant somaclones and the inheritance of resistance may depend on virus-host combinations, plant genotypes, starting material, in vitro techniques, type of regenerants, selection procedures, etc. Thus, to select virus-resistant somaclones, only the trial and error approach is valid.

## References


Bajaj YPS (ed) (1990) Biotechnology in agriculture and forestry, vol 11. Somaclonal variation in crop improvement I. Springer, Berlin Heidelberg New York

Barden KA, Schiller S, Murakishi NN (1986) Regeneration and screening of tomato somaclones for resistance to tobacco mosaic virus. Plant Sci 45: 209–213

Best RJ (1968) Tomato spotted wilt virus. Adv Virus Res 13: 65–146

Comeau A, Plourde A (1987) Cell, tissue culture and intergeneric hybridization for barley yellow dwarf virus resistance in wheat. Con J Plant Pathol 9: 188–192

Evans DA (1989) Somaclonal variation – genetic basis and breeding applications. Trends Genet 5: 46–50

Foroughi-Wehr B, Friedt W (1984) Rapid production of recombinant barley yellow mosaic virus-resistant *Hordeum vulgare* lines by anther culture. Theor Appl Genet 67: 377–382

Francki RIB, Fauquet CM, Knudson DL, Brown F (1991) Classification and nomenclature of virus. 5th Rep Int Comm on Taxonomy of Viruses. Arch Virol Suppl 2: 281–283

Gajos Z (1981) Przeniesienie odpornosci na virus brazowej plamistosci pomidora (tomato spotted wilt virus) z *Nicotiana alata* Link. et Otto do tytoniu szlachetnego przer skryzowanie obu gatunkow. Biul Inf Cent Lab Przem Tyton 1/2: 3–24

Gielen JJC, de Haan P, Kool AJ, Peters D, Van Grinsven MQVN, Goldbach RW (1991) Engineered resistance to tomato spotted wilt virus, a negative-strand RNA virus. Bio/Technology 9: 1363–1367
Jellis GJ, Gunn RE, Boulton RE (1984) Variation in disease resistance among potato somaclones. In: Winiger FA, Stockly A (eds) Abstr Conf Pap Triennal Conf EAPR, Interlaken. EAPR, Wageningen, pp 380–381
Kovalenko AG, Rud EA, Strelaeva NI, Oleschenko LT (1987), Responses of tobacco varieties, wild species and interspecific hybrids on artificial infection with tomato spotted wilt virus. Mikrobiol Zh 49: 85–89
Kovalenko AG, Scherbatenko IS, Oleschenko LT, Rud EA, Strelaeva NI (1989) The production of fertile somaclones of interspecific tobacco hybrids with high resistance to tomato spotted wilt virus. Cytol Genet 24: 59–65
Krishnamurthi M, Thaskal J (1974) Fiji disease resistant *Saccharum officinarum* var. Pindar subclones from tissue cultures. Proc Int Soc Sugarcane Technol 15: 130–137
Mackenzie DJ, Ellis PJ (1992) Resistance to tomato spotted wilt virus infection in transgenic tobacco expressing the viral nucleo capsid gene. Mol Plant-Microbe Interact 5: 34–40
Murakishi HH, Carlson PS (1976) Regeneration of virus-free plants from dark-green islands of tobacco mosaic virus-infected tobacco leaves. Phytopathology 66: 931–932
Murakishi HH, Carlson PS (1982) In vitro selection of *Nicotiana sylvestris* variants with limited resistance to TMV. Plant Cell Rep 1: 94–97
Murashige T, Skoog F (1962) A revisd medium for rapid growth and bioassays with tobacco tissue culture. Physiol Plant 15: 473–497
Nagy JI, Maliga P (1976) Callus induction and plant regeneration from mesophyll protoplasts of *Nicotiana sylvestris*. Z Pflanzen physiol 78: 453–455
Pang S-Z, Nagpala P, Wang M, Slighton JL, Gonsalves D (1992) Resistance to heterologous isolates of tomato spotted wilt virus in transgenic tobacco expressing its nucleocapsid protein gene. Phytopathology 82: 1223–1229
Reddy DVR, Wightman JA (1988) Tomato spotted wilt virus: thrips transmission and control. Adv Dis Vector Res 5: 203–220
Rice DJ, German TL, Mau FRL, Fujimoto FM (1990) Dot blot detection of tomato spotted wilt virus RNA in plant and thrips tissues by cDNA clones. Plant Dis 74: 274–276
Saha S, Gupta S (1989) Isolation of disease-free plants from tissue cultures of the TMV-infected leaf of tobacco var. Jayasri. Phytomorphology 38: 241–248
Schmidt H, Kleinhempel H (1986) Speziele Kulturen. In: Spaar D, Kleinhempel H (eds) Bekämpfung von Viruskrankheiten der Kultur pflanzen. Agropromizdat, Moskau pp 408–438
Shepard JF (1975) Regeneration of plants from protoplasts of potato virus X-infected tobacco leaves. Virology 66: 492–501
Scherbatenko IS, Oleschenko LT (1993) The production of tobacco androgenetic plants resistant to tomato spotted wilt virus. Cytol Genet 27: 48–52
Scherbatenko IS, Oleschenko LT (1995) The display of high resistance to tomato spotted wilt virus in cellular clones and hybrids of tobacco. Mikrobiol Zh 57: 65–71
Scherbatenko IS, Kovalenko AG, Oleschenko LT, Rud EA, Strelyaeva NI (1989) The production of somatic clones of tobacco resistant to tomato spotted wilt virus. Biol Nauki 6: 24–27
Scherbatenko IS, Kovalenko AG, Oleschenko LT, Olevinskaya ZM, Rud EA, Strelyaeva NI (1991a) Resistance of tobacco somaclones to tomato spotted wilt virus. Mikrobiol Zh 53: 75–80
Scherbatenko IS, Oleschenko LT, Olevinskaya ZM (1991b) Display of hypersensitivity and acquired resistance to TMV in tobacco regenerants. Mikrobiol Zh 53: 69–75
Ternowsky MF (1971) Interspecific hybridization and experimental mutagenesis in tobacco breeding. In: Control of viral diseases in plants. Nauka, Moscow, pp 260–312
Toyoda H, Oishi Y, Matsuda Y, Khatani K, Hirai T (1985) Resistance mechanism of cultured plant cells to tobacco mosaic virus. 1V. Changes in tobacco mosaic virus concentrations in somaclonal tobacco callus tissues and production of virus-free plantlets. Phytopathol Z 114: 126–133
Toyoda H, Khatani K, Matsuda Y, Ouchi S (1989) Multiplication of tobacco mosaic virus in tobacco callus tissues and in vitro selection for viral disease resistance. Plant Cell Rep 8: 433–436
Wang M, Gonsalves D (1990) ELISA detection of various tomato spotted wilt virus isolates using specific antisera to structural proteins of the virus. Plant Dis 74: 154–158
Wenzel G, Uhrig H (1981) Breeding for nematode and virus-resistance in potato via anther culture. Theor Appl Genet 59: 333–340

# II.6 Somaclonal Variation in Primula

M. KANDA[1]

## 1 Introduction

The genus *Primula*, a member of the Primulaceae family, consists of more than 500 species, which are naturally distributed in areas of the temperate zone in Europe, South America, North Africa, and Asia, including Japan (Smith and Forrest 1929). Most of them are perennial, hardy plants which bloom from early spring to summer. These plants usually have radical leaves and five-partite corollas. The plants are generally propagated by seeds or division of clones. Several species, such as *P. malacoides*, *P. obconica*, and *P. × polyantha*, are important ornamental plants which are grown commercially in many parts of the world for use as pot plants and garden flowers. Other species, such as *P. auricula* and *P. sieboldii*, are grown locally as pot plants. The chromosome numbers have been reported in the important wild species and some of the cultivated varieties of this genus, including *P. auricula* (2n = 62,), *P. malacoides* (2n = 18, 36), *P. obconica* (2n = 24, 48), and *P. sieboldii* (2n = 24, 36; Bolkhovskikh 1969).

Commercial cultivars of *Primula* need to improve the type of flower and its color continually. In addition to these factors, breeding work on *P. obconica* involves raising primine-free plants (Heyting and Toxopeus 1989). The present cultivars were developed by crossing and selection among spontaneous mutant plants. For instance, it is generally agreed that *P. × polyantha* arose as an interspecific hybrid among *P. elatior*, *P. veris*, and *P. vulgaris*; however, it is difficult to obtain interspecific hybrids in many combinations. Besides, seeds sometimes cannot be obtained by self-pollination because of the presence of a heteromorphic self-incompatibility system (Watts 1980).

Phenotypic or genetic changes in regenerated plants from in vitro cultures have been observed in many ornamental plants (Fujino 1990; Bajaj 1992). These somaclonal variations have contributed greatly to increasing variability in characters such as flower type and color. Therefore, somaclonal variation is expected to be applied to the genetic improvement of floral and marketable qualities in *Primula*.

## 2 In Vitro Culture Studies (Table 1)

Plant regeneration from anthers of *P. obconica* (Bajaj 1981) and from protoplasts of *P. malacoides* (Mii et al. 1990) has been studied. Propagation by in vitro culture

[1]Chiba Horticultural Experiment Station, 1762 Yamamoto, Tateyama, Chiba 294, Japan

Biotechnology in Agriculture and Forestry, Vol. 36
Somaclonal Variation in Crop Improvement II (ed. by Y.P.S. Bajaj)

**Table 1.** Summary of in vitro culture studies on *Primula*

| Species | Inoculum | Growth response | Medium (mg/l) | Reference |
|---|---|---|---|---|
| *P. malacoides* | Leaf | Callus | MS + 2,4-D (10) + BA (0.1) | Mii et al. (1990) |
| | Protoplast (Leaf callus) | Plant | MS + zeatin (0.1) → MS + zeatin (5 ~ 10) | |
| *P. obconica* | Inflorescence tip | Shoot | MS + NAA (1) + BA (1) | Coumans et al. (1979) |
| | Anther | Plant | MS + IAA (0.5) + 2,4-D (0.5) + zeatin (2) | Bajaj (1981) |
| | Leaf, flower bud | Shoot | MS + NAA (1) + BA (1) | Kanda (1992) |
| | Shoot tip | Plant | MS (growth regulator-free) | |
| *P. sieboldii* | Leaf, flower bud | Shoot | MS + NAA (0.1 ~ 0.5) + BA (1) | Matsumoto et al. (1986) |
| *P. tosaensis* | Petiole | Shoot | 1/2MS + NAA (0.1) + BA (1) | Kowase et al. (1993) |

has also been studied in *P. obconica* (Coumans et al. 1979), *P. sieboldii* (Matsumoto et al. 1986), and *P. tosaensis* (Kowase et al. 1993). The establishment of an in vitro propagation method is useful for the propagation of some variants with desirable characteristics, such as doubled flowers which do not normally produce seed. Another use of in vitro clonal propagation is the establishment of parental lines for the production of $F_1$ hybrids.

Our results on in vitro propagation of *P. obconica* indicated that shoots and roots developed on the growth regulator-free medium in shoot tip culture, whereas shoots were regenerated from callus tissues induced from leaf and flower bud explants on MS medium (Murashige and Skoog 1962) containing 1 mg/l α-naphthaleneacetic acid (NAA) and 1 mg/l 6-benzylamino purine (BA). The

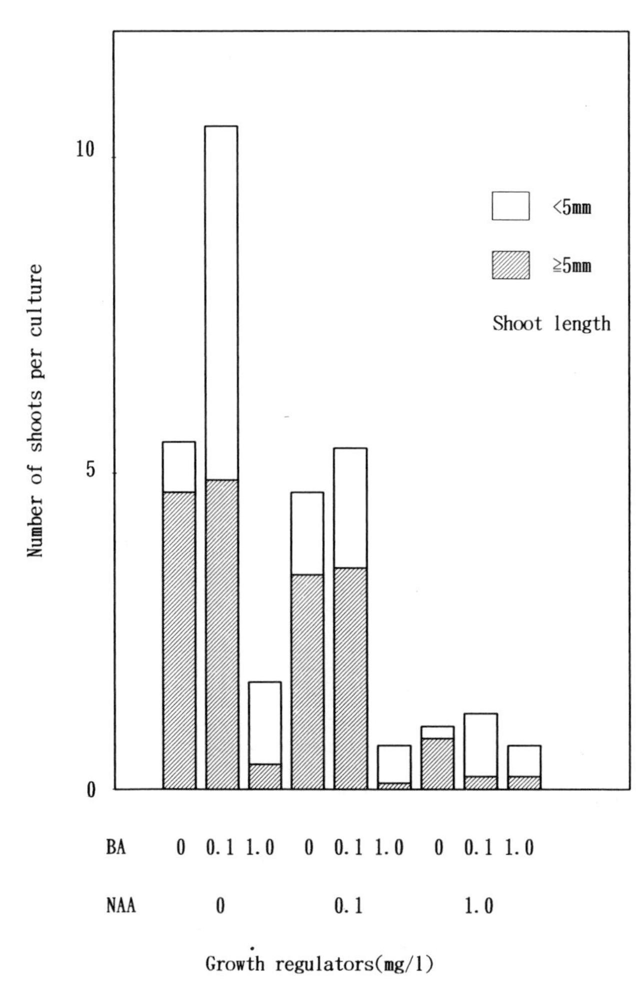


**Fig. 1.** Effects of NAA and BA combinations in subculture medium on shoot multiplication of *Primula obconica*

shoots developing from these explants were subcultured on MS media containing different concentrations of NAA and BA. Two months after transfer, about five axillary shoots ($\geqq$ 5 mm) developed from each shoot on MS media containing either 0.1 mg/l BA or no growth regulator (Fig. 1). For rooting, the propagated shoots were cultured for 1 month on growth regulator-free MS medium containing 0.2% Gellan Gum (Kelco, Division of Merck & Co. Inc.), as shown in Fig. 2. The rooted shoots in the test tubes were then transplanted to plastic pots containing Metro-mix 350 (W.R. Grace & Co.).

# 3 Somaclonal Variation

## 3.1 Somaclonal Variation in In Vitro-Propagated Plants

In *P. obconica*, somaclonal variation was investigated among the plants regenerated from shoot tips on growth regulator-free MS medium and from leaf segments and flower buds on MS medium containing 1 mg/l NAA and 1 mg/l BA, respectively (Kanda 1992). In this study, all explants were obtained from a double-flowered spontaneous mutant of *P. obconica* cultivar Apricot, which is

**Fig. 2.** Root formation from shoots cultured on growth regulator-free MS medium solidified with Gellan Gum

**Table 2.** Variants in flower color in *Primula obconica* plants propagated in vitro. (Kanda 1992)

| Explant | Culture period (weeks) | Times of subculture | No. of cultured lines[a] | No. of flowering plants | No. of cultured lines | |
|---|---|---|---|---|---|---|
| | | | | | Normal | Variant |
| Leaf | 70 | 6 | 3 | 134 | 1 | 2 |
| Shoot tip | 35 | 3 | 5 | 21 | 5 | 0 |
| | 70 | 6 | 5 | 110 | 5 | 0 |
| Flower bud | 35 | 3 | 2 | 8 | 2 | 0 |
| | 70 | 6 | 2 | 16 | 2 | 0 |

Explants of leaf and flower buds were cultured on MS medium containing 1 mg/l NAA and 1 mg/l BA. Shoot tips were cultured in growth regulator-free MS medium.
[a] A cultured line originated from an explant which produced a shoot.

a single-flowered cultivar. Table 2 shows that the variants in flower color occurred in all of the plants propagated for 70 weeks from two shoots originating from different leaf explants. These variants had a pale flower color compared to those of the mother plant, as shown in Fig. 3. According to the Japan Color Standard for Horticultural Plants, the color code number of the mother plant is in the range from 0705 (bright yellowish red) to 0702 (pale yellowish pink). All variants in flower color had a range of color from 0703 (yellowish pink) to 0701 (pinkish white).

No changes in chromosome number were found between the variants and normal plants. All the variants with altered flower color had the same diploid chromosome number (2n = 24) as the original double-flowered plants of *P. obconica* cultivar Apricot (Fig. 4).

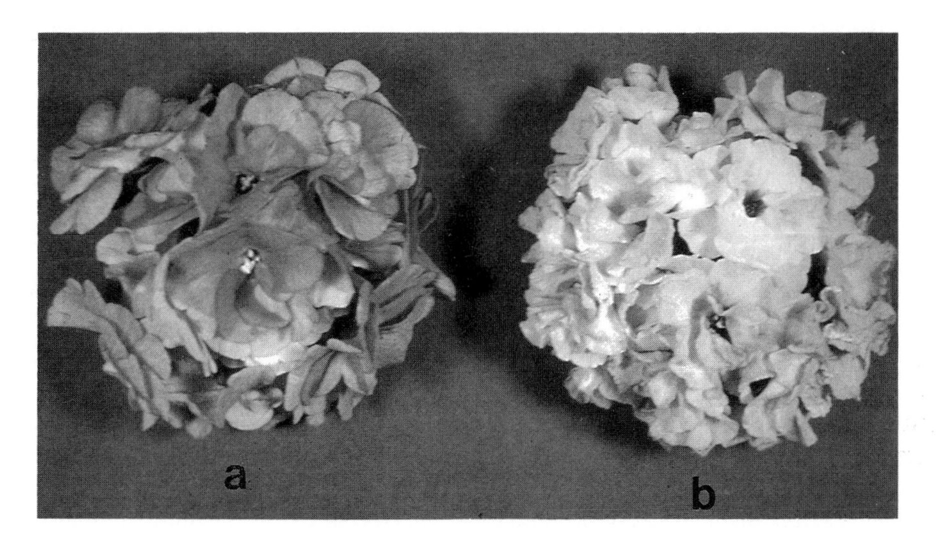


**Fig. 3a,b.** Flowers of *Primula obconica* plants propagated in vitro. **a** Normal plant. **b** Variant in flower color

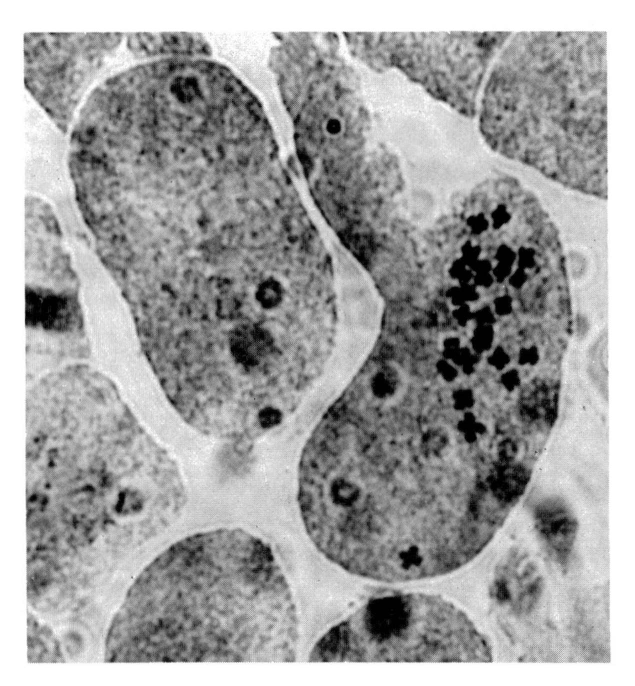

**Fig. 4.** Stable chromosome number (2n = 24) of the variant in flower color

In contrast to leaf explant culture, no variations were recorded among the plants which were propagated for 35 weeks from shoot tips and flower buds (Table 2). Prolonged subculture for 70 weeks induced no variations among the plants regenerated from those two sources. Flower color was normal in all of the investigated plants, and the color code number of the plants subcultured for 35 and 70 weeks was in the range from 0705 to 0702. Among the three lines derived from leaf tissues, only one line (L-1) produced no variations in flower color, and doubled flower character even after subculture for 113 weeks (Table 3). There were also no differences in the other morphological characters such as plant height, flower diameter, and leaf length between in vitro-propagated plants and mother plants which were propagated by division. Therefore, the period of subculture on growth regulator-free medium did not affect the frequency of variation.

In the present study, variation in flower color occurred in the plants regenerated from leaf explants of *Primula obconica*. A similar observation was also reported in *Anthurium scherzerianum* (Geier 1987) in which the frequency of variants was higher in leaf-derived plants than in spadix-derived plants. These results may suggest that the source of explants influences the frequency of variation in in vitro propagation of these species.

Factors such as growth regulator composition of the medium may also influence the frequency of variation (Oono 1984). In the present study, shoots were subcultured on growth regulator-free MS medium. Therefore, variation

**Table 3.** Morphological characters of in vitro-propagated plants of *Primula obconica.* (Kanda 1992)

| Line | Weeks of culture | Times of subcultures | No. of plants investigated | Plant height (cm) | Flower diameter (cm) | Leaf length (cm) | Flower color | |
|---|---|---|---|---|---|---|---|---|
| | | | | | | | Normal (%) | Variant (%) |
| L-1[a] | 113 | 9 | 30 | 19.6 | 4.2 | 9.6 | 100 | 0 |
| Mother plants (division) | | | 5 | 19.6 | 4.1 | 9.7 | 100 | 0 |

[a] This line is one of the lines derived from leaf explants and produced no variations after subculture for 70 weeks.

among the plants propagated from leaf explants should have occurred at the very early stage of culture during which growth regulators might have affected induction of the variation. Both source of explants and growth regulator composition may relate to the high frequency of somaclonal variations in in vitro culture.

### 3.2 Gamma-Ray Irradiation on In Vitro Plants

In order to increase the range of variation, the effect of gamma-ray ($^{60}Co$) irradiation on in vitro plants of *P. obconica* was examined (Kanda et al. 1990). All explants were obtained from a pale flower color variant plant which was regenerated from leaf explants of a double-flowered natural mutant of *P. obconica* cultivar Apricot. The plantlets in test tubes were irradiated by gamma-ray from a $^{60}C$ source at the exposure doses of 0, 5, 7.5, 10, 15, 20, and 30 kR and cultured on MS medium containing no growth regulator. Thirty days postirradiation, all irradiated plantlets survived in the test tubes. However, the growth parameters of the plantlets, such as plant height, number of leaves, root length, and number of roots, were inferior to those of the control plantlets (data not shown). Shoots produced on each explants were transferred, and then subcultured on MS medium without growth regulator.

Plants derived from the shoots were transplanted to plastic pots containing Metro-mix 350 (W.R. Grace & Co.), and grown under greenhouse conditions. Table 4 shows that one variant in flower color was found among the plants transferred from in vitro plants irradiated at 7.5 kR gamma-ray. This variant had a flower color different from those of the control plants. According to the Japan color standard for horticultural plants, the color code number of this variant was in the range from 3102 (pale yellow green) to 3101 (yellowish white). In contrast to the variant plant, the range of control plants in flower color is from 0703 (yellowish pink) to 0701 (pinkish white). Flower color mutants induced by

**Table 4.** Variant in flower color induced by gamma-ray irradiation on in vitro plants of *Primula obconica*. (Kanda et al. 1990)

| Treatment[a] | | No. of plants flowering[b] | No. of plants variant in flower color |
|---|---|---|---|
| Dose (kR) | Dose rate (R/h) | | |
| Control | 0 | 52 | 0 |
| 5 | 500 | 39 | 0 |
| 7.5 | 750 | 66 | 1 |
| 10 | 1000 | 22 | 0 |
| 15 | 1500 | 4 | 0 |
| 20 | 2000 | 0 | – |
| 30 | 3000 | 0 | – |

[a] Each treatment consists of five in vitro plants.
[b] Plants propagated from in vitro plants irradiated by gamma-ray.

irradiation were observed in many ornamental plants (Yamaguchi 1987). Therefore, it is expected that the range of variations could be spread by using gamma-ray irradiation on in vitro plants.

## 4 Summary and Conclusions

Somaclonal variation was investigated in *P. obconica* plants regenerated from shoot tips on growth regulator-free MS medium and from leaf segments and flower buds on MS medium containing 1 mg/l NAA and 1 mg/l BA, respectively. After repeating the subculture for 70 weeks on growth regulator-free medium, all the plants regenerated from two out of three leaf segments showed a variation in flower color. However, no variations were recorded among the plants regenerated from shoot tips and flower buds. In addition, variation in flower color was increased by using gamma-ray irradiation from a $^{60}$C source at the exposure doses of 7.5 kR on in vitro plant. After being cultured in the greenhouse for 2 years, both these variants continuously had flower colors different from the mother plant. The present study showed the possibility of induced somaclonal variation being applied to the genetic improvement of floral and marketable qualities of *Primula* cultivars in which it is difficult to obtain seeds.

## 5 Protocol

Explant. Leaf segments (rectangular pieces with 1 cm wide). Medium. Basal medium; Murashige and Skoog (1962) medium containing 20 g/l sucrose. The pH was adjusted to 5.8 before autoclaving. Callus and shoot formation from leaf segment on basal medium containing 1 mg/l NAA, 1 mg/l BA, and 8 g/l agar. Shoot multiplication and root formation on basal medium without growth regulator and 0.2% Gellan Gum.

Culture condition 23 °C, continuous illumination with fluorescent lamps (3000 lx).

*Acknowledgments.* The author wishes to thank Dr. M. Mii, Chiba University, for his advice in chromosome analysis and critical reading of the manuscript. Thanks are also due to Dr. S. Nagatomi, Institute of Radiation Breeding, National Institute of Agrobiological Resources, for his valuable suggestions in gamma-ray irradiation.

## References

Bajaj YPS (1981) Regeneration of plants from ultra-low frozen anthers of *Primula obconica*. Sci Hortic 14: 93–95

Bajaj YPS (ed) (1992) Biotechnology in agriculture and forestry, vol 20. High-Tech and micropropagation IV. Springer, Berlin Heidelberg New York

Bolkhovskikh Z (1969) Chromosome numbers of flowering plants. Acad Sci USSR, pp 591–595 (Reprint by Otto Koeltz Science Publishers, Koenigstein, Germany, 1974)

Coumans M, Coumans-Gilles M.-F, Delhez J, Gaspar T (1979) Mass propagation of *Primula obconica*. Acta Hortic 91: 287–293

Fujino M (1990) Occurrence of variation in tissue cultures of ornamental plants. Bio Horti 4: 15–22

Geier T (1987) Micropropagation of *Anthurium scherzerianum*: propagation schemes and plant conformity. Acta Hortic 212: 439–443

Heyting J, Toxopeus SJ (1989) Breeding primine-free *Primula obconica*. Neth J Agric Sci 37(4): 371–378

Kanda M (1992) Variation in in vitro propagated plants of *Primula obconica*. In: Plant tissue culture and gene manipulation for breeding and formation of phytochemicals. NIAR, Tsukuba, Japan, pp 179–183

Kanda M, Nagatomi S, Yamaguchi M (1990) Mutant induced by gamma-ray irradiation on in vitro plants of *Primula obconica*. Jpn J Breed 40 (Suppl 2): 256–257

Kowase M, Matsumoto T, Kato S, Sakakibara I (1993) Propagation of *Primula tosaensis* var. *rhodotricha* Ohwi by tissue culture. Abstr 13th Annu Meet Japanese Assoc Plant Tissue Culture Kyoto, Japan, 34 pp

Matsumoto T, Kowase M, Itami K (1986) In vitro propagation of *Primula sieboldii*. Abstr Jpn Soc Hortic Sci (Suppl 1): 410–411

Mii M, Kadowaki S, Ohashi H, Nemoto K, Yamada K (1990) Plant regeneration from protoplasts of *Primula malacoides*. Jpn J Breed 40 (Suppl 1): 16–17

Murashige T, Skoog F (1962) A revised medium for rapid growth and bioassays with tobacco tissue cultures. Physiol Plant 15: 473–497

Oono K (1984) Genetic variability in cultured cells and their use for breeding. Plant Tissue Cult Lett 1(1): 2–7

Smith WW, Forrest G (1929) The sections of the genus *Primula*. J R Hortic Soc 54: 4–50

Watts L (1980) Flower and vegetable plant breeding, breeding systems. Grower Books, London

Yamaguchi T (1987) Mutation breeding of ornamental plants. Bull Inst Radiat Breed 7: 49–67

# II.7 Somaclonal Variation in *Rauwolfia*

V.A. Kunakh[1]

## 1 Introduction

*Rauwolfia* (Apocynaceae family) species are distributed in tropical regions of Africa, Central and South America, Australia, and Southeast Asia. Altogether, the genus covers 140 species, most of which are shrubs, less frequently trees up to 18–21 m in height (*R. macrophylla* Stapf and *R. caffra* Sond), or intermediate forms up to 6 m high (*R. vomitoria* Afzel).

*Rauwolfia* species are known to accumulate considerable amounts of alkaloids. Some species contained more than 80 alkaloids, many of which were biologically active (Kaul 1963–1964; Martinez 1983; Stockigt 1988). Some species, *R. serpentina* in particular, have been widely used in popular medicine in India for more than 3000 years. A number of species (*R. serpentina*, *R. vomitoria*, *R. canescens*, *R. verticillata*, etc.) are used in the pharmaceutic industry, mainly for production of sedative and cardiovascular medicines.

World demand for *Rauwolfia* raw material is increasing. In India alone, the demand for dry *Rauwolfia* roots exceeds 650 t per year. Because of the severe depletion of natural *Rauwolfia* thickets, extensive efforts have been undertaken to raise *Rauwolfia* as an agricultural crop in India, Vietnam, China, Cuba, Georgia, etc. (Granda 1987; Antipova et al. 1988; Singh et al. 1990). Prospects for improvement in plant productivity are under study (Singh and Nand 1989); novel, more productive plant varieties have been both singled out from wild types and obtained experimentally (Janaki 1982; Kaul 1963–1964).

Two possibilities exist to increase the yield and quality of *Rauwolfia* raw material. One is the production of novel plant varieties by conventional methods, while the other involves the generation of cell strains differing in their potential due to induced somaclonal variation followed by regeneration of plants with promising properties.

## 2 In Vitro Culture Studies

Since 1964, studies on the initiation and examination of *R. serpentina* tissue culture have been repeatedly reported (Butenko 1964; Mitra and Kaul 1964;

[1] Institute of Molecular Biology and Genetics, National Academy of Sciences of Ukraine, Kiev 252143, Ukraine

Biotechnology in Agriculture and Forestry, Vol. 36
Somaclonal Variation in Crop Improvement II (ed. by Y.P.S. Bajaj)

Vollosovich and Butenko 1970; Ohta and Yatazawa 1979; Stockigt et al. 1981; Yamamoto and Yamada 1986, etc). Callus tissues were exposed to mutagens that resulted in mutant strains accumulating increased amounts of indole alkaloids (Kovaleva et al. 1972; Vollosovich et al. 1976; Vollosovich 1989). Further improvement of cell strains and conditions for their maintenance increased indoline alkaloid accumulation by as much as 6–8%, and in selected cases up to 20% per dry tissue (Vollosovich 1989; Kunakh 1994). This is nearly 100% of that found in nature. These strains were examined both cytologically and genetically (Kunakh and Alkhimova 1989; Kunakh 1994).

Additionally, other *Rauwolfia* species were introduced into culture in vitro, and the productivity of cultured cells is being investigated (Vollosovich et al. 1972; Kunakh and Gubar 1993). Studies on production and characterization of cell hybrids between *R. serpentina* and other species of the Apocynaceae family were initiated (Kostenyuk et al. 1991a, b); multiple shoot cultures of *R. serpentina* were generated (Roja et al. 1990); and plant regeneration from callus of *R. canescens* (Upadhyay and Batygina 1992) and *R. serpentina* (Akram et al. 1990) was reported to occur. The latter showed normal phenotype and appeared to be diploid. Recently, somatic embryogenesis has also been reported in protoplast-derived callus of *R. vomitoria* (Tremouillaux-Guiller and Chenieux 1995).

# 3 Somaclonal Variation

## 3.1 Experimental Material

Plants of several species, i.e., *R. serpentina*, *R. verticillata*, *R. canescens*, *R. vomitoria*, and *R. chinensis*, were raised in the greenhouse, and callus and suspension culture were generated from them. Cells recovered from these plants were studied after 2 to 3 years of maintenance in vitro. In addition, cell cultures of *R. serpentina* subcultured in vitro for more than 20 years were examined. These were:

*Cell Line A*. Strain establishment records and results of its examination were reported previously (Vollosovich et al. 1976; Kunakh and Alkhimova 1989). It was maintained as a callus tissue on agar nutrient 5S medium containing 5% sucrose (Vollosovich et al. 1979).
*Strain K-20*. This was generated as a result of line A selection on medium containing 5-methyltriptophan, and maintained on 6S agar medium containing 6% sucrose (Vollosovich et al. 1982).
*Strain K-27*. This resulted from cell line A exposed to ethylenimine following its selection for increased productivity on agar 10S medium containing 10% sucrose.
*Strain A-10S*. This is cell line A, conditioned to growth on agar 10S medium containing 10% sucrose.
*Strain F*. This was derived from line A by exposure to the ethylenimine mutagen followed by selection for an increased biomass increment on 10S medium. It accumulates ca. two to three times more biomass than line A.

*Strain R-III*. This originated from cell line A as a result of its prolonged selection in liquid nutrient medium. It is maintained as a suspension in RL medium (Kaukhova et al. 1981b).

Cell line A itself and the strains derived from it are auxin- and cytokinin-independent. They differ from each other in rate of growth, tissue morphology, and the number of cytological and biochemical traits, including spectrum and amount of alkaloids accumulated (Table 1). The strains examined share not only a common origin but are distinguished by the same major alkaloid, ajmaline. Under certain, rather complicated maintenance conditions, the amount of ajmaline accumulated can account for up to 20% of this dry tissue mass. More detailed evaluations of these strains' productivity, with some peculiarities of their genome and biochemical variability are given in Kunakh (1994).

## 3.2 Karyological Variability

Populations of cultured *Rauwolfia* cells were distinguished by their high level of variability (Figs. 1,2). An especially high level of karyological variability was found to occur in the long-term cultured (since 1964) highly productive cell strain A and its derivatives. These strains showed cell nuclei differing in size and shape. Fragmented and lobed, presumably budding nuclei, as well as spindle- and band-shaped nuclei of up to 28 $\mu$ long to carry 4–6 nucleoli were encountered. In highly productive strains like A, K-20, and especially K-27, the proportion of cells with such nuclei could amount to 50%.

Analysis of cells dividing on the 7th to 15th days of growth revealed the occurrence of various abnormalities, expressed both as chromosome aberrations (bridges and fragments in anaphases) and as spindle disturbances (in both anaphases and metaphases of mitosis). However, the incidence of chromosome aberrations was not comparatively high. In young cultures, it was at the level of 1–2%, about 3% in cell line A and strain R-III, while in strain K-27 it amounted to 13%. The majority of the aberrations (60–75%) presented the bridges without fragments. This suggests a considerable contribution by the "bridge-break-reassociation" cycle to the formation of chromosome aberrations, and low

**Table 1.** Alkaloid content in examined *R. serpentina* strains as a percentage relative to dry weight. (Data obtained jointly with Dr. S. I. Gubar)

| Strain | Total alkaloids | Ajmaline | Vomilenine | Reserpine | Serpentine | Ajmalicine | Rescinamine |
|---|---|---|---|---|---|---|---|
| Line A | 1.8–1.9 | 0.4–0.5 | 0.2–0.3 | 0.02–0.04 | × | × | 0.003 |
| K-20 | 2.5–3.3 | 0.9–1.2 | 0.7–0.8 | 0.03–0.04 | × | × | × |
| K-27 | 2.4–2.9 | 0.9–1.4 | 0.2–0.3 | 0.01 | 0.05 | 0.05 | × |
| A-10S | 1.9–2.1 | 0.5–0.6 | 0.3–0.4 | 0.05 | × | × | × |
| F | 2.9–3.1 | 0.1–0.2 | – | × | – | – | × |
| R-III | 1.4–1.8 | 0.4–0.6 | – | – | – | – | – |

× = Values beyond the resolution power.
– = measurements were not performed.

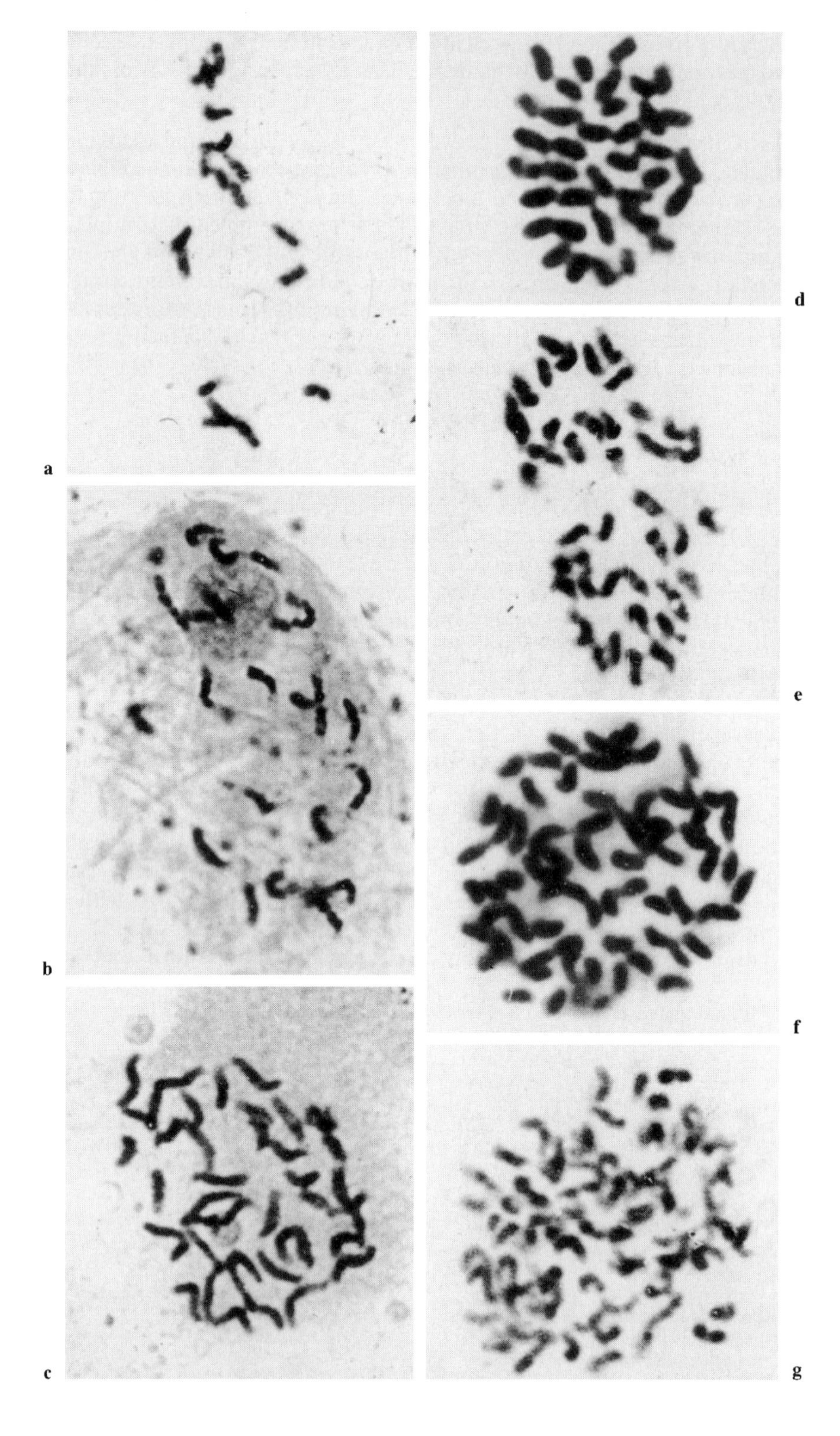
a
b
c
d
e
f
g

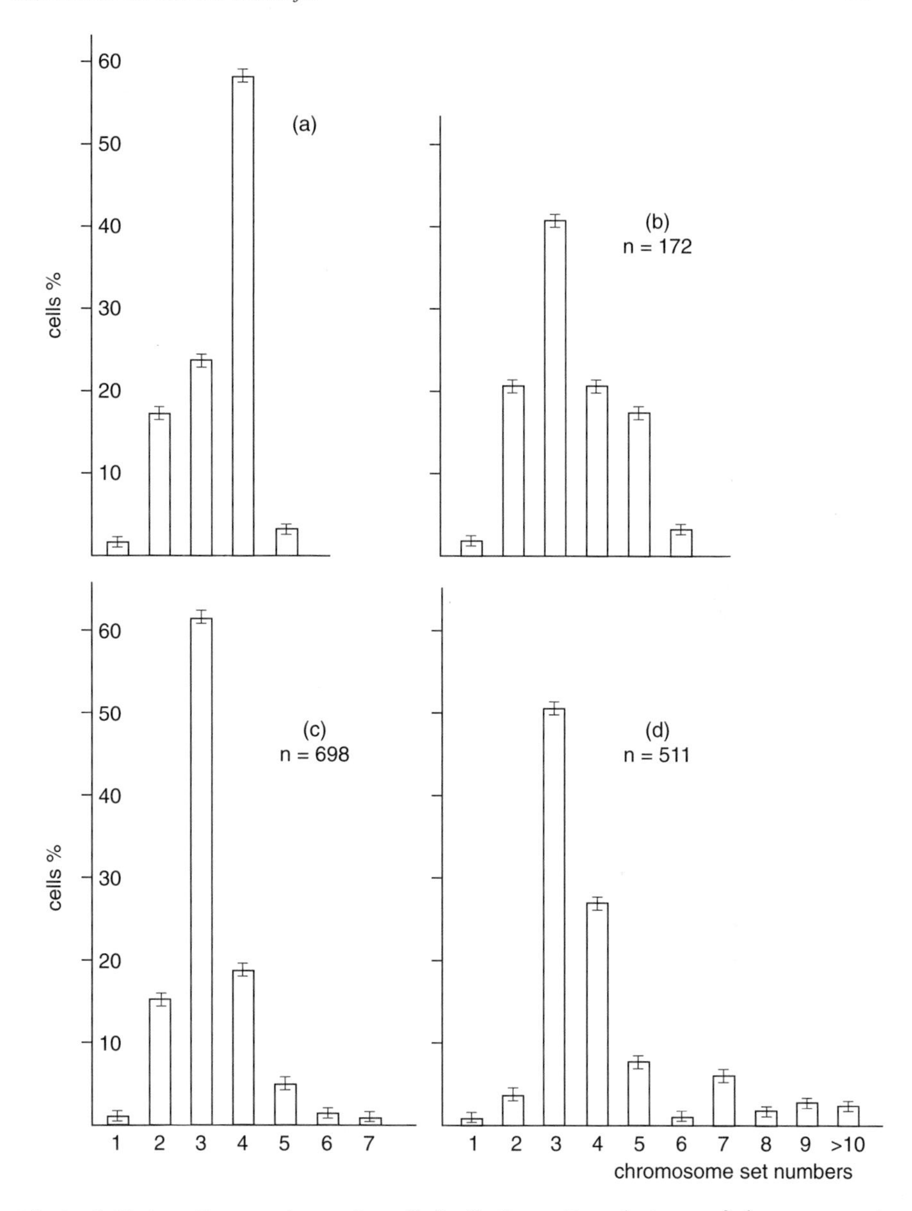


**Fig. 2a–d.** Various *R. serpentina* strains cell distribution patterns in terms of chromosome set number. **a** Cell line A. **b** Suspension R-III strain. **c** K-20 strain. **d** K-27 strain. *n* Number of metaphases studied

**Fig. 1a–g.** Number of chromosomes in metaphase plates of *R. serpentina* cultured strains ($2n = 2 \times = 22$) **a** 16 chromosomes = hypodiploid. **b** 22 chromosomes = diploid. **c** 33 chromosomes = triploid. **d** 36 = chromosomes = hypertriploid. **e** 44 chromosomes = tetraploid. **f** 55 chromosomes = pentaploid. **g** Around 66 chromosomes = near hexaploid

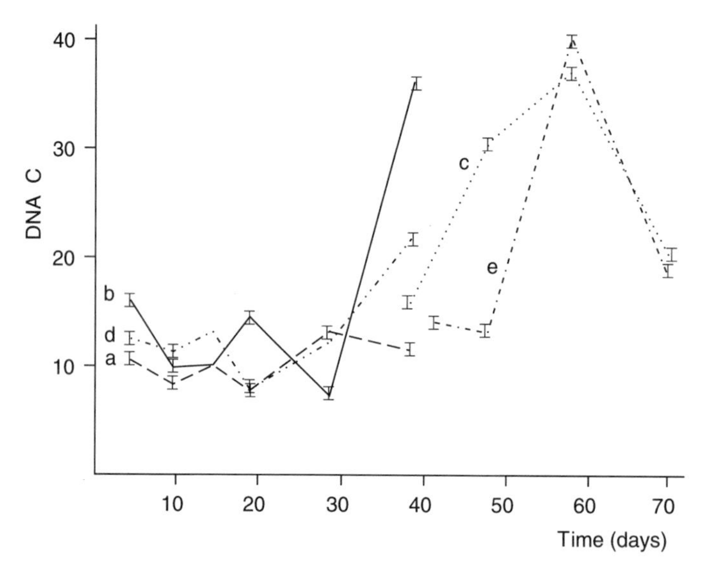


**Fig. 3.** Dynamics of mean DNA content per nucleus in *R. serpentina* cell line A (*a*) and its K-20 (*b*), K-27 (*c*), R-III (*d*), and A-10S (*e*) derivative strains

incidence rates for "fresh" chromosome breaks (see Kunakh 1984). The incidence of other abnormalities was somewhat lower, i.e., spindle irregularities, endomitoses, cytomixis, and extrusion of chromosome material. Occasionally, it exceeded 3%, but was usually within the range of 0.3–0.8%.

The cultures examined were also fairly heterogenous in terms of chromosome number, the cell line A providing a mixploid cell population to involve the tetraploid modal class. Other cell strains, were maintained on different nutrient media and originating from cell line A both through selection of spontaneous mutations (strain R-III and K-20) and as a result of chemical mutagenesis, proved to be stable mixploid populations. However, in each cell line A derivative strain mainly triploid cells contributed to its modal class, while the occurrence of polyploid cells containing more than the tetraploid chromosome number was markedly higher. More detailed results of cytological studies are provided by Kunakh et al. (1982), Kunakh (1984, 1994), and Kunakh and Alkhimova (1989).

### 3.3 Variability in DNA Amount

Examination of a 4-year-old *R. serpentina* plant demonstrated the diploid nature of the apical meristem and juvenile leaf cells; about 85% of the nuclei to be tested carried from 2C to 4C DNA. Cultured cells showed considerable variability in this trait. They carried from 1C to 27–30C DNA. Some individual nuclei of the K-20 strain carried 56C or more DNA on the 40th day of growth, while maximum DNA content (115C) was documented in a few nuclei of K-27 strain on the 60th day of growth.

Mean DNA content per nucleus during passage proved to be unsteady (Fig. 3). Although every individual strain displayed its variation in a distinct manner, on the whole, DNA content variations resembled the time course of highlyploid mitosis numbers during passage, as was described above for line A (see Kunakh and Alkhimova 1989). In the latter case, however, this process appeared to be shifted in time, occurring against the background of an almost complete absence of mitoses, and was more drastically pronounced.

Attention is called to the fact that a discrepancy exists between the results based on chromosome number count and those dealing with DNA content per nucleus in cultured cells. In particular, cells distinguished by maximum ploidy level carried 6 to 8, occasionally as many as 15 chromosome sets, the model class of the studied strains being composed of triploid and tetraploid cells. DNA amount per nucleus, however, comprised on average more than 10C, a significant proportion of nuclei containing 36–115C DNA.

To discover the reasons underlying this mismatch, line A, taken at the peak of its mitotic activities (on the 5th day of growth), was scrutinized for DNA amount in both interphase and prophase nuclei separately, and a chromosome number count was performed in metaphases of the same preparations. Results obtained so far suggest that the modal class among dividing cells at this time period was determined by triploid cells (Fig. 4a). Mean DNA content within the prophase nuclei was equal to $12.6 \pm 0.8$C (instead of the expected 6C), while that of the interphase nuclei was estimated to be $7.1 \pm 0.6$C (Fig. 4b, c). These data indicate that in cultured cells, as compared with the intact plant, amount of DNA per haploid genome appears to be ca. twofold (Kunakh et al. 1986).

It was assumed that one of the reasons for the increase in DNA amount per genome of cultured cells may be the magnification of certain sequences, as was found to be the case experimentally (Solovyan et al. 1987). Results of this work

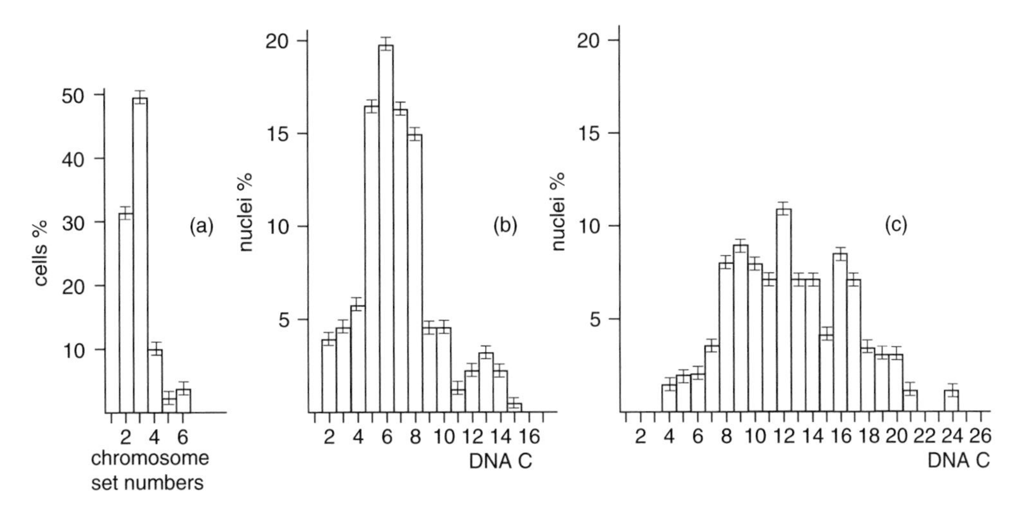


**Fig. 4a–c.** Line A distribution pattern on the 5th day of growth in terms of chromosome numbers (**a**) and DNA content per nucleus in interphase (**b**) and prophase (**c**) cells

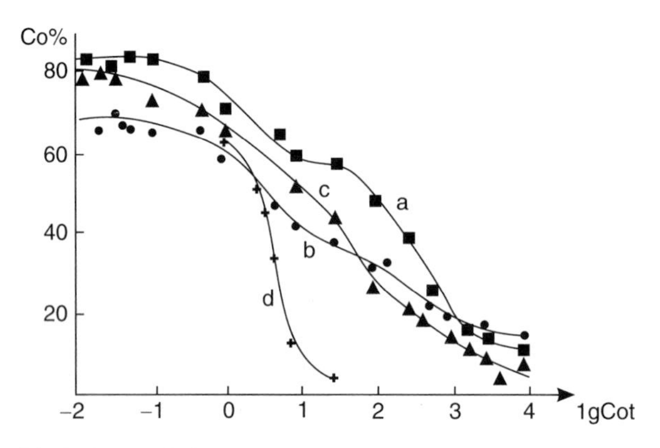


**Fig. 5.** Reassociation kinetics of *R. serpentina* total DNA intact plant. *b* K-20 strain; *c* cell line A; *d E. coli* DNA, taken for comparison

are summarized in Fig. 5 to show that DNA of intact plant and that of *Rauwolfia* cell strains seems to differ in reassociation rate, with callus DNA reassociation occurring more rapidly within the wide range of Cot variable values. This suggests a total increase in proportion shared by repeated sequences within the genomes of cultured cells.

Data given in Table 2 indicate that the proportion of sequences (Cot not more than 100) within the genomes of cultured cells increased as the ploidy level rose. Hence, in the case examined, genome multiplication as a whole (polyploidization) was accompanied by differential multiplication (endoreduplication or amplification) of repetitive DNA sequences.

## 3.4 Variability of DNA Sequences

Dot hybridizational analysis of DNA repeated sequences derived from the intact plant and cultured *R. serpentina* cells showed that only an insignificant proportion of intact plant sequences would hybridize with DNA of long-term-passaged in vitro cells (Fig. 6). At the same time, repeated sequences of primary callus genome and intact plant of *R. serpentina*, as well as DNA sequences from intact plant of other species, like *R. verticillata*, exhibit a rather high degree of hybridization between each other. Available evidence suggests that considerable

**Table 2.** The proportion of repeated sequences in the genomes of intact plant and cultured cells of *Rauwolfia serpentina*

| Genome | Ploidy modal class | Sequences (%) | |
|---|---|---|---|
| | | Cot up to 0.01 | Cot up to 100 |
| Intact plant | 2n | 18 ± 3 | 50 ± 2 |
| K-20 strain | 3n | 34 ± 5 | 68 ± 1 |
| Line A | 4n | 20 ± 1 | 73 ± 1 |

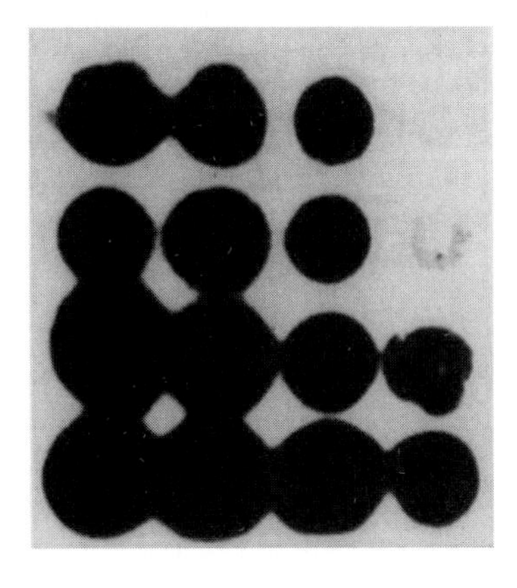

**Fig. 6.** Dot hybridization of total DNA from intact plant and cultured cells to involve fraction of 32P-labeled sequences from *R. serpentina* intact plant *a R. verticillata*; *b R. serpentina*; *c* primary callus of *R. serpentina*; *d* long-term–passaged *R. serpentina* cells (K-20 strain)

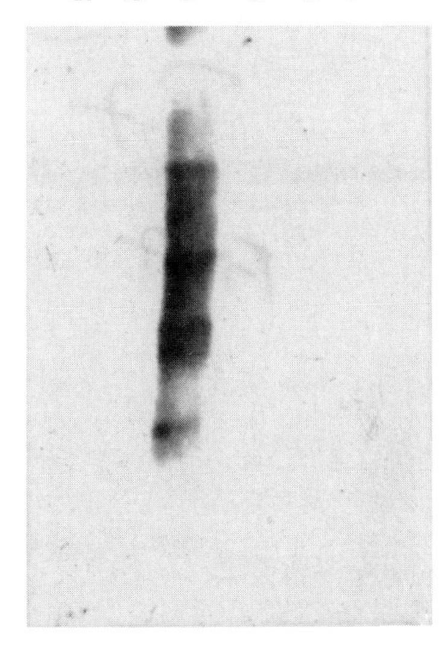


**Fig. 7.** Blot hybridization of DNA with 32-labeled fraction of immediately reassociating DNA (0-fraction), resulting from *R. serpentina* cell line A. *a*, *b*, *d–f* DNA from intact plants, respectively: *R. verticillata*, *R. serpentina*, *R. caffra*, *R. vomitoria*, *R. canescens*; *c* DNA from *R. serpentina* K-20 strain

rearrangements within the fraction of long-term-passaged *R. serpentina* cell genome presumably involve alterations of various repeated sequences. These changes seem to be multiple in nature, and in terms of magnitude, may exceed interspecies variability. To verify the validity of this suggestion, comparative blot-hybridizational analysis of *R. serpentina* cultured cell genomes and those of intact plants of other *Rauwolfia* species was carried out.

It was established that in the genomes of cultured *R. serpentina* cells, as opposed to intact plants of various *Rauwolfia* species, amplification of the so-called 0-fraction occurred, i.e., the fraction to be reassociated at zero Cot values and supposedly involving self-complementary DNA sequences (Fig. 7).

From the results presented in Fig. 8, it appears that in the genomes of cultured *R. serpentina* cells, a dramatic copy number depletion of the major class of rDNA sequences takes place, while the latter found in the genomes of intact plants of various *Rauwolfia* species remains relatively conservative. Data provided in Fig. 9 demonstrate another kind of variability that may not be due to amplification/deamplification of the genome sequences in cultured *R. serpentina* cells. Of all the *Rauwolfia* species examined, the rearrangements within a "genus-

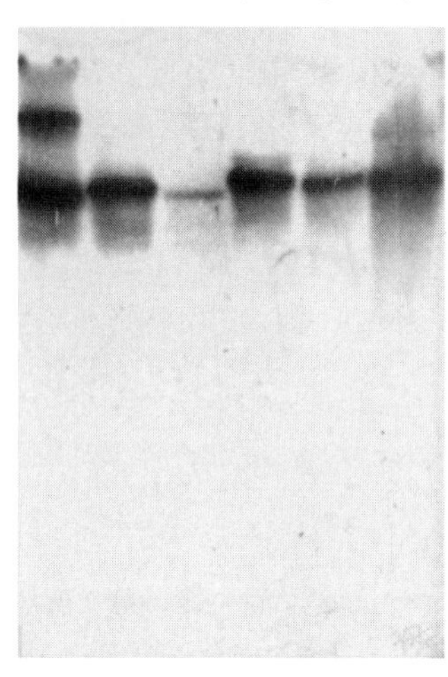


**Fig. 8.** Blot hybridization of DNA with 32P-labeled lemon ribosomal gene fragment. For explanations see Fig. 7

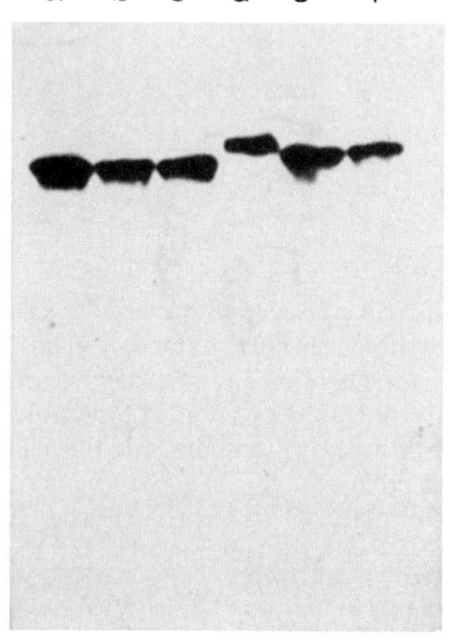


**Fig. 9.** Blot hybridization of DNA with 32P-labeled genus-specific AluI genome fragment derived from intact *R. caffra* plant *a–c*, *e–f* DNA from intact plants, respectively: *R. chinensis*, *R. caffra*, *R. serpentina*, *R. verticillata*, *R. canescens*; *d* DNA of *R. serpentina* K-20 strain

specific" fragment, resulting in the restriction fragment length polymorphism (RFLP), were restricted to *R. serpentina* callus genomes.

Thus, genome variability in cultured *R. serpentina* cells seems to be diverse in nature to affect various sequences and in terms of magnitude may exceed interspecies variability. These results and similar data, available for another plant, *Crepis capillaris*, are outlined by Solovyan et al. (1989, 1990, 1994a).

### 3.5 Relationship with Interspecies Genome Variations

Blot-hybridizational analysis of the genomes of various *Rauwolfia* species involving cytochromoxidase gene fragment as 32P-labeled probe revealed the RFLP among the species examined (Fig. 10), to be comparable with the alteration pattern found in the genomes of intact plants of *R. vomitoria* and *R. caffra*. Hence, it seems likely that a certain correlation exists between genome variability under in vitro conditions and naturally occurring genome variations among various *Rauwolfia* species. The probability of such a correlation was examined in detail (Solovyan et al. 1994b).

Endonuclease restriction analysis demonstrated the availability of two types of Alu-1-fragments within the genomes of intact plants (Fig. 11). One type of fragment (Fig. 11 thin arrow) remains unaffected among the species examined, while the other (Fig. 11 thick arrow) is responsible for the interspecies RFLP. The results of blot-hybridizational analysis to involve as a probe "species-related" Alu-1-fragment, derived from the intact *R. caffra* plant, support the data from restrictional analysis by showing changes not only in the genomes of intact plants of various *Rauwolfia* species, but also in the genome of cultured *R. serpentina* cells (Fig. 12). The changing pattern of the sequence in question, as seen in the genome of cultured *R. serpentina* cells, may be likened to its changes in the genome of intact *R. verticillata* plant.

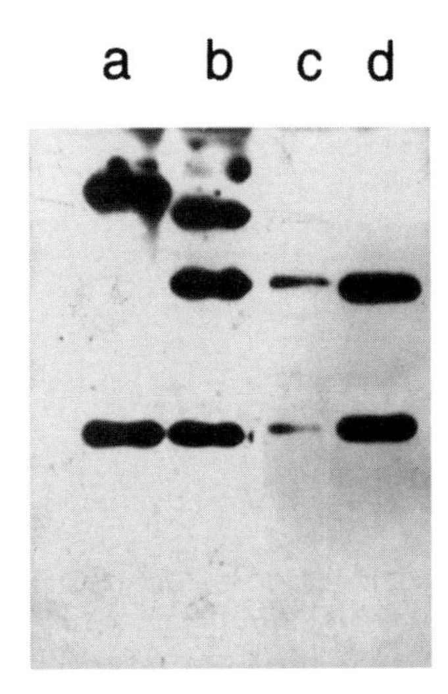


**Fig. 10.** Blot hybridization of DNA with 32P-labeled cytochromeoxidase gene fragment: *a, c, d* DNA from intact plants, respectively: *R. serpentina, R. caffra, R. vomitoria*; *b* DNA from *R. serpentina* K-27 strain

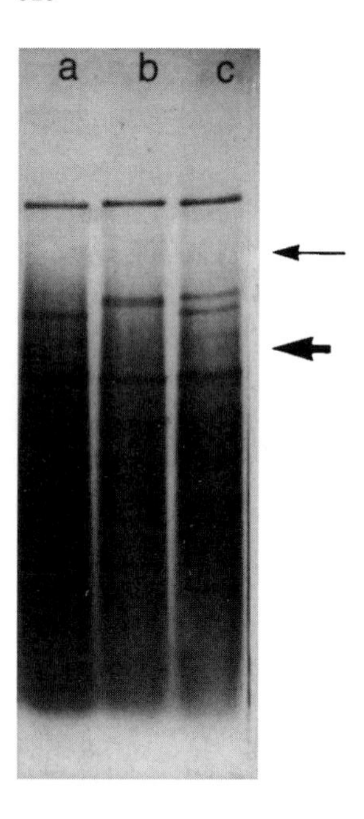


**Fig. 11.** AluI-restriction analysis of DNA from intact *R. verticillata* (*a*), *R. caffra* (*b*) and *R. serpentina* (*c*) intact plants. Genus-specific and species-related restriction fragments are marked by *thin* and *thick arrows*, respectively

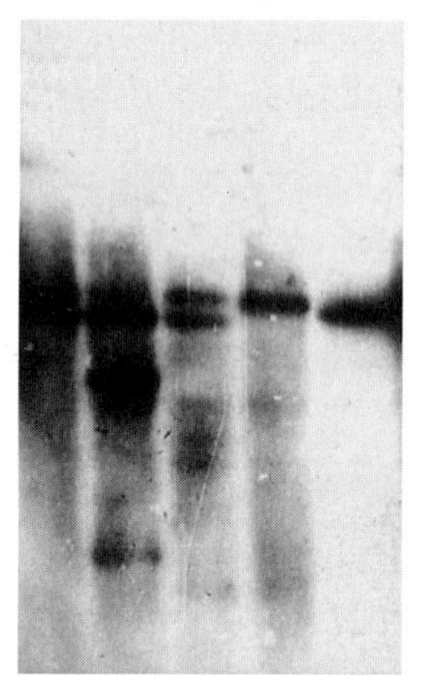


**Fig. 12.** Blot-hybridization of DNA with 32P-labeled species-related AluI-fragment derived from intact *R. caffra* plant; *a, b, d, e* DNA from intact *R. verticillata*, *R. serpentina*, *R. caffra*, *R. vomitoria* plants, respectively; *c* DNA from *R. serpentina* K-27 strain

As a result of cultured cell genome studies, the "callus-specific" E-coR1-restriction fragment was discovered. Blot-hybridizational analysis showed that genomic sequences bearing homology to the E-coR-1-fragment undergo changes not only in cultured *R. serpentina* cells but in the genomes of intact plants of other *Rauwolfia* species as well (Fig. 13), with the changes in pattern of the above

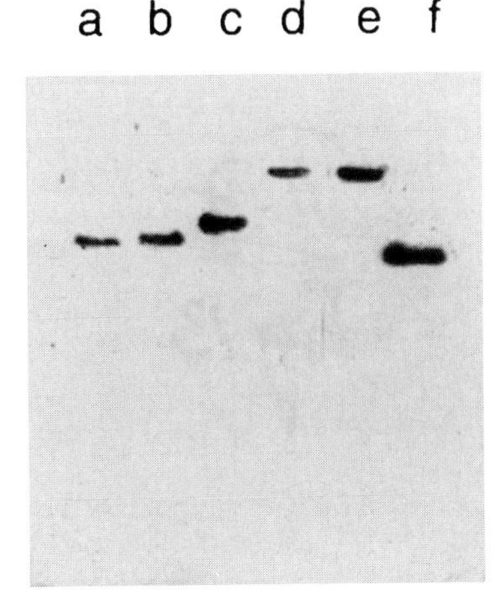


**Fig. 13.** Blot hybridization of DNA with 32P-labeled callus-specific EcoR1 fragment, derived from *R. serpentina* cell line A (EcoR1 digest). *a, b, c, d, e* DNA from *R. vomitoria, R. canescens, R. caffra, R. verticillata, R. serpentin* intact plants respectively; *f* DNA from *R. serpentina* K-27 strain

sequences in the callus genome of *R. serpentina* and *R. canescence* as intact plants being comparable.

These results, coupled with those involving other restrictases, are given in detail by Solovyan et al. (1994b).

### 3.6 Somaclonal Variation for Indoline Alkaloids

Based on a high level of *Rauwolfia* cell population heterogeneity, and taking into account the positive value of the heritability coefficient of the trait "indoline alkaloid contents" (see Kunakh, and Alkhimova 1989), generation of some clonal variants, i.e., cell clones differing by their productivity is theoretically conceivable. Such clones were established and characterized by Kunakh (1994).

Repeated attempts at deriving viable passaged clones from isolated protoplasts of callus and suspension cultures, originating from cell line A, failed. Clones were obtained from individual cells and cell aggregates using the suspension strain R-III.

The strains of cell suspension representing individual cells and small aggregates containing up to 20–25 cells were plated into Petri-dishes and grown in the darkness at 27 °C. Altogether, 16 variants of nutrient media were tested, with B5 medium being most successful. However, this medium also exhibited a low level of colony-forming capacity. It failed to exceed values of $10^{-5}$–$10^{-6}$. The colonies produced in the B5 medium were transferred to the phytohormone-free 5S medium (Vollosovich et al. 1979). A total of 32 clones capable of passaged growth in this medium were obtained, and were monitored for more than 5 years. The clones showed regular growth, but differed from each other by type and rate of growth, morphology, and the ability for alkaloid biosynthesis.

Table 3 gives some results from studies of total alkaloids and indoline alkaloids in ten different clones. It is obvious that alkaloid accumulation was low. Continuous selection for the rate of growth and that of intermittent supporting

**Table 3.** Alkaloid content in the clones derived from individual cells of *R. serpentina* R-III strain suspension culture

| Clone no. | Total alkaloids[a] | | | Indoline alkaloids | | |
|---|---|---|---|---|---|---|
| | 6th[b] | 21st | 46th | 6th | 21st | 46th |
| 4 | 1.94 | 1.23 | 0.80 | 0.16 | 0.26 | 0.21 |
| 10 | 1.17 | 1.17 | 0.90 | 0.09 | 0.08 | 0.39 |
| 14 | 1.25 | 0.95 | 1.21 | 0.20 | 0.33 | 0.22 |
| 16 | – | – | 0.90 | 0.19 | 0.07 | 0.04 |
| 20 | 1.27 | 1.30 | 0.90 | 0.07 | 0.09 | 0.10 |
| 27 | 1.03 | 1.01 | – | 0.05 | 0.08 | 0.05 |
| 28 | 1.10 | 1.13 | 1.04 | 0.14 | 0.03 | 0.04 |
| 29 | 1.08 | 1.02 | 0.80 | 0.07 | 0.35 | 0.30 |
| 30 | 1.01 | 0.95 | 0.93 | 0.38 | 0.24 | 0.38 |
| 31 | 1.35 | 1.40 | 1.34 | 0.10 | 0.40 | 0.71 |

[a] Total alkaloids and indoline alkaloids in starting R-III strain, respectively, 1.40 and 0.42%.
[b] 6th–46th = Passage numbers.

selection, in every five to six passages, for the trait "indoline alkaloid contents" proved ineffective. The selection somewhat increased productivity in only three of ten samples examined in detail. As a result, clones No. 10 and No. 29 brought their productivity closer to the level of a starting strain, while clone No. 31, showing a similar total alkaloid level, towards the completion of the investigation accumulated twice the amount of indoline alkaloids as the original R-III strain.

The low productivity of resultant clones may be explained on the basis that within the rather heterogenous cell population only cells showing low potential capacity for alkaloid biosynthesis were able to produce colonies. The proportion of such cells in the population, as observed earlier, was not large, somewhere at the level of spontaneous mutations ($10^{-5}$–$10^{-6}$).

The apparent contrast between the data on cloning and previously obtained results from the study of heterogeneity and heritability for indoline alkaloid contents (Kunakh and Alkhimova 1989) can be interpreted in this way. In the latter case, sections of the callus tissue under 1 g in weight were used; these involved a large number of cells. Such a number of cells available appeared sufficient to retain innate population features, and mechanisms for its genetic and physiological homeostasis. The antagonism between the processes of proliferation and those of secondary metabolism in this case was less pronounced than in clones generated from individual cells. Alternatively, the latter may restore the initial level of alkaloid biosynthesis only after prolonged selection, and then occasionally even exceed it.

Such an interpretation of the data available seems to contradict the French workers' view, who conducted similar studies on cultured *Choisya ternata* cells (Tremouillaux-Guiller et al. 1987). These authors believe that to ensure effective cell line selection for increased alkaloid contents, the approach based on protoplast cloning is more advantageous than that using cell aggregates for cloning, since cell aggregates were distinguished by an unstable alkaloid accumulation pattern. Based on personal experience, we find it possible to suppose that the structural unit of the examined culture as a biological system retaining its capacity for an increased biosynthesis of indoline alkaloids may not be an individual cell, but a multicellular aggregate with a living mass of at least 0.1 g.

Thus, the cloning of the cell line A and its derivatives in an effort to obtain more productive somaclonal variants in the above experiments proved to be rather elaborate. This may be due to the observation that line A represents a more complicated biological system than an ordinary cell cluster (Kunakh et al. 1982). The process of restoration of some innate features of the system and above all its ability for ajmaline oversynthesis following detachment of individual cells is sufficiently durable.

Another in vitro approach for increasing alkaloid contents is through somatic hybridization by protoplast fusion (Kostenyuk et al. 1994).

### 3.7 Induction of Regenerants

In callus tissues, especially towards the end of the passage, a great number of cell differentiation elements appeared—tracheids and sieve tubes (Kaukhova et al. 1981a). Conventional manipulation proved unsuccessful in inducing regeneration with these strains, derived from callus established in 1964. Previously, we had succeeded in inducing regeneration from long-term-passaged calli of various cultivars and genetic lines of *Pisum sativum*. The method was based on callus preconditioning to organogenesis that involved its maintenance on growth substance-deficient medium followed by alternating maintenance on deficient and normal medium. The essential factor of organogenesis with *P. sativum* was the occurrence of gibberellin in the medium, while abscisic acid was the prerequisite for organogenesis in long-term culture (Kunakh et al. 1984).

Following this approach, experiments were conducted to induce organogenesis in cell line A, K-20, and K-27 strains. Cells were maintained on media supplemented with kinetin (kin), IAA, gibberellic acid (GA), and abscisic acid (AA). Concentrations of each phytohormone were tested within the range of 0 and 30 mg/l, as was described earlier when cell line A productivity was examined (Kunakh and Alkhimova 1989). Using the Malyshe method of factorial experiment, a total of 25 versions of phytohormone combinations were screened. Shoot regeneration was induced as follows:

Tissue transfer on media, containing (mg/l); kin-3, IAA-3, GA-1, AA-30 or kin-1, IAA-0, AA-10. Maintain on these media in darkness for five passages followed by transfer on to initial medium with phytohormones omitted.

During the second to fourth passages, organogenesis developed as a sporadic shoot formation on phytohormones lacking media. The shoots were able at most to achieve a height of 3 to 5 cm. Any attempt at inducing or propagating plant root formation proved ineffective. In this regard, induction of somatic embryogenesis would be worthwhile (Tremouillaux-Guiller and Chenieux 1995).

## 4 Summary and Conclusions

*Rauwolfia* cells maintained in vitro result in a high level of structural and functional genome variability. In particular, this is shown by alterations in genome structure (chromosome number and morphology, DNA amount, variability of DNA sequences), changes in the protein spectrum, the alkaloid spectrum, and its amount. Taking advantage of this variability concomitant with mutagen application enabled the establishment of a wide range of cell strains resistant to antimetabolites and differing in terms of productivity, and cell strains among which overproducers accumulating up to 20% of ajmaline per dry mass, were singled out (Kunakh 1994).

Long-term passaging in vitro may result in rearrangements whose magnitude seems to exceed a naturally occurring interspecies variability. Alterations appear to be multiple in nature, affect various DNA sequences, and may include amplification and copy number depletion, as well as other kinds of genome rearrangements contributing to RFLP occurrence. Alternatively, the changes in some sequences within the genome of cultured *R. serpentina* cells appear to follow those that are characteristic of other *Rauwolfia* species found in nature. It seems likely that there is a certain parallelism between changes arising in nature as a result of species evolution and those occurring in vitro. What is the incidence of "parallel" changes, and to what extent they may be expressed by plant regenerants, is the question for future investigations.

*Acknowledgments.* The author thanks Dr. Victor Solovyan for kindly providing some illustrations and Dr. Vladimir Adonin for assistance in translating the text into English.

## References

Akram M, Ilahi I, Mirza MA (1990) In vitro regeneration and field transfer of Rauwolfia plants. Pak J Sci Ind Res 33: 270–274

Antipova EA, Nikolaeva LA, Bajelidze ASh (1988) Comparative characteristics of alkaloid content and composition among some *Rauwolfia* L. species. Restit Resur 24: 575–578

Butenko RG (1964) Isolated tissue culture and physiology of plant morphogenesis. Nauka, Moscow

Granda LMM (1987) Informe sobre la introduccion de plantas medicinales exoticas: *Rauwolfia vomitoria* Afzel. Rev Cubana farm 21: 187–193

Janaki A (1962) Tetraploidy in *Rauwolfia serpentina*. Curr Sci 31: 520–521

Kaukhova IE, Kunakh VA, Legeida VS, Vollosovich AG (1981a) Cytological study of high-productive cell line of *Rauwolfia serpentina* Benth in submerged culture. Tsitol Genet 3: 33–37

Kaukhova IE, Vollosovich AG, Tsigankov VA (1981b) Nutrient medium selection for submerged culturing in *Rauwolfia serpentina* Benth tissues. Rastit Resur 17: 217–224

Kaul MLH (1963–1964) Natural polyploidy in *Rauwolfia serpentina*. J Sci Res Banaras Hindu Univ 14: 100–102

Kostenyuk I, Lubaretz O, Borisyuk N, Voronin V, Stockigt J, Gleba Y (1991a) Isolation and characterization of intergeneric somatic hybrids in the Apocynaceae family. Theor Appl Genet 82: 713–716

Kostenyuk IA, Lubaretz OF, Voronin VV, Gleba YY (1991b) Cell engineering developed for Apocynaceae plant biotechnology. Biopolymeri Kletka 7: 26–34

Kostenyuk IA, Lubaretz OF, Endress S, Stockigt J, Gleba YY (1994) Somatic hybridization in the family Apocynaceae (*Catharanthus*, *Rauwolfia*, *Rhazya*, and *Vinca* species). In: Bajaj YPS (ed) Biotechnology in agriculture and forestry, vol 27. Somatic hybridization in crop improvement I. Springer, Berlin Heidelberg New York, pp 405–424

Kovaleva TA, Shamina ZB, Butenko RG (1972) Effect of nitrogen mustard on isolated tissue culture of *Rauwolfia*. Genetika 8: 46–54
Kunakh VA (1984) Peculiarities of structural mutagenesis in populations of cultured plant cells. In: Dubinin NP (ed) Uspekhi Sovrem Genetiki, vol 12, Nauka, Moscow, pp 30–62
Kunakh VA (1994) Genome variability and accumulation of indoline alkaloids in *Rauwolfia serpentina* Benth cell culture. Biopolymeri Kletka 10: 3–30
Kunakh VA, Alkhimova EG (1989) *Rauwolfia serpentina*: in vitro culture and the production of ajmaline. In: Bajaj YPS (ed) Biotechnology in agriculture and forestry, vol 7. Medicinal and aromatic plants II. Springer, Berlin Heidelberg New York, pp 398–416
Kunakh VA, Gubar SI (1993) Rauwolfia: strategy for generation of cell strains-producents of alkaloids. In: Skryabin KG (ed) Plant biotechnology and molecular biology. Abst 2nd Symp, Pushchino, Russia, 382 pp
Kunakh VA, Kaukhova IG, Alpatova LK, Vollosovich AG (1982) Peculiarities of cell behavior in tissue culture of *Rauwolfia serpentina* Benth. Tsitol Genet 5: 6–10
Kunakh VA, Voityuk LI, Alkhimova EG, Alpatova LK (1984) Callus tissue formation and induction of organogenesis in *Pisum sativum* L. Fiziol Rast 31: 542–543
Kunakh VA, Kostenyuk IA, Vollosovich AG (1986) An increase in the amount of nuclear DNA during biosynthesis of alkaloids in the culture of *Rauwolfia serpentina* Benth tissue. Dokl Acad Nauk Ukr SSR Ser B 7: 62–65
Martinez PJA (1983) Distribucion de alcaloides indolicos en species del genero *Rauwolfia*. Boll Resen CIDA Plant Med 7: 3–58
Mitra GC, Kaul KN (1964) In vitro culture of root and stem callus of *Rauwolfia serpentina* Benth for reserpine. Indian J Exp Biol 2: 49–51
Ohta S, Yatazawa M (1979) Growth and alkaloid production in callus tissues of *Rauwolfia serpentina*. Agric Biol Chem 43: 2297–2303
Roja G, Benjamin BD, Heble MR, Patankar AV, Sipahimalani AT (1990) The effect of plant growth regulators and nutrient conditions on growth and alkaloid production in multiple shoot cultures of *Rauwolfia serpentina*. Phytother Res 4: 49–52
Singh JN, Nand K (1989) Effect of nitrogen and phosphorus on yield and quality of sarpaganha (*Rauwolfia serpentina* Benth). Indian Drugs 26: 461–464
Singh M, Singh A, Singh DV (1990) Effect of length and thickness of *Rauwolfia serpentina* root-cuttings on plant stand and root yield. Indian Drugs 27: 483–485
Solovyan VT, Kostenyuk IA, Kunakh VA (1987) Genome changes in *Rauwolfia serpentina* Benth cells cultivated in vitro. Genetika 23: 1200–1208
Solovyan VT, Popovich VA, Kunakh VA (1989) Genome rearrangements in *Crepis capillaris* L (Wallr) cultured cells. Genetika 25: 1768–1775
Solovyan VT, Zakhlenjuk OV, Kunakh VA (1990) *Rauwolfia* genome rearrangements during culturing in vitro. Biopolymeri Kletka 6: 103–106
Solovyan VT, Spiridonova EV, Kunakh VA (1994a) Genome rearrangements in cell cultures of *Rauwolfia serpentina*: diverse pattern of genome variations. Genetika 30: 250–254
Solovyan VT, Spiridonova EV, Kunakh VA (1994b) Genome rearrangements in cultured *Rauwolfia serpentina* cells. 11. Relation to interspecific genome variation. Genetika 30: 399–403
Stockigt J (1988) Alkaloidbiosynthese in *Rauwolfia*. Heilmittel mit 3000-jahriger Tradition. GIT 32: 608–615
Stockigt J, Pfitzner A, Firl J (1981) Indoline alkaloids from cell suspension cultures of *Rauwolfia serpentina* Benth. Plant Cell Rep 1: 36–39
Tremouillaux-Guiller J, Chenieux JC (1995) Somatic embryogenesis from leaf protoplasts of *Rauwolfia vomitoria* Afz. In: Bajaj YPS (ed) Biotechnology in agriculture and forestry, vol 31. Somatic embryogenesis and synthetic seed II. Springer, Berlin Heidelberg New York, pp 357–370
Tremouillaux-Guiller J, Andreu F, Creche J (1987) Variability in tissue cultures of *Choisya ternata*. Alkaloid accumulation in protoclones and aggregate clones obtained from established strains. Plant Cell Rep 6: 375–378
Upadhyay N, Batygina TB (1992) Development of plantlets from cultured tissues of *Rauwolfia canescens* L. In: Proc 11 Int Symp Embryol and Seed Repord, St Petersburg, pp 578–579
Vollosovich AG (1989) Some peculiarities of alkaloid accumulation in tissue culture of *Rauwolfia*

*serpentina*. In: Highlights mod Biochem. Proc 14th Int Congr Biochem (Prague, 10–15 July 1988). Utzecht, Tokyo, 1989, Vol 2, pp 1177–1182

Vollosovich AG, Butenko RG (1970) Tissue culture of *Rauwolfia serpentina* as a resource of alkaloids. In: Butenko RG (ed) Culture of isolated organs, tissues and cells of plant. Nauka, Moscow, pp 253–257

Vollosovich AG, Nikolaeva LA, Zharko GR (1972) Tissue culture of some medicinal plants from *Rauwolfia* genus. Rastit Resur 8: 331–338

Vollosovich AG, Puchinina TN, Nikolaeva LA (1979) Optimization of the composition of macrosalts for a *Rauwolfia serpentina* Benth tissue culture. Rastit Resur 15: 516–528

Vollosovich AG, Puchinina TN, Lisunova NA (1982) Optimization of the composition of macrosalts for a Rauwolfia tissue culture. Restit Resur 18: 239–243

Vollosovich NE, Vollosovich AG, Kovaleva TA, Shamina ZB, Butenko RG (1976) Strains of a *Rauwolfia serpentina* Benth tissue culture and their productivity. Rastit Resur 12: 578–583

Yamamoto O, Yamada Y (1986) Production of reserpine and its optimization in cultured *Rauwolfia serpentina* Benth cells. Plant Cell Rep 5: 50–63

# II.8 Somaclonal Variation in *Scilla scilloides* Complex

J.W. Bang[1] and H.W. Choi[1]

## 1 Introduction

### 1.1 Botany and Distribution of the Plant

*Scilla scilloides* Complex of the family Liliaceae, a bulbous perennial plant with horticultural value, is widely distributed in Far East Asia: Korea, mainland China, and Japan (Noda 1976; Araki 1985; Yu and Araki 1991). It can be used as plant material for studying genomic stability in cultures, since its chromosome complement consists of two kinds of genomes, A ($x = 8$) and B ($x = 9$). The largest metacentric chromosome $a_1$ is the marker of the AA genome.

In Korean natural populations, ten cytogenetic types have been found within a single species of *S. scilloides* Complex (Araki 1972, 1985; Noda and Lee 1980; Choi and Bang 1990; Bang and Choi 1991, 1993). They are: AA ($2n = 16$), BB ($2n = 18$), ABB ($2n = 26$), AAAB ($2n = 33$), AABB ($2n = 34$), ABBB ($2n = 35$), BBBB ($2n = 36$), AABBB ($2n = 43$), AAABBB ($2n = 51$), and AAAABBBB ($2n = 68$).

B-chromosomes have been described in natural populations of *S. scilloides* Complex (Haga 1961). They are classified into two groups, F, which is metacentric, and f, which is telocentric B-chromosome (Haga 1961). The frequencies of B-chromosomes among Korean natural populations differ depending upon the cytogenetic types (Choi and Bang 1990; Bang and Choi 1991).

### 1.2 Genetic Variability and Objectives

Many bulbous plants such as lily, hyacinth, and *Ornithogalum* seldom mutate after adventitious shoot formation (Pierik 1987), and mutation often appears during callus, cell, and protoplast cultures. It has been known for some time that growth in a callus phase can be associated with chromosome stability (D'Amato 1977). In the genus *Scilla*, chromosomal changes in cultured cells were reported in diploid *S. indica* (Charkravarty and Sen 1983), and *S. siberica* (Deumbling and Clermont 1989), but there are no reports showing the chromosomal stability in the callus cells derived from different cytogenetic type plants. Furthermore, these results were obtained in autosomes not including the B chromosome. The

[1]Department of Biology, College of Natural Sciences, Chungnam National University, Daejon 305–764, Korea

Biotechnology in Agriculture and Forestry, Vol. 36
Somaclonal Variation in Crop Improvement II (ed. by Y.P.S. Bajaj)

variation of B chromosomes in callus cells was reported only in *Secale cereale*, where the selective function of B chromosomes in cultures was proposed (Asami et al. 1976). B chromosome frequency is highly maintained in the natural populations of *S. scilloides*. When plants with B chromosomes are applied to cultures, the behavior and effect of the B chromosome can be elucidated.

However, little is known about the behavior of different explants within a given genotype. *S. scilloides* Complex has been shown to have various cytogenetic types, which are composed of two well-differentiated genomes, A $(x = 8)$ and B $(x = 9)$; (Araki 1971; Haga and Noda 1976). The stability of the A and B genomes in cultured cells may differ depending on the cytogenetic type. Therefore in this chapter in vitro culture, plant regeneration, and chromosomal variation in cultured cells of *S. scilloides* Complex with different genome compositions are summarized. The genetic effect of B chromosomes on autosomes during cultures is also discussed.

## 2 In Vitro Culture and Plant Regeneration

### 2.1 In Vitro Culture

Cytogenetic types were analyzed from somatic cells of the root-tips of *S. scilloides* Complex (Choi and Bang 1990; Bang and Choi 1991), in which types AA $(2n = 16)$, BB $(2n = 18 + 2F + 2f)$, AABB $(2n = 34)$ and BBBB $(2n = 36 + 1F + 1f)$ plants were selected for culture. Plants of diploid BB and eutetraploid BBBB were chosen to clarify the behavior and effect of the B-chromosome during cultures.

For tissue culture, the bulbs were washed with 70% ethanol for 30 s, rinsed in sterilized distilled water (SDW) three times, sterilized in 3% sodium hypochlorite solution for 10 min followed by washing with SDW. Bulb scale leaves were dissected into $5 \times 5$ mm segments and placed on MS medium (Murashige and Skoog 1962) containing 2 ml/l 2,4-D, 2 ml/l NAA, 2 ml/l BAP and 3% (w/v) sucrose.

The cultures were maintained in the growth room at a temperature of $25 \pm 1$ °C under 4000 lx for a 16-h photoperiod followed by 8 h dark.

### 2.2 Plant Regeneration from Embryogenic Callus

Callus initiated from the surfaces of the bulb segments after 2 weeks of culture. The first change noted after the bulb segments were placed on the medium was swelling. Two kinds of calli were found; a callus yellow in color and compact in shape, and a white and friable one. Embryos initiated from the former (Fig. 1A), no embryos were formed from the latter. Calli containing somatic embryos were transferred to fresh MS medium within 3 weeks of initiation. Shoots were developed on the same medium used for callus induction (Fig. 1B) and multiple shoots in dense clusters were found in over 90% of the cultures (Fig. 1C). Shoots were transferred to hormone-free medium. This subculture resulted in root

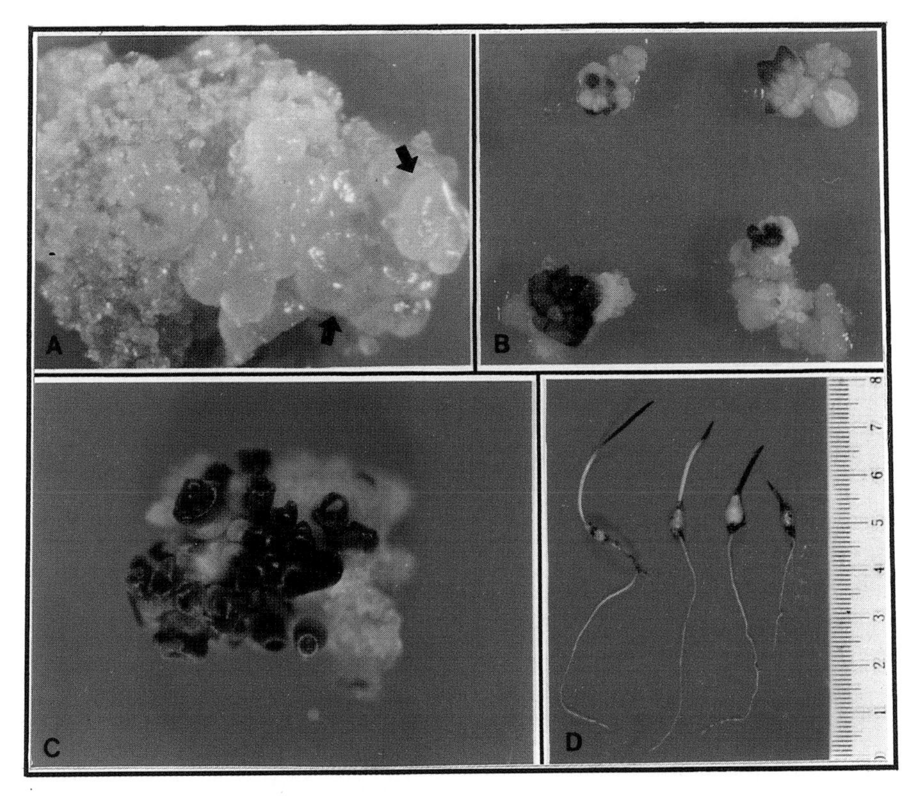


**Fig. 1A–D.** Plant regeneration via somatic embryogenesis from bulb segment culture of *S. scilloides* Complex. **A** Embryogenic callus initiated from bulb segment culture. *Arrows* indicate somatic embryos. **B, C** Subcultured embryogenic calli and developing shoots. **D** Regenerated plantlets

development within 2 weeks. Regenerated plants were successfully transferred to soil and greenhouse conditions. New bulbs were developed from the bases of the plantlets in soil (Fig. 1D). Whole plantlets were obtained in 3 months of culture.

# 3 Somaclonal Variation

## 3.1 Morphological and Chromosomal Variation in Regenerated Plants

Regenerated plantlets grown in culture media were very different in size and morphology (Fig. 2A). They had different numbers of leaves (one to five) and bulb shapes after habituation in soil (Fig. 2B). In cytogenetic analysis, all the regenerated plants except one aneuploid in type AABB had the autosome numbers as the original explants (Table 1).

All regenerated plants from the type AA plant carried normal chromosome complements and normal morphology (Fig. 3). It is suggested that regenerated

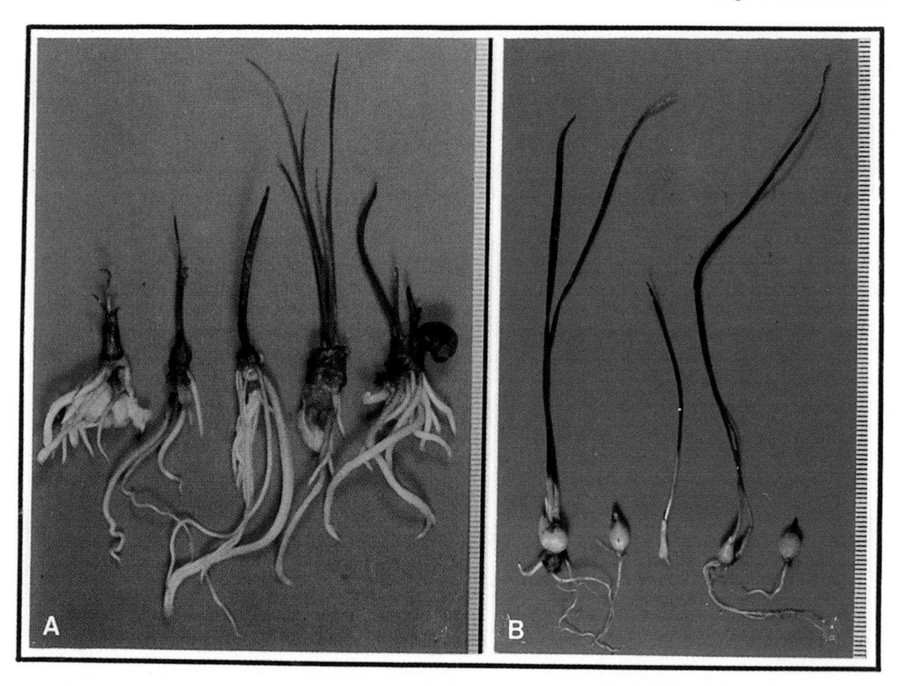


**Fig. 2A, B.** Phenotypic variation of regenerated plants developed from the bulb segment culture of type AABB (2n = 34). **A** Regenerated plants grown in MS medium. **B** Whole plantlets habituated in soil

**Table 1.** Chromosome variation in regenerated plants derived from embryogenic callus of *S. scilloides* Complex

| Cytogenetic types of original explants | Chromosome nos. in regenerants | No. of regenerants investigated |
|---|---|---|
| AA (2n = 16) | 16 | 24 |
| | | Total 24 |
| | 18 | 2 |
| BB (2n = 18 + 2F + 2f) | 18 + 2F + 2f | 56 |
| | | Total 58 |
| | 34 | 30 |
| | 34 + 1s | 30 |
| AABB (2n = 34) | 34 + 2s | 1 |
| | 34 + 1s | 4 |
| | 43 + 1s | 1 |
| | | Total 66 |
| | 36 | 48 |
| BBBB (2n = 36 + 1F + 1f) | 36 + 1F + 1f | 8 |
| | | Total 56 |

plantlets develop from cells which are cytogenetically normal and stable. Cells with chromosomal changes are not competent for regeneration (D'Amato 1977; Deumbling and Clermont 1989).

Autosomes in type BB regenerants were quite stable (2n = 18), while the elimination of B chromosomes were found in two plants (Fig. 4).

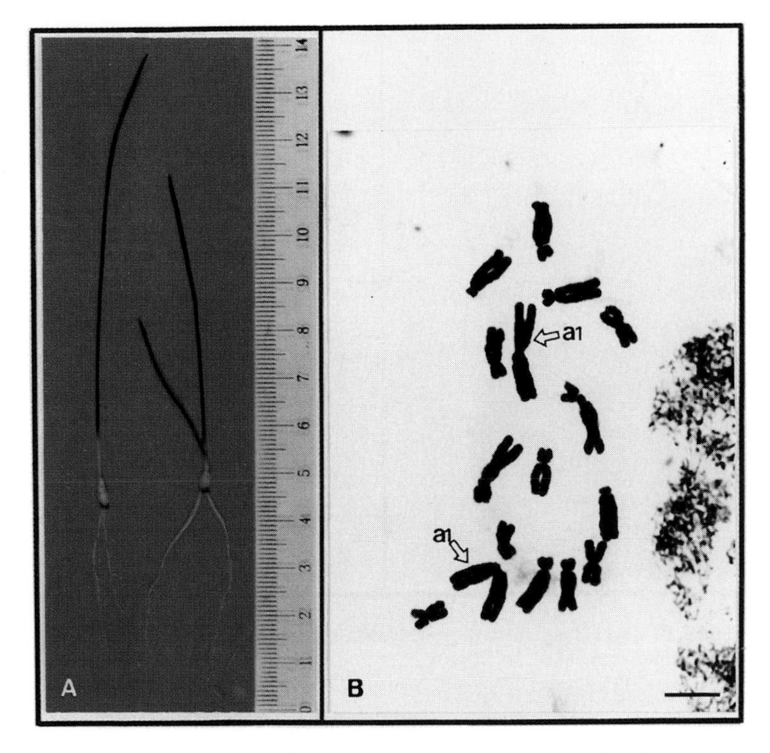


**Fig. 3A, B.** Regenerated plants developed from bulb segment culture of type AA and its chromosome complement (2n = 16). $a_1$ Marker chromosome of AA genome; *bar* 10 µm

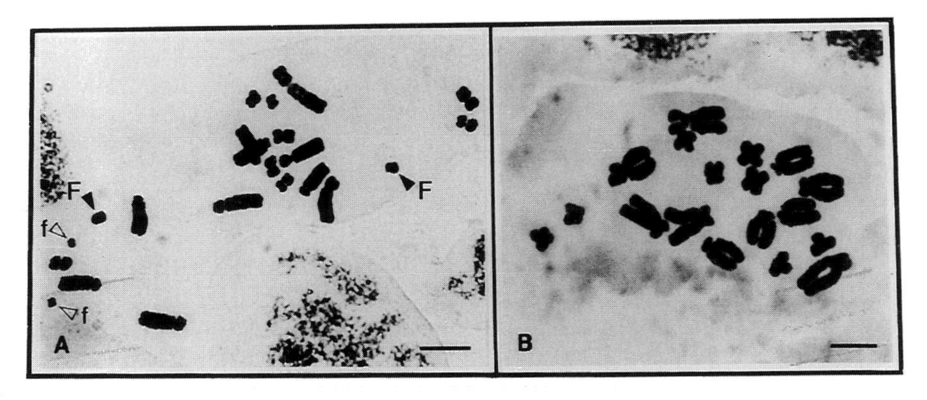


**Fig. 4A, B.** Chromosome complements in regenerated plants developed from bulb segment culture of type BB (2n = 18 + 2F + 2f). **A** Normal chromosome complement (2n = 18 + 2F + 2f). **B** 2n = 18 showing loss of B chromosomes. *F* metacentric, *f* telocentric B chromosomes; *bars* 10 µm

It has been described that the karyotype in alloploidy might be unstable during culture and that aneuploidy frequently occurred (Armstrong et al. 1983). In *S. scilloides*, autosomes were well maintained in allotetraploid AABB regenerants. Chromosome segments which were not observed in the original explants

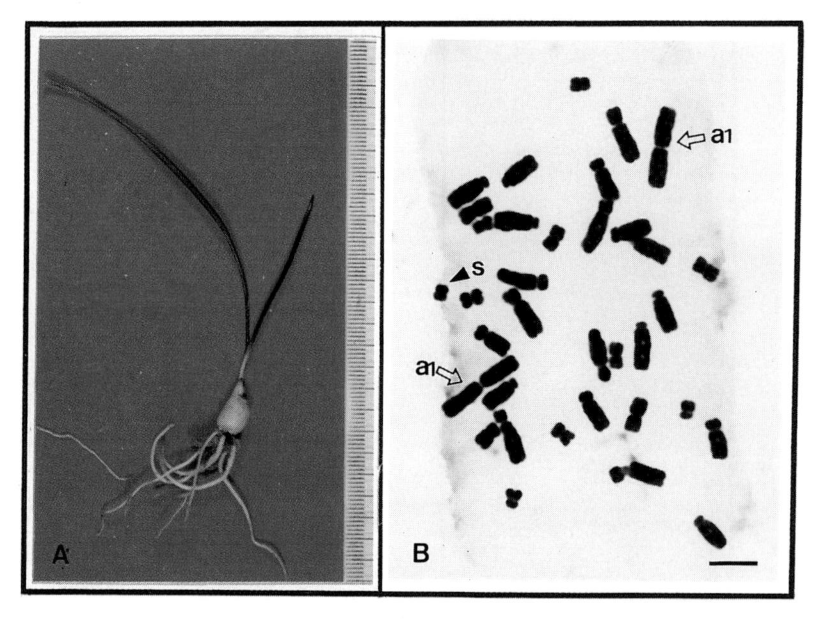


**Fig. 5A, B.** Regenerated plant of type AABB (**A**) and its chromosome complement (**B**) showing one chromosome segment (*s*); *bar* 10 μm

became visible (Fig. 5). This might supply a clue for the origin of the B chromosome. Plants with extra chromosome segments were indistinguishable from regenerants with comparable chromosome numbers. One aneuploid was found in type AABB regenerants that carried one chromosome segment (Table 1). Chromosome segments in regenerated plants were reported in celery (Orton 1985). The frequency of chromosomal variations in regenerated plants is lower than that in callus cells (Table 2; Bang et al. 1994).

Phenotypic variation was found in regenerated plants. In wild plants of *S. scilloides*, flower stalks emerged from the base after leaf shedding, while plants with flowers accompanied by leaves were found in the regenerated plants. A difference in anthesis was also observed (Larkin et al. 1984).

All the regenerated plants developed from the eutetraploid BBBB carried normal autosome numbers. However, a considerable loss of B chromosome was observed. Only 8 of 56 plants had the same number of B chromosomes as the original explants (Table 1).

Morphological variation was observed amongst regenerated plants which could, in part, be related to chromosome number. Some of this variation is due to changes in the chromosomes (Creissen and Karp 1985). These changes are largely undesirable, but not all variation in chromosomes is undesirable. Morphogenetic potential is related to the degree of chromosome variation (Murashige and Nakano 1967). In *S. scilloides*, autosome numbers were normal in all of the regenerant plants except one aneuploid, though variation in phenotype has been found.

**Table 2.** Chromosome variation in callus cells derived from different cytogenetic type plants of *S. scilloides* Complex. (Bang et al. 1994)

| Cytogenetic types of original explants | Chromosome nos. in callus cells | No. of cells investigated |
|---|---|---|
| | 7 | 2 |
| | 12 | 2 |
| | 13 | 4 |
| | 14 | 7 |
| | 15 | 13 |
| AA (2n = 16) | | |
| | 16 | 245 |
| | 16 + 1s | 1 |
| | 16 + 2s | 2 |
| | 16 + 3s | 1 |
| | | Total 277 |
| | 18 | 1 |
| | 18 + 1F + 3f | 1 |
| | 18 + 2F | 8 |
| BB (2n = 18 + 2F + 2f) | 18 + 2F + 1f | 4 |
| | 18 + 2F + 2f | 57 |
| | 18 + 2F + 3f | 6 |
| | 18 + 3F + 1f | 1 |
| | | Total 78 |
| | 19 | 2 |
| | 22 | 1 |
| | 31 | 1 |
| | 32 | 1 |
| AABB (2n = 34) | 33 | 1 |
| | 34 | 58 |
| | 34 + 1s | 2 |
| | 34 + 2s | 1 |
| | | Total 67 |
| | 29 | 1 |
| | 36 | 50 |
| | 36 + 1f | 3 |
| BBBB (2n = 36 + 1F + 1f) | 36 + 1F + 1f | 8 |
| | 48 | 1 |
| | 68 | 1 |
| | 77 | 1 |
| | | Total 65 |

## 3.2 Karyotypic Changes in Cultured Cells

The cytological examination of callus cells followed Bang's method (1990). Actively growing calli were subcultured on fresh medium. For cytological analysis, a small amount of subcultured callus was placed in the refrigerator (5 °C) overnight, soaked in saturated 1-bromonaphthalene for 5–6 h, then fixed in acetic alcohol (1:3). The squash method was adopted for chromosome preparation.

Considerable chromosome loss and karyotypic rearrangement occurred in callus cells derived from type AA plants (Table 2; Bang et al. (1994). The chromosome complement of AA consists of one pair of V-shaped large metacen-

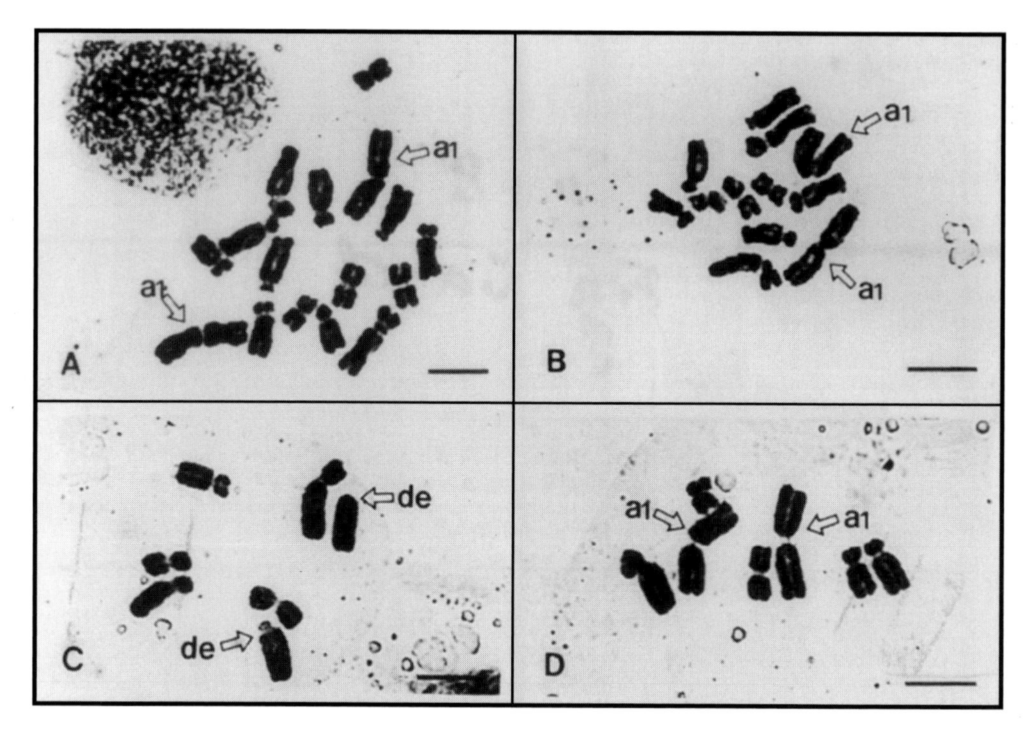


**Fig. 6A–D.** Chromosome variation in the callus cells derived from type AA plant (2n = 16). **A** 2n = 16. **B** 2n = 12. **C** 2n = 7. **D** 2n = 7. $a_1$ Marker chromosome of AA genome; *de* deficient chromosome; *bars* 10 μm

tric ($a_1$), five pairs of I-shaped subtelocentric ($a_2$ to $a_6$) and two pairs of v-shaped small metacentric ($a_7$ to $a_8$). The $a_2$ chromosome has secondary constriction (Araki 1971; Choi and Bang 1990). Elimination of chromosomes occurred sporadically in all homologous pairs of type AA callus (Fig. 6). Loss of one homologous chromosome was a common phenomenon. Loss of a whole homologous pair was found in chromosomes $a_1$ (Fig. 6B) and $a_4$ (Fig. 6D). A deficiency in chromosome arms was also found (Fig. 6C). Structural changes such as dicentric due to centromeric shift and breakage of the centromeric region of $a_1$ were observed (Fig. 7). The autosomal stability in callus cells of type AA was much lower than that in any other types (Table 2; Bang et al. 1994).

Original explants of type AA applied to culture carried no B-chromosome. However, chromosome segments (s) were observed in 4 amongst 277 cells (Table 2). Occurrence of chromosome segments, which were not found in the mother plants, was reported in *Vicia faba* (Jha and Roy 1982). Chromosome segments newly appearing in cultured cells might supply a clue for the origin of the B-chromosome.

No autosomal variation was detected in type BB, while numerical variation in B-chromosomes was found (Table 2). The chromosome complement of BB consists of five pairs subtelocentric ($b_1$ to $b_5$), four pairs metacentric ($b_6$ to $b_9$),

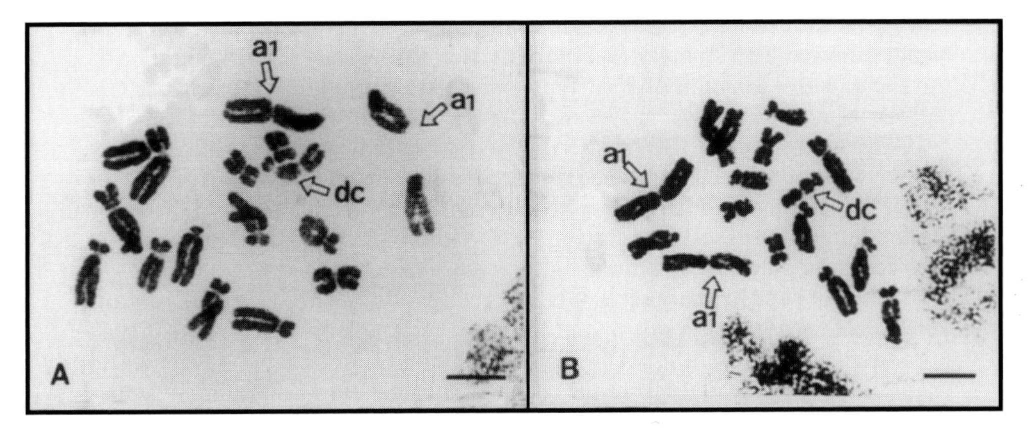


**Fig. 7A, B.** Structural changes of chromosome in the callus cells of type AA. Breakage in chromosome $a_1$(**A**), and dicentric chromosomes are showing (**A, B**). *dc* Dicentric; *bars* 10 μm

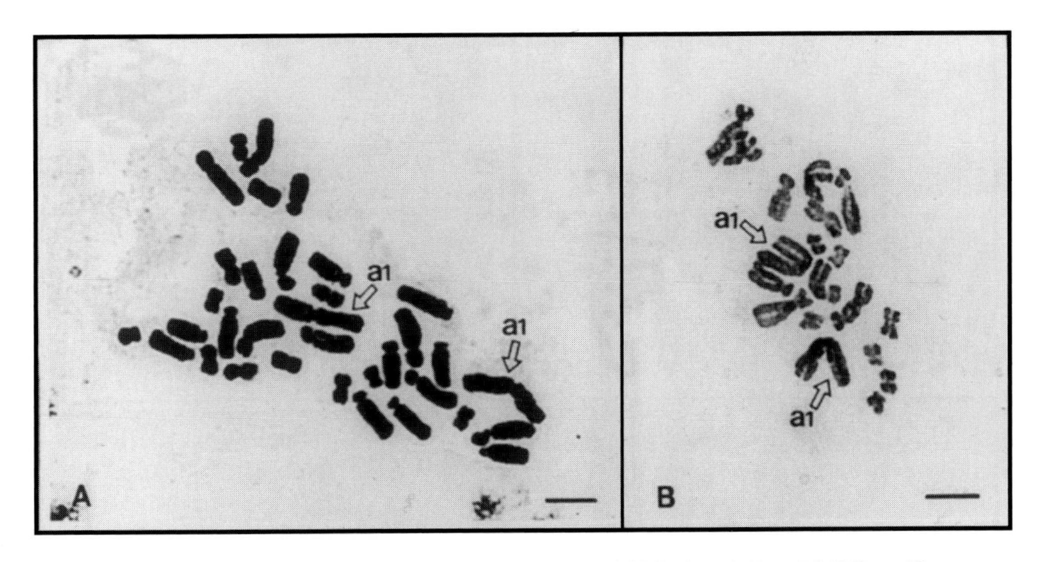


**Fig. 8A, B.** Karyotypes of the callus cells derived from type AABB plant. **A** (2n = 34). **B** (2n = 19). $a_1$ Marker chromosome of AA genome; *bars* 10 μm. (Bang et al. 1994)

and the $b_1$ chromosome has secondary constriction (Araki 1971; Bang and Choi 1993).

In allotetraploid AABB, hypoploid cells caused by an elimination of chromosomes were observed (Fig. 8; Bang et al. 1994), while a hypoploid cell and three hyperploid cells were found in eutetraploid cells of BBBB (Table 2; Bang et al. 1994). Chromosomal stability in suspension cells derived from diploid and tetraploid wild wheat was higher than that in hexaploid common wheat (Bang 1990). *S. scilloides* showed autosomal stability in diploid; however, type AA was on a level with allotetraploid AABB, while autosomes in type BB with B chromosomes showed high stability. The stability of autosomes in type BB might be due to the selective function of B chromosomes (Asami et al. 1976; Bang and Choi 1991).

In eutetraploid BBBB, numerical and structural variations of autosomes which were not detected in type BB were found. The multiplication of chromosome complements which was not observed in other cytogenetic types was also found in type BBBB callus cells (Fig. 9). The elimination of B chromosomes was the common phenomenon in the cultured cells of eutetraploid.

It is noteworthy that polyploids and aneuploids arose during callus cultures of *Daucus carota*, *Oryza sativa*, and *Triticum*, although normal diploids may be produced after regeneration (D' Amato 1978). This means that plant regeneration can selectively take place from the cells with normal autosome complements.

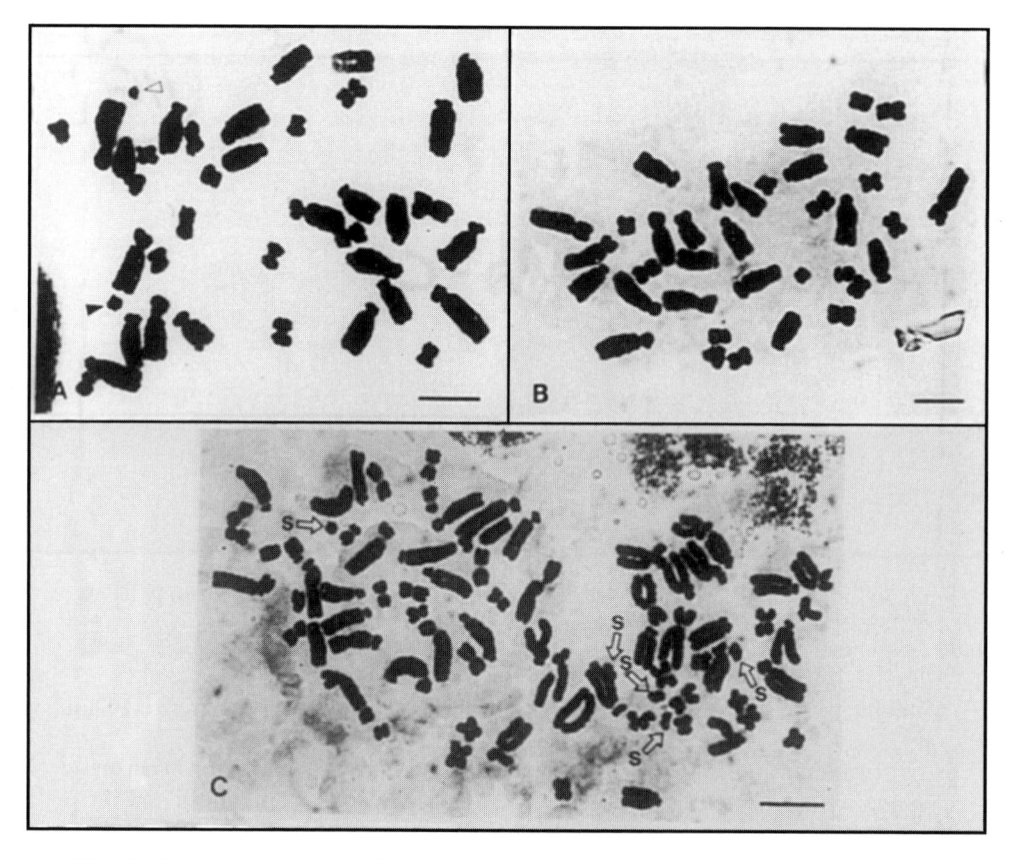


**Fig. 9A–C.** Karyotypes of the callus cells derived from type BBBB plant (2n = 36 + 1F + 1f). **A** Normal chromosome (2n = 36 + 1F + 1f). **B** 2n = 36 showing loss of B chromosomes. **C** hyperploid cell (2n = 71 + 5s). *s* Chromosome segments; *bars* 10 μm

**Table 3.** Frequency of B chromosome variation in callus cells and regenerants derived from type BB and BBBB plants of *S. scilloides* Complex

| Cytogenetic types of | Variation rate (%) | |
|---|---|---|
| original explants | Callus cells | Regenerated plants |
| BB (2n = 18 + 2F + 2f) | 26.9 (21/78)[a] | 3.4 (2/56)[a] |
| BBBB (2n = 36 + 1F + 1f) | 87.7 (57/65)[a] | 85.7 (48/56)[a] |

[a] Varigates/total numbers investigated.

### 3.3 B-Chromosome in Somaclonal Variation

Cultures of type BB and BBBB were derived from plants with 4B (2F + 2f) and 2B (1F + 1f), respectively (Tables 1, 2). Stability of B chromosomes in type BB was higher than that in BBBB. No numerical variation except whole B chromosome loss was found in regenerated plants. B chromosomes incidentally increase crossing-over in autosomes (Carlson et al. 1993) and may carry a gene regulating the expression of a structural gene (Ruiz et al. 1980). B chromosomes might have an effective role in stabilizing of autosomes in cultures of *S. scilloides* complex. It is also suggested that autosomes and B chromosomes are closely related in chromosome stability during culture.

In type BB, the frequency of B chromosome variation in callus cells was 26.9%, and one out of 78 cells carried no B chromosome, while the frequency was 3.4% in type BBBB (Table 3). It is interesting that the variation rate in B chromosomes in eutetraploid cells with 4B was much higher than in diploid BB cells with 2B. This trend was more elevated in regenerated plants (Table 3), while autosomes in type BB were more stable than in BBBB. Fusion of small B chromosomes was found in type BB.

In the case of type BBBB, only 8 amongst 65 cells carried the same number of B chromosomes as the original explant. However, the frequency of cells with the same number of B chromosomes as the mother plant in diploid BB was much higher than in eutetraploid BBBB (Table 2). It is suggested that the stability of B chromosomes in cultured cells differs depending on ploidy level.

Numerous factors such as ploidy level, genotype, regeneration procedure, source of tissue applied, hormonal composition in culture medium, endomitosis, and nondisjuction affect the chromosome variation in cultured cells (Karp 1988; Oh and Kim 1988; Bajaj 1990). In *S. scilloides*, different cytogenetic type, ploidy level, and the B chromosome might have affected the chromosome variation in the callus cells.

## 4 Summary and Conclusions

Somaclonal variation in cultured cells and regenerated plants developed from the different cytogenetic type plants of *Scilla scilloides* complex was investigated.

In cytogenetical analysis, all the regenerated plants except one aneuploid had the same autosome numbers as the original explants. It is suggested that plants derived from embryogenic callus are cytogenetically normal and stable. Autosomes were well maintained in allotetraploid AABB regenerants, though phenotypic variations such as leaf number, bulb shape, life form, and anthesis were found. Chromosome segments which were not observed in the original explants appeared anew. This might supply a clue for the origin of the B chromosome.

The autosomal stability in callus cells of type AA was much lower than in any other types. Chromosome stability in type AA was on a level with allotetraploid AABB, while autosomes in type BB with B chromosome showed higher stability. This stability of autosomes in type BB might be due to the selective function of the B chromosome. The elimination of B chromosomes was the common phenomenon in the cultured cells of eutetraploid BBBB. B chromosomes might play an effective role in the stability of autosomes in culture.

Different cytogenetic type, ploidy level and the B chromosome might affect the chromosomal variation in the callus cells of *S. scilloides* Complex.

# 5 Protocol

The following protocols are used for regeneration of plants from cultures of *S. scilloides* Complex:

1. Bulb scale segments are cultured on agar-solidified MS medium supplemented with 2 mg/l 2,4-D, 2 mg/l NAA and 2 mg/l BAP for plant regeneration. Calli appear after 2 weeks of incubation at $25 \pm 1$ °C under 16 h photoperiod.
2. Calli containing somatic embryos are transferred to fresh MS medium within 3 weeks of initiation for shoot development.
3. The obtained shoots are transplanted to hormone-free MS medium for rooting. Rooted plantlets are subsequently acclimated to greenhouse conditions after gradual hardening-off.

*Acknowledgment.* The authors thank Professor S.M. Boo for critically reading and Ms. M.K. Lee for help in the preparation of the manuscript.

# References

Araki H (1971) Cytogenetics of *Scilla scilloides* Complex III. Homoeology between genomes A ($x = 8$) and B ($x = 9$). Jpn J Genet 46: 265–275

Araki H (1972) Cytogenetic study of *Scilla scilloides* complex from Korea. Jpn J Genet 47: 147–150

Araki H (1985) The distribution of diploids and polyploids of the *Scilla scilloides* complex in Korea. Genetica 66: 3–10

Armstrong KC, Nakamura C, Keller K (1983) Karyotype instability in tissue culture regenerants of Triticale (*Triticosecale* Wittmack) cv. Welsh from 6-month-old callus cultures. Z Pflanzenzuecht 91: 233–245

Asami H, Inomata N, Okamoto M (1976) Chromosome variation in callus cells derived from *Secale cereale* L. with and without B-chromosome. Jpn J Genet 51: 297–303

Bajaj YPS (1990) Somaclonal variation—origin, induction, cryopreseryation, and implications in plant breeding. In: Bajaj YPS (ed) Biotechnology in agriculture and forestry, vol 11. Somaclonal variation in crop improvement I. Springer, Berlin Heidelberg New York, pp 3–48

Bang JW (1990) Chromosome variation in suspension cells derived from cultured immature embryo of *Triticum* spp. Korean J Bot 33: 189–196
Bang JW, Choi HW (1991) Cytogenetic studies of *Scilla scilloides* Complex from Korea II. Genome distribution in Chejudo populations. Korean J Bot 34: 145–150
Bang JW, Choi HW (1993) Cytogenetic studies of *Scilla scilloides* Complex III. Karyotype of cytotype BB and B-chromosome composition. Korean J Bot 36: 281–284
Bang JW, Park JH, Choi EY (1994) Chromosome variation in callus cells derived from different cytogenetic type plants of *Scilla scilloides* Complex. Korean J Plant Tissue Cult 21: 59–63
Carlson WR, Roseman R, Cheng Y (1993) Localizing a region on the B-chromosome that influences crossing over. Maydica 38: 107–113
Charkravarty B, Sen S (1983) Chromosomal changes in the scale leaf callus of diploid *Scilla indica*. Proc Indian Natl Sci Acad Part B Biol Sci 49: 120–124
Choi HW, Bang JW (1990) Cytogenetic studies of *Scilla scilloides* Complex from Korea I. Distribution of genomes and composition and frequencies of B-chromosome. Korean J Bot 33: 237–242
Creissen GP, Karp A (1985) Karyotypic changes in potato plants regenerated from protoplasts. Plant Cell Tissue Organ Cult 4: 171–182
D'Amato F (1977) Cytogenetics of differentiation in tissue and cell cultures. In: Reinert J, Bajaj YPS (eds) Applied and fundamental aspects of plant cell, tissue, and organ culture. Springer, Berlin Heidelberg New York, pp 343–357
D'Amato F (1978) Chromosome number variation in cultured cells and regenerated plants. In: Thorpe TA (ed) Frontiers of plant tissue culture. IAPTC/Univ of Calgary, Calgary, pp 287–295
Deumbling B, Clermont L (1989) Changes in DNA content and chromosomal size during cell culture and plant regeneration of *Scilla siberica*: Selective chromatin diminution in response to environmental conditions. Chromosoma 97: 439–448
Haga T (1961) Intra-individual variation in number and linear patterning of chromosomes. I. B-chromosomes in *Rumex*, *Paris* and *Scilla*. Proc Jpn Acad 37: 627–632
Haga T, Noda S (1976) Cytogenetics of the *Scilla scilloides* complex I. Karyotype, genome and population. Genetica 46: 161–176
Jha TB, Roy SC (1982) Chromosomal behaviour in cultures of *Vicia faba*. Cytologia 47: 465–470
Karp A (1988) Origins and causes of chromosome instability in plant tissue culture and regeneration. In: Brandham PE (ed) Kew Chromosome Conf III. Her Majesty's Stationery Office, London, pp 185–191
Larkin PJ, Ryan SA, Brettell RIS, Scowcroft WR (1984) Heritable somaclonal variation in wheat. Theor Appl Genet 67: 443–456
Murashige T, Nakano R (1967) Chromosome complement as a determinant of the morphogenic potential of tobacco cells. Am J Bot 54: 963–970
Murashige T, Skoog F (1962) A revised medium for rapid growth and bioassays with tobacco tissue culture. Physiol Plant 15: 473–497
Noda S (1976) Outline of natural habitats and geographical distribution of *Scilla scilloides* in East Asia. Bull Cult Nat Sci Osaka Gakuin Univ 2: 77–102
Noda S, Lee HS (1980) Relationship between chromosome constitution of 3 species of Liliaceae and human activities. Rep Sci Res, Ministry of Education, Japan, pp 33–55
Oh MH, Kim SK (1988) Chromosomal and morphological variation in protoplast-derived petunia (*Petunia hybrida*) plants. Korean J Genet 10: 265–271
Orton TJ (1985) Genetic instability during embryogenic cloning of celery. Plant Cell Tissue Organ Cult 4: 159–169
Pierik RLM (1987) In vitro culture of higher plants. Martinus Nijhoff, Dordrecht, pp 231–279
Ruiz RM, Posse F, Oliver JL (1980) The B-chromosome system of *Scilla autumnalis* (Liliaceae): effects at isozyme level. Chromosoma 79: 341–348
Yu Z, Araki H (1991) The distribution of diploids and polyploids of the *Scilla scilloides* Complex in the Northeastern District of China. Bot Mag Tokyo 104: 183–190

# II. 9 Somaclonal Variation in *Zinnia*

S.M. Stieve[1] and D.P. Stimart[1]

## 1 Introduction

### 1.1 Importance of *Zinnia*

The genus *Zinnia* (Asteraceae, Helianthus tribe) consists of about 17 species of annuals, perennial herbs, or small shrubs native to the southwestern United States and Central and South America. *Zinnia elegans* Jacq., native to Mexico, is a commercially important herbaceous annual grown as a bedding plant or cut flower. The wide variety of flower types, flower colors, and plant heights combined with drought tolerance and long bloom all contribute to make zinnias popular. Zinnias have accounted for more seed packet sales than any other garden flower sold by mail or garden center display. However, its susceptibility to *Alternaria zinnia* Pape (causal agent of alternaria blight), *Erysiphe cichoracearum* DC. ex Merat (causal agent of powdery mildew), and *Xanthomonas campestris* pv. *zinniae* Hopkins and Dowson (causal agent of bacterial leaf and flower spot) has in recent years posed an economic threat to commercial seed producers. Most zinnias available on today's market are $F_1$ hybrids derived from inbred hybridization. Some open-pollinated cultivars remain.

### 1.2 Genetic Variability and Breeding Objectives

One source of potentially valuable genes for resistance to the three major pathogens of *Z. elegans* and subsequent improvement of cultivars is *Zinnia angustifolia* HBK. Successful hybridizations between *Z. elegans* (2n = 24; Torres 1963; Ramalingam et al. 1971; Gupta and Koak 1976) and *Z. angustifolia* HBK (2n = 22; Olorode 1970; Ramalingam et al. 1971) have been reported (Ramalingam et al. 1971; Boyle and Stimart 1982), although partial to complete sterility in the $F_1$ generation placed serious limitations on their subsequent use. Boyle and Stimart (1982) restored partial fertility by colchicine treatment of axillary buds, resulting in *Z. marylandica* (Spooner et al. 1991), a disomic polyploid (2n = 46).

Phytopathological studies demonstrated that these induced amphiploids possess high levels of resistance to *A. zinniae*, *E. cichoracearum*, and *X. campestris* pv. *zinniae* (Terry-Lewandowski and Stimart 1983). In addition, all advanced

[1] Department of Horticulture, University of Wisconsin, Madison, WI 53706, USA

Biotechnology in Agriculture and Forestry, Vol. 36
Somaclonal Variation in Crop Improvement II (ed. by Y.P.S. Bajaj)

generations of amphiploids failed to segregate for resistance to *E. cichoracearum*; there was a very high degree of genetic uniformity for this trait.

### 1.3 Need to Produce Somaclonal Variants

Cytogenetic analysis showed lagging univalents and irregular distribution of chromosomes to the gametes to be the major contributing factors in $F_1$ hybrid sterility (Terry-Lewandowski et al. 1984). Bivalent associations were observed in the $F_1$ hybrid which indicated partial homology between parental genomes. In contrast, the induced amphiploids formed predominantly bivalents at metaphase I due to suppression of pairing between homoeologous chromosomes. Terry-Lewandowski et al. (1984) concluded that absence of multivalent associations in colchicine-induced amphiploids implied a genetic control of chromosome pairing. Observed cytological behavior is substantiated further by the lack of segregation for resistance to *E. cichoracearum* (Terry-Lewandowski and Stimart 1983) and for morphological traits among amphiploid families (Terry-Lewandowski et al. 1984).

Self-pollination of the amphiploids essentially produces seed with clonal uniformity, and sterile progeny results from backcrossing to diploid *Z. elegans* or *Z. angustifolia*, or crossing with the tetraploid *Z. elegans* 'State Fair' (Boyle 1990, pers. comm.). Since these traditional methods of breeding for improvement are ineffective, somaclonal variation was examined as a method of introducing genetic variation.

## 2 In Vitro Culture

In vitro culture of zinnias has been used to clone plants and recover interspecific hybrids. The production of male sterile lines for hybrid seed production was achieved by in vitro culture of axillary buds on MS salts and organics (Murashige and Skoog 1962) supplemented with 1 μM benzyladenine (Rogers et al. 1992). Interspecific hybrids of *Zinnia peruviana* and *Z. elegans* have been obtained through embryo culture (Shahin et al. 1971). Embryos appeared normal early in development; however, by the second week breakdown occurred and culture on White's standard medium (White 1943) supplemented with indole-3-acetic acid was necessary to recover hybrid plants.

Research has focused recently on propagation of zinnias by adventitious shoot formation (Stieve 1991). Results demonstrated adventitious shoots formed on cultured *Z. marylandica* embryo tissue. Optimum conditions for adventitious shoot formation, based on number of adventitious shoots formed and percent embryos forming adventitious shoots, were 16-day-old cotyledons oriented adaxial surface down on MS salts and organics supplemented with 0.2 μM thidiazuron (N-phenyl-N′-1,2,3-thidiazol-5-ylurea, TDZ). Embryos cultured on 22.2 μM TDZ produced more callus and took longer to form adventitious shoots.

# 3 Somaclonal Variation

## 3.1 Morphological Variation in Regenerated Plants

A seed-derived orange-flowered *Zinnia marylandica* plant derived from *Z. angustifolia* × *Z. elegans* 'Thumbelina Mini-Salmon' was propagated asexually; flower heads on clonal plants were self-pollinated, cotyledons of 16-day-old embryos were excised and cultured in vitro, and seed-derived control plants of the amphiploid parent were grown as described (Stieve et al. 1992; Stieve and Stimart 1993). One hundred and forty-nine of 510 adventitious shoots and 23 of 31 adventitious shoots formed on 0.2 and 22.2 μM TDZ, respectively, rooted and flowered. There were 86 control plants grown to flowering.

Leaf length and width, internode length, ray petal number, and seed set were on average largest in control plants (Table 1) when compared to plants regenerated from culture ($R_0$). Control plant means for peduncle length and petal width were less than TDZ-derived plants. Flower diameter and petal length control means were similar to 0.2 μM TDZ-derived plants, but larger than 22.2 μM TDZ means. On average, 0.2 μM TDZ means were largest for leaf length, flower diameter, petal length, and seed set when compared to 22.2 μM TDZ values. Mean values for leaf width, internode length, petal width, and ray petal number were similar between TDZ treatments. Peduncle length was largest on 22.2 μM TDZ.

Variances of $R_0$ plants derived from 0.2 μM TDZ were significantly greater than control plants for leaf length and width, peduncle length, flower diameter, and ray petal length, width and number (Table 1). Internode length and seed set variances were similar to control plants. Plants derived from 22.2 μM TDZ had variances significantly greater than control plants for internode and peduncle length, petal number, and seed set, and greater variance than 0.2 μM TDZ-derived plants for four of nine parameters observed.

Twelve variant $R_0$ plants were identified from 0.2 μM TDZ (8% of the population) and three from 22.2 μM TDZ (13% of the population). Aberrant characteristics identified in 0.2 μM TDZ-derived plants included tallness (line 3-1-2); dwarfness (lines 2-1-10, 5-1-5, and 5-1-6); increased seed set (line 17-2-4 and 17-2-6); fasciated flower heads (line 7-1-2; Fig. 1), upwardly curved ray petals (line 5-1-7; Fig. 2a), striped ray petals (line 16-2-2; Fig. 2b); and darker red-orange (line 17-2-5; Fig. 1) or muted orange ray (line 16-2-3) petals. Additionally, two regenerated plants had variegated leaf sectors. These areas were confined to plant sectors not near flower stems. Variegated areas became less distinct as plants aged and normal flowers were produced, so variegation was not studied further.

Aberrant characteristics identified in 22.2 μM TDZ-derived plants included large disc diameter of flower heads with short ray petals (line BxT3-1; Fig. 2c) and green spots on ray petals (lines BxT3–14 and BxT3–15; Fig. 1). Two plants producing green ray petal spots had nonspotted flowers until approximately five flowers opened, after which 75% had spotted rays.

Three examples were observed where the same variant arose in cotyledons of one embryo: dwarfness in lines 5-1-6 and 5-1-7, high seed set in lines 17-2-4 and

**Table 1.** Summary of *Zinnia marylandica* characteristics in seed-propagated (control) or adventitious shoots ($R_0$) derived from 0.2 or 22.2 μM thidiazuron. (Stieve et al. 1992)

| Population/analysis | Size (mm) | | | | | | | | Seed set (%) |
|---|---|---|---|---|---|---|---|---|---|
| | Leaf | | Internode | Peduncle | Flower | Ray petal | | | |
| | Length | Width | length | length | diameter | Length | Width | No. | |
| Control[a] | | | | | | | | | |
| Mean | 53.1 | 24.4 | 65.2 | 42.2 | 48.8 | 18.0 | 10.4 | 14.5 | 26.3 |
| Range | 36.7–69.0 | 15.7–33.2 | 38.3–99.7 | 19.3–79.0 | 34.3–57.0 | 12.2–21.3 | 8.2–12.5 | 12.5–18.7 | 0.0–71.7 |
| Variance | 56.1 | 14.8 | 140.5 | 154.7 | 17.9 | 2.8 | 0.8 | 1.6 | 0.1[d] |
| $R_0$ 0.2 μM TDZ[b] | | | | | | | | | |
| Mean | 49.4 | 20.4 | 46.6 | 48.9 | 48.7 | 18.5 | 11.5 | 12.2 | 17.9 |
| Range | 25.0–72.5 | 10.2–35.7 | 14.0–82.3 | 14.0–99.7 | 32.3–60.7 | 11.3–23.3 | 6.3–16.7 | 7.7–21.0 | 0.0–80.0 |
| Variance | 90.9 | 24.0 | 124.5 | 281.9 | 33.2 | 5.6 | 2.8 | 6.2 | 0.1[e] |
| Fisher's Test[g] | | | | | | | | | |
| Versus control | ** | ** | NS | ** | ** | ** | ** | ** | NS |
| $R_0$ 22.2 μM TDZ[c] | | | | | | | | | |
| Mean | 46.2 | 21.0 | 48.9 | 61.1 | 46.7 | 17.3 | 11.4 | 12.1 | 1.1 |
| Range | 37.0–72.5 | 16.0–28.2 | 31.0–111.0 | 20.7–110.0 | 39.7–54.3 | 13.3–21.2 | 10.0–12.7 | 8.0–16.7 | 0.0–6.6 |
| Variance | 64.2 | 10.8 | 377.0 | 361.2 | 13.8 | 2.9 | 0.6 | 6.3 | 3.4[f] |
| Fisher's Test | | | | | | | | | |
| Versus control | NS | NS | ** | * | NS | NS | NS | ** | ** |
| Versus 0.2 μM TDZ | NS | NS | ** | NS | NS | NS | NS | NS | ** |

[a,b,c] Based on 86, 141, and 23 plants, respectively.
[d,e,f] Based on 84, 138, and 22 plants, respectively.
[g] Comparison by Fisher's Test for equal variance.

**Fig. 1.** *Zinnia marylandica* 8411-2c seed-derived control plants (*row 1*; rows numbered from top to bottom) and $R_1$ adventitious derived progeny showing line 17-2-5 dark red-orange rays (*row 2, right*), line BxT3–14 and BxT3–15 green spotted rays (*row 3; left, center*) and short dentate rays (*row 3, right*), line 17-2-5 crested disk florets (*row 5, left*), line 16-1-4 tubular ray florets (*row 5, center*), and line 7-1-2 fasciated flower heads (*row 6, right*)

17-2-6, and green spots on ray petals in lines BxT3–14 and BxT3–15. Two seedlings from lines 14-1-4 (unreported) and 7-1-2 produced adventitious shoots on cotyledons. Shoot apical meristems of these seedlings never developed and plants were lost.

### 3.2 Heritability of Regenerated Plants

Control plants and the previously described 12 $R_0$ plants with variant characteristics were self-pollinated as described previously (Stieve et al. 1992; Stieve and Stimart 1993). Data were taken on seed-derived progeny ($R_1$) for the variant characteristic for which the parent was selected. No variants of either control plants or seed derived progeny of control plants were observed.

Tallness was transmitted sexually to line 3-1-2 progeny and dwarfness to line 2-1-10, but not 5-1-5 or 5-1-6 (Table 2). Tallness and dwarfness were due to

**Fig. 2a–c.** Upward-curved ray petals of *Zinnia marylandica* line 5-1-7 (**a**) and striped ray petals of line 16-2-2 (**b**) adventitiously derived on MS medium supplemented with 0.2 μM thidiazuron; large disk area and short ray petals of line BxT3-1 adventitiously derived from 22.2 μM thidiazuron (**c**)

increased or decreased internode and peduncle lengths. Increased seed set was transmitted sexually in lines 5-1-6, 17-2-4, and 17-2-6. $R_1$ plants had lower seed set than $R_0$ plants, but exceeded control seed set by 2.3-fold (Table 2). Disk area diameter and short petal length were transmitted sexually to progeny of line BxT3–1. $R_1$ ranges for internode length, peduncle length, plant height, disk diameter, petal length, and seed set were larger than $R_0$ plants, reflecting the larger number of observations going into them (Table 2). Flower diameter ranges of $R_0$ and $R_1$ generations were similar.

Fasciated flower heads were transmitted sexually and observed in 40% of line 7-1-2 (Table 3). Upwardly curved ray petals were transmitted to all progeny of line 5-1-7. Distorted ray petals of line 16-1-4 were transmitted to 33% of progeny. Ray petals of one plant were fused and tubular near the flower head and flared at the tip (Fig. 1). Ray petal striping was not inherited by progeny of line

**Table 2.** Heritability of *Zinnia marylandica* characteristics from tissue culture-derived adventitious plants ($R_0$) to self-pollinated progeny ($R_1$). (Stieve et al. 1992)

| Trait | Line | Generation | No. of plants observed | Internode length (mm) | Peduncle length (mm) | Plant height (mm) | Flower diameter (mm) | Disk diameter (mm) | Petal length (mm) | Seed set (%) |
|---|---|---|---|---|---|---|---|---|---|---|
| | Control | Unknown | 45 | 58.8 + 2.5[a]<br>28.7 – 93.0 | 42.4 + 2.1<br>18.7 – 74.3 | 55.0 + 1.8<br>31.0 – 69.0 | 47.1 + 0.58<br>36.0 – 55.5 | 12.5 + 0.03<br>9.0 – 15.5 | 18.1 + 0.30<br>13.3 – 23.0 | 17.5 + 3.8<br>0.0 – 39.6 |
| Tallness | 3-1-2 | $R_0$ | 1 | 75.0 + 6.9<br>63.0 – 87.0 | 88.7 + 11.3<br>67.0 – 105.0 | | | | | |
| | | $R_1$ | 111 | 93.7 + 2.4<br>33.3 + 152.7 | 83.0 + 3.8<br>10.0 – 181.7 | 60.4 + 1.1<br>35.0 – 84.5 | | | | |
| Dwarfness | 2-1-10 | $R_0$ | 1 | 41.2 + 4.2<br>34.0 – 48.5 | 29.7 + 3.0<br>24.0 – 31.0 | | | | | |
| | | $R_1$ | 10 | 34.8 + 2.6<br>21.7 – 48.0 | 22.0 + 1.8<br>13.7 – 29.3 | 32.6 + 2.6<br>22.0 – 43.5 | | | | |
| Dwarfness | 5-1-5 | $R_0$ | 1 | 45.0 + 8.0<br>37.0 – 61.0 | | | | | | |
| | | $R_1$ | 14 | 74.7 + 2.2<br>59.7 – 88.3 | | 54.9 + 1.0<br>49.5 – 58.5 | | | | |
| Dwarfness, high seed set | 5-1-6 | $R_0$ | 1 | 36.3 + 1.8<br>34.0 – 40.0 | | | | | | 65.3 + 2.2<br>60.9 – 69.1 |
| | | $R_1$ | 14 | 64.3 + 3.3<br>45.3 – 94.0 | | 50.8 + 1.8<br>39.0 – 63.0 | | | | 41.7 + 4.6<br>16.0 – 55.8 |
| High seed set | 17-2-4 | $R_0$ | 1 | | | | | | | 48.4 + 5.7<br>35.0 – 58.0 |
| | | $R_1$ | 9 | | | | | | | 39.2 + 2.5<br>27.2 – 52.9 |
| High seed set | 17-2-6 | $R_0$ | 1 | | | | | | | 65.3 + 1.9<br>61.7 – 68.3 |
| | | $R_1$ | 9 | | | | | | | 42.2 + 3.3<br>29.4 – 64.6 |
| Disk area diameter and petal length | BxT3-1 | $R_0$ | 1 | | | | 39.7 + 2.4<br>35.0 – 59.3 | 15.9 + 1.5<br>11.0 – 19.5 | 13.3 + 1.0<br>11.5 – 15.0 | |
| | | $R_0$ | 16 | | | | 42.6 + 1.4<br>35.5 – 59.3 | 17.0 + 0.3<br>15.5 – 20.0 | 15.6 + 0.4<br>13.0 – 19.3 | |

[a] Numerator represents mean and standard error, respectively; denominator represents range.

**Table 3.** Heritability of *Zinnia marylandica* characteristics from tissue culture-derived adventitious plants ($R_0$) to self-pollinated progeny ($R_1$). (Stieve et al. 1992)

| $R_0$[b] trait observed | Line | $R_1$ generation[a] | | |
|---|---|---|---|---|
| | | No. observed | Trait observed | % With character |
| Flower head fasciation | 7-1-2 | 15 | Flower head fasciation | 40 |
| Upward ray petal curvature | 5-1-7 | 22 | Upward ray petal curvature | 100 |
| Distorted rays | 16-1-4 | 6 | Distorted rays | 33 |
| Striped rays | 16-2-2 | 10 | Striped rays | 0 |
| Dark red-orange rays | 17-2-5 | 50 | Dark red-orange rays | 0 |
| | | | Upward ray petal curvature | 100 |
| | | | Crested disks | 2 |
| Muted orange rays | 16-2-3 | 48 | Muted orange rays | 81 |
| | | | Dark red-orange rays | 2 |
| | | | Light yellow-orange rays | 10 |
| | | | Pink rays | 4 |
| | | | Green ray petal spots | 2 |
| Green ray petal spots | BxT3-14 | 7 | Green ray petal spots | 29 |
| Green ray petal spots | BxT3-15 | 6 | Green ray petal spots | 0 |

[a] Self-pollinated progeny of tissue culture-derived plants.
[b] Plants regenerated from tissue culture by adventitious shoots.

16-2-2 (Table 3), but appeared to segregate into solid-colored lighter orange or darker red-orange ray petals. Darker red-orange rays of line 17-2-5 were not inherited by progeny (Table 3), but a new aberration of crested disk florets was observed in one plant (Fig. 1). All progeny except crested had upward-curved ray petals. Muted orange rays of line 16-2-3 were transmitted sexually to 81% of $R_1$ plants (Table 3). Additionally, 2% displayed darker red-orange ray petals and were taller, 10% lighter yellow-orange ray petals, 2% solid orange ray petals, 50% of which had two green spots near the tip, and 4% had pink rays and were shorter. One pink-flowered plant quit flowering after 1 month and new growth ceased. Green ray petal spots were inherited by two of seven progeny for BxT3–14 and none of the six progeny observed in line BxT3–15 (Table 3). A new aberrant flower type with very short dentate ray petals arose in progeny of both lines (Fig. 1). All progeny of these two lines grew slowly and took up to 3 months to flower.

## 4 Summary and Conclusions

Research on *Zinnia marylandica* showed that regeneration of plants from tissue culture results in many variations in plant height, fertility, and flower color and morphology. Variant characteristics included tallness, dwarfness, high seed set, large disk area diameter, fasciated flower heads, and ray petal curvature, distortion, striping, spotting, and color variations. All variants except striped ray petals and dark and red-orange ray petal color were transmitted sexually from $R_0$ to $R_1$

plants, indicating that genetic rather than epigenetic changes had occurred. New aberrant characteristics of pink ray petals, crested disk florets, green spots on ray petals, upwardly curved ray petals, and tubular ray petals, not observed in the $R_0$ generation, arose in the $R_1$ ($R_0$ selfed) progeny. This suggests that recessive genetic changes occurred in culture, genetic change occurred in $R_0$ plants, or that $R_0$ genomes are unstable and genetic rearrangement continues out of culture. The appearance of new variants in self-pollinated progeny of regenerated plants may demonstrate the need to study future generations of phenotypically normal $R_0$ plants to detect segregating characteristics. Seed-derived control plants displayed normal and uniform growth.

The variation in tissue culture observed with *Z. marylandica* may be due to a mutated cell early in culture or a preexisting mutation which gave rise to one or a few vegetative daughter cells that formed shoots (Broertjes and Keen 1980). Also, variation might have originated from the tissue culture environment allowing for increased genetic recombination (Skirvin and Janick 1976) through homoeologous pairing, allowing pre-existing genetic differences of *Z. elegans* and *Z. angustifolia* parental species to be expressed.

Fisher's Test for Equal Variance showed 0.2 μM TDZ-derived plants had more variation than control or 22.2 μM-derived plants for most characteristics observed. This suggests that higher TDZ levels do not induce more genetic variation, therefore 0.2 μM TDZ is adequate for obtaining somaclonal variations of *Z. marylandica.*

*Zinnia marylandica* was selected for study because it was resistant to three major zinnia pathogens but displayed no morphological variation. Somaclonal variants of commercial interest, including high seed set and novel flower pigmentation patterns and shapes, were obtained and should be tested for continued disease resistance, which may have been lost or mutated as a result of genetic change in culture. Future research with variant plants may be hindered by reduced fertility of variant plants, since many appear to set little or no seed. One solution to this may be crossing morphological variants with high seed-setting variants, which may promote increased seed set. Placing mutations of interest into the amphiploid genotype by backcrossing may prove useful in the development of commercially viable cultivars. Also of interest would be putting variant plants through another tissue culture regeneration cycle to see if they are genetically unstable, perhaps reverting to the parental phenotype or to an even more aberrant phenotype.

## References

Boyle TH, Stimart DP (1982) Interspecific hybrids of *Zinnia elegans* Jacq. and *Z. angustifolia* HBK: embryology, morphology, and powdery mildew resistance. Euphytica 31: 857–867

Broertjes C, Keen A (1980) Adventitious shoots: do they develop from one cell? Euphytica 29: 73–87

Gupta PK, Koak R (1976) Induced autotetraploidy in *Zinnia elegans* Jacq. Cytologia 41: 187–191

Murashige T, Skoog F (1962) A revised medium for rapid growth and bioassays with tobacco tissue cultures. Physiol Plant 15: 473–497

Olorode O (1970) The evolutionary implications of interspecific hybridization among four species of *Zinnia* sect. Mendezia (Compositae). Brittonia 22: 207–216

Ramalingam RS, Rangasamy SRS, Raman VS (1971) The cytology of an interspecific hybrid in *Zinnia*. Cytologia 36: 522–528
Rogers RB, Smith MAL, Cowen, RKD (1992) In vitro production of male sterile *Zinnia elegans*. Euphytica 61: 217–223
Shahin SS, Campbell WF, Pollard LH, Hamson AR (1971) Interspecific hybrids of *Zinnia peruviana* and *Z. elegans* through embryo culture. J Am Soc Hortic Sci 96: 365–367
Skirvin RM, Janick J (1976) 'Velvet Rose' *Pelargonium*, a scented geranium. HortScience 11(1): 61–62
Spooner DM, Stimart DP, Boyle TH (1991) *Zinnia marylandica* (Asteraceae: Heliantheae), a new disease resistant ornamental hybrid. Brittonia 43(1): 7–10
Stieve SM (1991) Adventitious shoot formation and somaclonal variation in *Zinnia marylandica*. MS Thesis, University of Wisconsin, Madison
Stieve SM, Stimart DP (1993) *Zinnia marylandica* tissue culture-induced variation and heritability. Acta Hortic 336: 389–395
Stieve SM, Stimart DP, Yandell BS (1992) Heritable tissue culture induced variation in *Zinnia marylandica*. Euphytica 64: 81–89
Terry-Lewandowski VM, Stimart DP (1983) Multiple resistance in induced amphiploids of *Zinnia elegans* and *Zinnia angustifolia* to three major pathogens. Plant Dis 67: 1387–1389
Terry-Lewandowski VM, Bauchan GR, Stimart DP (1984) Cytology and breeding behavior of interspecific hybrids and induced amphiploids of *Zinnia elegans* and *Zinnia angustifolia*. Can J Genet Crystal 26: 40–45
Torres AM (1963) Taxonomy of *Zinnia*. Brittonia 15: 1–25
White PR (1943) A handbook of plant tissue culture. The Jaques Cattel Press, New York

# Subject Index

# Springer-Verlag and the Environment